Digital Library Use

Digital Libraries and Electronic Publishing
William Y. Arms, series editor

Digital Library Use

Social Practice in Design and Evaluation

Edited by Ann Peterson Bishop, Nancy A. Van House,
and Barbara P. Buttenfield

The MIT Press
Cambridge, Massachusetts
London, England

This book was set in Sabon on 3B2 by Asco Typesetters, Hong Kong.

Library of Congress Cataloging-in-Publication Data

Digital library use : social practice in design and evaluation / edited by Ann Peterson
 Bishop, Nancy A. Van House, and Barbara P. Buttenfield.
    p.   cm. — (Digital libraries and electronic publishing)
 Includes bibliographical references and index.
  ISBN 978-0-262-02544-7 (hc. : alk. paper) — 978-0-262-52785-9 (pb.)
  1. Digital libraries—Social aspects. 2. Digital libraries—Planning. 3. Information
technology—Social aspects. I. Bishop, Ann P. II. Van House, Nancy A. III. Buttenfield,
Barbara Pfeil. IV. Series.
ZA4080.D546   2003
025′.00285—dc21   2002045248

The MIT Press is pleased to keep this title available in print by manufacturing single copies, on demand, via digital printing technology.

Contents

Part III

Foreword

Bruce Schatz

The world has changed radically with the emergence of the Internet. Information retrieval used to be a specialized topic known by a few experts and practiced by a few librarians. Today, millions of ordinary people all over the world routinely search the Internet in an attempt to find useful information to solve their problems.

This emergence has made the process of organizing and searching digital collections a critical international need. As the Internet itself becomes increasingly part of the structure of the world, so will the process of creating useful digital libraries become a critical part of society (Schatz 1997). Previous generations of the Internet were focused largely on the technology itself. When the Internet was originally developed in the 1960s, the focus was on transmitting packets of data correctly from one machine to another. Such transmitting could be engineered in a value-free fashion in the abstract world of bits.

Today the focus has shifted dramatically to searching documents usefully across many collections over the Internet. Such searching must be engineered to meet the needs of users in the concrete world of people. There are not correct answers to most queries in information retrieval, merely useful ones.

This shift from correct to useful has correspondingly created a shift in the focus of projects needed to develop infrastructure for the Internet. Advances in technology remain important, but considerations in sociology become equally important. The development of an information-retrieval system is determined largely by technology. But the deployment is determined largely by sociology.

A successful digital library is a place where a group of users (people) can effectively search a group of documents (collection) via an information system (technology). These three components must be in harmony, and all must be effective for the digital library to be useful.

Shortly after the Internet began to be widely used in the mid-1990s, the first generation of digital library projects began. Since the technology was still quite new, these projects were largely research projects by government-sponsored universities or major library organizations. One major catalyst was the National Science Foundation (NSF), National Aeronautics and Space Administration (NASA), and Defense Advanced Research and Projects Agency (DARPA)'s Digital Libraries Initiative (DLI) sponsored from 1994 to 1998.

The first generation of systems tends to be where the major styles are set. This first generation was technology focused but generally had a goal of fielding some sort of useful system as well. The tension between cool new technologies and mundane daily needs was never resolved but was significantly noted by these projects (Schatz and Chen 1996, 1999).

This volume contains thoughtful descriptions of the sociology research conducted in several of these projects as well as work that represents the growing community of researchers investigating social aspects of digital libraries. Fostering the development of this community is one of the lasting contributions of the DLI. As principal investigator of one of the DLI projects, I am pleased to see the final result at last and glad to have been able to encourage its production with words and monies.

In this volume, we see evidence of the struggle to determine which research methods to use for which stages of digital library development and deployment. The process of dealing with conflicting goals over the course of multiyear projects is also described. These descriptions allow readers to gain some feeling for the balancing act in information systems between technological and sociological factors.

Some of the descriptions may seem theoretical in nature. Although these first-generation projects tended to be rather academic, the problems and solutions considered are much the same as the commercial projects of later generations. As the initial foray into critical infrastructure, these descriptions are significant practically in addition to being valuable historically.

Digital libraries will form a major part of the structure of everyday life in the future. Stakeholders of all types, from system builders to policy makers, will be forced to deal with their successes and failures.

It is hoped that all will find useful guidance from this book and move closer to the dream of usefully providing access to all the world's knowledge.

References

Schatz, B. 1997. Information Retrieval in Digital Libraries: Bringing Search to the Net. *Science*, 275 (January 7) (special issue on Bioinformatics), 327–334.

Schatz, B., and H. Chen. 1996. Building Large-Scale Digital Libraries (Guest Editor's Introduction to Special Issue on Digital Libraries). *IEEE Computer*, 29(5), 23–27.

Schatz, B., and H. Chen. 1999. Digital Libraries: Technological Advancements and Social Impacts. IEEE Computer, 32(2) (special issue on Digital Libraries), 45–50.

Digital Library Use

1

Introduction: Digital Libraries as Sociotechnical Systems

Nancy A. Van House, Ann Peterson Bishop, and Barbara P. Buttenfield

This book is about digital libraries as *sociotechnical systems*—networks of technology, information, documents, people, and practices. It is about digital libraries' interactions with the larger world of work, institutions, knowledge, and society, as well as with the production of knowledge. And it is about creating, managing, and evaluating DLs.

The term *digital library* (DL) encompasses a wide range of working systems and research prototypes, collections of information and documents, and technologies.[1] Much of the discussion about DLs is about technology or about specific applications (e.g., Joint Conference on Digital Libraries 2001). This book takes a different approach. We are interested in understanding the social aspects of DLs—not just social impacts but the web of social and material relations within which DLs operate.

This book originated in the work of its editors and authors in designing, evaluating, and simply trying to understand DLs and their uses. Initially, this book was to be about DL evaluation, but it rapidly became apparent to the authors and editors that we were concerned with much more. We began with the belief that a good DL is useful. Like a traditional library, a useful DL fits the needs, activities, and contexts of the people who use it, as well as those of the people who create it, operate it, and contribute to its content. The more we delved into DLs and their social worlds, the more we found ourselves drawn into questions not just about DLs but about documents, collections, and classification; activity, work, and knowledge; politics and values; institutions; and identity, organizations, and communities.

The contributors to this volume see technology as "embedded in the social world in complicated ways, and this is particularly true for digital libraries, which are intertwined with the cognitive processes of a complex society" (Philip E. Agre, chapter 9). DLs form part of a long history of the mutual constitution of knowledge, documents, technology, and the social (David M. Levy, chapter 2).

This book has two goals. One is to inform policy and professional practice in DLs with socially grounded understanding of DLs as part of a web of social relations and practices. Another is to perform "technically informed social analysis" (Bowker, Star, Turner, and Gasser 1997, p. xiii) of phenomena of interest to social scientists that are highlighted by digital libraries, specifically issues of work, groups, and knowledge.

The chapters in this volume are unified by a sociotechnical approach. In this context, this phrase has two meanings: the first, already introduced, views digital libraries as composed of people, activity, artifacts, and technology. The second is an analytical stance that "privileges neither the social nor the technological and in which neither is reducible to the other" (Levy, chapter 2). Technology and the social are instead mutually constituted; the ongoing dynamic of their relationship is one of the themes of this book.

In this introduction, we consider socially grounded research in digital libraries generally and discuss why this kind of research is needed. We describe the varied domains and methods that come together in these chapters and identify major themes. We outline the book, summarize chapters, and end with some reflections on the implications of the book and of our approach to DL research.

Interconnections

Computers have escaped from the laboratories that once contained them. They pervade offices. They have settled into dining rooms, third-grade art rooms, and botanists' knapsacks. Information circulates among desktop computers, hand-held organizers, and mobile phones. With information technology operating in such a wide sphere of human activity, the consequences of problems in such areas as usability and access become significant. Information technology and systems simply have more power to influence our lives, for good or ill. And as the users of information technology have widened from professionals to everyone, the gap between users and designers has widened. So to understand and design for use, we need to know something about what people are doing at their desks and in the field and what else rests on those desks and dining room tables.

As information technology becomes more embedded in everyday activities, we become more aware of its role in social worlds. At the simplest level, people rely on friends, relatives, and passers-by to learn how to use information technology. Increasingly, social protocols develop around various kinds of information systems, such as those serving stock traders and auctions (Clifford Lynch, chapter 8). Technology creates linkages among information resources, groups, and individuals that have never existed and could not have existed before.

Our systems of collective cognition and the artifacts, technologies, and practices that support them are central to knowledge, activity, identity, community, and order. Information technology supports, shapes, and transforms our individual and collective knowledge processes: "The DL is not simply a new technology or organizational form but a change in the social and material bases of knowledge work and the relations among people who use and produce information artifacts and knowledge" (Nancy A. Van House, chapter 11).

To understand, use, plan for, and evaluate digital libraries, we need to attend to *social practice*, which we define as people's routine activities that are learned, shaped, and performed individually and together. As Vicki L. O'Day and Bonnie A. Nardi (chapter 4) put it:

Design problems get harder—and more realistic—as more interconnections between people, tools, and practices are revealed. A technological innovation may look good in isolation and yet turn out to be problematic or incomplete in actual settings of use.... When people look only at technical features when they make decisions about how to apply new technologies, they are likely to miss some of the interconnections that shape successful practice.

In other words, we need to uncover "the practical everyday reality" (Catherine C. Marshall, chapter 3) of workplaces, libraries, and other settings in which DL use happens and the "network of social and material relations" (Van House, chapter 11) in which DLs are embedded.

Multiple Research Trajectories

Socially informed research on digital library design, use, and evaluation sits at the convergence of several research trajectories. These intersections help to explain the multifaceted (some would say bewildering) state of DL research. We can categorize these approaches to research according to their levels of analysis, sponsors and institutions, and disciplinary bases.

Philip E. Agre (chapter 9) describes DL design, evaluation, and analysis as deriving from three disciplinary levels of analysis. The first, the physical and cognitive mechanics of work, is the subject of research in human-computer interaction (HCI) and ergonomics. Library and information science (LIS) seeks to understand the organization of information and the search habits of individual users. The third and highest level of analysis draws from social theory to examine the embedding of DLs in the larger social world. The chapters in this book address, at various times and in various ways, all three of these levels.

Clifford Lynch (chapter 8) categorizes the kinds of DLs currently being deployed according to their sponsors and institutional needs and goals. Each has its own research needs, as well. Traditional libraries, especially academic libraries, are moving into the digital distribution of materials to their primary clientele. Drawing primarily on library and information science, they rely largely on user surveys to ascertain needs and satisfaction and on performance measures to assess success. In contrast, commercial systems emphasize coherent collections to support targeted domains. Their interest lies in maximizing market share and profitability. These providers typically draw on marketing and HCI, often using focus groups, usability testing, and surveys of user satisfaction. Federal research and development agencies support technological innovation and investigation of broad social impact. They customarily sponsor prototypes as opposed to full-scale functional systems. This research is often multifaceted in goals and methods. The chapters in this volume address all these different categories of DLs but report most often on prototypes, not real-world DLs.

Another way to look at DL research is in terms of the disciplines or professions represented. Computer science is of course heavily represented in the DL world. Among the contributors to this volume, Agre, Christine L. Borgman, Levy, Lynch, Marshall, O'Day, and Schatz have backgrounds in computer science. Many computer scientists adopt an attitude of "build it and they will come" (Lynch, chapter 8; Gary Marchionini, Catherine Plaisant, and Anita Komlodi, chapter 6). However, a growing emphasis on usability assessment and user-centered design has led computer scientists and systems designers to adopt some of the methods and perspectives of the social sciences—ethnographic methods (Blomberg, Giacomi, Mosher, and Swenton-Wall 1993), ethnographically informed methods like contextual design (Beyer and Holtzblatt 1998), user-centered design (Vredenberg et al. 2001), and, more generally, design grounded in a better understanding of actual users at work in their own settings (Badre 2002; Hackos and Redish 1998).

Library and information science (LIS) is concerned with information, documents, information systems, with users and uses, and with technologies ranging from books and three-by-five-inch cards to computers and telecommunications and, of course, digital libraries. Among the contributors to this book, Agre, Ann Bishop, Borgman, Geoffrey C. Bowker, Komlodi, Levy, Marchionini, Laura J. Neumann, Plaisant,

Mark A. Spasser, Susan Leigh Star, and Van House all have been associated with LIS to varying degrees. Within the LIS research community, Patrick Wilson (1996) identifies two distinct enterprises. One, closely related to computer science, is concerned with the technology of computer-based information systems. Wilson (1996, p. 319) describes the second as "a field of social, behavioral, and humanistic studies …" a branch of what Europeans call the human sciences, which, he states, is difficult to delineate but has to do with information and users. The second is closely aligned with the research reported here.

A strength of LIS has been its long emphasis on user needs as a basis for design and evaluation (e.g., Bishop and Star 1996; Dervin and Nilan 1986; Paisley 1968; Van House, Weil, and McClure 1990; Van House et al. 1987). However, simply asking users directly about potential uses of new technology, resources, or services yields limited information. Users often have trouble predicting how they will incorporate new capabilities into existing practices and how needs and activities may change. As a consequence, LIS research has tended to look at characteristics of user groups, on the one hand, and at use of libraries and information systems, on the other (e.g., Dillon 1994; Marchionini 1995; Paepcke 1996; Savolainen 1998).

Other disciplines are represented in this book and in the DL world. Barbara Buttenfield is a cartographer. Her work in geographic information systems (GIS) led her to DLs containing georeferenced data. Vicki O'Day trained in computer science and is now a Ph.D. student in anthropology; Bonnie Nardi's educational background is also in anthropology. Geoffrey Bowker is trained as an historian, Susan Leigh Star as a sociologist, and Christine Borgman in communications. David Levy trained as a calligrapher after finishing a Ph.D. in computer science.

Whatever their training, contributors' thinking has become thoroughly hybrid, formally or informally, drawing on multiple disciplines and methods. Socially grounded DL research is defined more by the phenomena in which it is interested and its sociotechnical orientation than by specific methods, theories, or approaches.

Multiple Methods

Given the eclectic nature of DL research, an important issue addressed by contributors to this book is how to, methodologically and conceptually, "cross the great divide" (Bowker et al. 1997) between social science and computer science. DL research provides one instance of developing methods to "examine the social and cultural structures within which technologies are embedded" (Lyman and Wakeford

1999, p. 359). Which analytical stands and methods are appropriate for understanding DLs as sociotechnical systems? In empirical work, how do we choose an application or a setting and investigate it in ways that are helpful for a wide spectrum of DL research methods and practices? How do we do research that speaks to technologists, system developers, managers, funders, and users?

The chapters in this volume are by turns analytical and empirical. A number of authors take a primarily analytical approach, raising issues and questions or suggesting frameworks for understanding DLs as sociotechnical systems. Most of these—including Agre (chapter 9), Levy (chapter 2), Spasser (chapter 12), Star, Bowker, and Neumann (chapter 10), and Van House (chapter 11)—draw on contemporary social theory to illuminate how DLs are mutually constituted with social practice, structures, and values. A primary contribution of the book may be to model how social theory can expand our understanding of the processes of knowledge production in ways that inform DL design and evaluation, as part of a research literature utilizing social theory to improve understanding and design of information systems (see, for example, Bowker and Star 1999; Brown and Duguid 2000; Coyne 1995; Hakken 1999; Lyman and Wakeford 1999; Nardi and O'Day 1999; Suchman, Blomberg, Orr, and Trigg 1999).

Many chapters are grounded empirically and reflect some of the various systems that may be called digital libraries. They include DLs that have been implemented, are in the design stage, or have been merely proposed. They include collections of published and unpublished documents, images, and even records of plant observations. Some are available to the world at large via the Internet; others only to small workgroups. Some serve experts, others schoolchildren, yet others anyone and everyone.

This book illustrates the strengths of methodological pluralism, both across and within studies. The methods represented are "rigorously eclectic" (Spasser, chapter 12). Multiple methods are sometimes needed to suit different DLs' goals and circumstances or to study DLs at various stages of development and evaluation. Furthermore, different analytical perspectives mandate different—and sometimes multiple—methods. Marchionini, Plaisant, and Komlodi (chapter 6) use the analogy of medical imaging techniques in describing how "a plethora of data slices" from multifaceted approaches can be integrated in a manner that is not algorithmic but "systematic, interpretive, and driven by high-level goals."

Understanding DLs as embedded in complex social systems tends to promote the use of multiple methods. For example, Spasser (chapter 12) notes that "[DLs] are always embedded in a range of attitudinal, individual, institutional, and societal processes, and thus observable outcomes are always generated by a range of micro and macro forces that together produce observed situated activities." He describes his use of multiple methods as the desire to "take as many 'cuts' at the data from as many angles as is feasible to maximize the strength, density, and validity of theoretical ideas that emerge from data collection and analysis."

Particularly notable is the extent to which the chapters in this volume illustrate the usefulness of naturalistic methods—e.g., Ann Peterson Bishop, Bharat Mehra, Imani Bazzell, and Cynthia Smith (chapter 7), Marshall (chapter 3), O'Day and Nardi (chapter 4), Spasser (chapter 12), Star, Bowker, and Neumann (chapter 10), Van House (chapter 11)—including interviews, ethnographic fieldwork, and participant observation. Investigating people's understanding of their work, their actual practices, and their interpretations of their circumstances requires the flexibility, depth of inquiry, and long-term engagement with a field site typical of naturalistic methods.

Many of these chapters rely on multiple cases, including Marchionini, Plaisant, and Komlodi (chapter 6) and Star, Bowker, and Neumann (chapter 10). Marshall asks in chapter 3, "Why so many cases?" and answers that "the breadth of sources allows me to see things from a variety of use and maintenance perspectives." She demonstrates how looking across instances of DLs allows researchers to get a handle on common and important phenomena such as boundaries and metadata. Quoting Ray Pawson and Nick Tilley (1997), Spasser (chapter 12) notes that "you move from one case to another not because they are descriptively similar but because you have ideas that can encompass them both."

Themes in Socially Grounded Digital Library Research

A number of themes run through this book and socially grounded digital library research more generally. We have already discussed the common thread of a sociotechnical perspective. Another is the priority of users and the need for design and evaluation methods that emphasize users' experience (although Lynch in chapter 8 astutely questions the extent to which real-world DLs will be influenced by user-centered approaches when users aren't the ones developing or paying for a DL). A number of other themes are worth highlighting in this introduction, with some examples from the chapters that follow.

Content

Content is a key issue that is often overlooked in more technology-oriented digital library research: "At the core of effective digital library design is the relationship between the content to be provided and the user community to be served" (Borgman, chapter 5). The usefulness of a DL depends critically on its content being relevant to and usable by its clientele. Spasser (chapter 12) studies a little-examined aspect of DL design—the assembly and vetting of content in a DL that integrates content from hundreds of individual contributors. Bishop, Mehra, Bazzell, and Smith (chapter 7) describe a method of involving users in decisions about health information for an underserved community. Van House (chapter 11) highlights the importance of the trust in the content of the library in people's willingness to use the DL and to contribute their own work to its content.

Transparency

In a book about evaluating information systems and technology, usability is, of course, a major topic (Borgman, chapter 5). However, the argument of this book is that usability, as it is generally understood in HCI, is too limited a concept for assessing how well DLs serve their intended users. Star, Bowker, and Neumann (chapter 10) focus on transparency. When a DL is transparent, users don't have to know about the underlying machinery or software. Transparency cannot be assumed (Agre, chapter 9). Rather, it is achieved as a "product of a shifting alignment of information resources and social practices" (Star, Bowker, and Neumann, chapter 10). Major questions are for whom and under what circumstances a DL is transparent and what happens to transparency when we move from a single user to a larger community (Star, Bowker, and Neumann, chapter 10) or across user communities (Borgman, chapter 5; Van House, chapter 11; and Marchionini, Plaisant, and Komlodi, chapter 6).

Work Practice, Communities of Practice, and Mutual Constitution and Convergence

A major contention of this book is that DL design needs to be based on an understanding of users and their work: "A deep understanding of work is needed to make an artifact useful; an elegant design is no guarantee of utility" (Marshall, chapter 3). (Lynch, in chapter 8, suggests that *work* is perhaps too narrow a term, since DLs support decision making and behavior more broadly.) This is a shift away from traditional perspectives in LIS and in HCI that look at people as users of libraries and information systems and toward a more holistic understanding of how people make decisions, form opinions, and draw on information resources. This shift toward understanding the entire constellation of goals, activities, and resources of DL users—a shift that Lynch (chapter 8) describes as radical (largely because he sees it as unpopular with the institutions that oversee specific systems)—is described by most of authors in this book as necessary.

Work and the production and use of knowledge are social activities. Star, Bowker, and Neumann (chapter 10) and Van House (chapter 11) focus most explicitly on communities of practice, which share work practices, understandings, language, values, and orientations as well as information and which shape their members' understandings and even identity. Van House argues that DL communities of practice are not just those that use and contribute content but also those that build and operate a DL.

Star, Bowker, and Neumann (chapter 10) take as a major theme how transparency results from the convergence or mutual constitution that occurs when use and practice fit design and access. Many other chapters address in various ways the convergence of technology, practices, artifacts, and communities. For example, Bishop et al. (chapter 7) describe a project aimed at developing accessible and appropriate digital health information for black women.

The work that needs to be understood is not just the users'. Levy (chapter 2) encourages us to look at the work that documents do for us and how we delegate work to them as "talking things." Marshall (chapter 3) looks at the work that collections do, and Star, Bowker, and Neumann (chapter 10) look at classification systems.

Access, Equity, and Multiplicity

Access is a long-standing issue for traditional libraries. This concept is particularly sticky for DLs. Some are accessible only to specific audiences. Some are designed for specific audiences but, because of the openness of the Internet, are available to everyone. In at least one case report here (Van House, chapter 11), this gives rise to fears of misuse of the data. Some are designed for anyone—or, in some cases, for everyone who will pay (personally or institutionally).

Access is partly technical, consisting of access to the technology or usability. Access is also cognitive and may refer, for example, to what a user needs to know about a subject area or the DL. Finally, it may be social, relying on the user's participation in a community of practice or on social class or economics.

Because DLs operate in a variety of complex worlds and increasingly serve a wide range of users, many chapters—including Marchionini, Pleaisant, and Komlodi (chapter 6) and Star, Bowker, and Neumann (chapter 10)—argue that multiple communities and views need to be incorporated in their design. Equity of access and multiplicity of voices are especially important issues for traditionally marginalized groups (Bishop et al., chapter 7). However, as collections go digital, blurring the boundaries between published and unpublished, public and private, questions arise about the inclusion of information from a variety of sources and the issues of expertise, authority, and quality (Van House, chapter 11).

Scale

Scale includes the sheer size of the repository, the number of collections incorporated, and the size and number of targeted users and user communities. Spasser's Flora of North America project (chapter 12) must coordinate the work of thousands of contributors. The Library of Congress has never before had to serve such a large and varied a user community as it does, at least potentially, via the Internet (Marchionini, Pleaisant, and Komlodi, chapter 6). Star, Bowker, and Neumann (chapter 10) focus on how transparency is achieved as scale increases.

One issue related to scale is the question of universality versus locality or customization. A DL designed to serve the entire range of the U.S. citizenry (Marchionini, Plaisant, and Komlodi and the National Digital Library in chapter 6) has very different demands from, say, one serving a group of teachers in a single locale (also Marchionini, Plaisant, and Komlodi). Although small, customized DLs might seem easier to design, customization raises other questions of defining the audience and avoiding overfragmentation of the information world.

Scale may also refer to the rate at which a collection is expected to grow and to sometimes-paradoxical decisions, such as when to cull the collection to support additional growth. Allometric models have been taken from evolutionary biology to analyze paradoxes of scale in digital library collections, with mixed success (Buttenfield 1995).

Boundaries

Increasing scale often means crossing boundaries. Digital information crosses boundaries easily, but Marshall (chapter 3) demonstrates that many assumptions about the seamlessness of the DL, the "library without walls," are inaccurate. Boundaries are a major theme of Marshall's chapter, but they appear in many others—boundaries among collections (Levy, chapter 2), between organizations (Van House, chapter 11), between groups of users (Van House, chapter 11; Star, Bowker, and Neumann, chapter 10), and even between documents (Levy, chapter 2; Marshall, chapter 3).

Some boundaries need to be bridged, such as those of the "digital divide" (Bishop, Mehra, Bazzell, and Smith, chapter 7). Others perform useful functions. Van House (chapter 11) notes that crossing boundaries, including from private to public realms or across knowledge communities, sometimes throws into question the trustworthiness and credibility of information and sources. The point is not that boundaries are desirable or undesirable but that they have desired and undesired effects. As Marshall (chapter 3) says, they are "potential sites for new kinds of sociotechnical intervention" as well as places where we may want to tread carefully before intervening.

Place

Boundaries often imply place. Lynch (chapter 8) notes that DLs resist geography and institutional boundaries; they "dismiss place in favor of intellectual and nature-of-work coherence."

Yet the digital library remains a place in the view of many contributors, albeit less a physical concept than metaphorical or conceptual. Agre (chapter 9) says, "A library, even when it is digital, is still a place—the place where a scholarly community or a social movement can conduct its collective cognition with a reasonable degree of autonomy." O'Day and Nardi's (chapter 4) information ecologies play on the metaphor of place. Agre (chapter 9) talks about collections as workspaces for workgroups: "We still know little about the construction of such places, but perhaps we can renew our appreciation of the need for them." Bishop, Mehra, Bazzell, and Smith (chapter 7) describe a project that intentionally crafts a new public space where health professionals mingle with local community members.

Digital and Traditional Libraries

The relationship between digital and traditional libraries—conceptual, organizational, and functional—is addressed in many chapters. Some DLs are outgrowths of traditional libraries; other DLs relate to traditional libraries mostly metaphorically. Metaphors are both fruitful and constraining as we think about new uses of information technology, as O'Day and Nardi (chapter 4) demonstrate.

Lynch (chapter 8) notes that many of the values of traditional libraries are at odds with the market orientation of commercial DLs. He also points out DLs often go beyond traditional libraries by "provid[ing] an environment for actually doing active work rather than just locating and reviewing information that can support work processes." Yet many are missing some of the components of traditional libraries. Van House (chapter 11) notes that the publishing system and librarians' selection procedures and standards provided a form of quality control over library collections that is missing in some DLs.

Observing that librarians are often noticeably absent from DLs, several chapters ask specifically about the role of librarians in digital libraries. And when librarians are absent, what is lost? O'Day and Nardi (chapter 4) identify librarians as a "key-stone species" and note the "missing safety net of human assistance" (in Borgman's phrase, chapter 5) in computerized systems. Van House (chapter 11) asks who does the articulation work in DLs. Agre (chapter 9) argues that librarians "retain a considerable role in ensuring that libraries continue to encourage ... values.... This role is centrally one of design, not the command-and-control style of design from which computers first emerged but a participatory style in which the well-being of social institutions and their participants cannot be separated from the construction of technical systems."

Stability and Change

One inevitable theme in any book about information technology is change. These authors don't simply note (or celebrate or bemoan) its prevalence but grapple with ways of understanding it. Both stability and change need to be explained, not accepted as a matter of course.

Levy (chapter 2) takes change as one of his major themes. He notes our anxiety in the face of change and our continual collective efforts to create (temporary) stability. He notes that while one major function of documents is to be stable or repeatable, in practice both paper and digital documents are both fixed and fluid.

Some chapters focus on processes of evolution and coevolution. Levy (chapter 2) frames his study of documents over time by saying that "we will fail to see the current transformation correctly unless we also see the ways in which current developments are deeply continuous with the past." O'Day and Nardi (chapter 4), prompted by their ecological metaphor, look at the mutual adaptation among tools and social practices and ask what opportunities these create.

Another approach focuses on dynamics—on the factors tending toward both stability and instability. Spasser (chapter 12) looks at the contradictions and tensions that continually threatened the stability of a specific DL, a flora-collecting project that organizes the contributions from hundreds of participants, and the organizational strategies that are adopted to keep the project going. Lynch (chapter 8) describes DL development over the last twenty years, including the continual emergence of new forms and the tension between preexisting models and frameworks and the emerging digital realm.

Some view DLs as agents of social change. Bishop, Mehra, Bazzell, and Smith (chapter 7) are the most assertive in their promotion of participatory action research to empower people in designing DLs that will foster new community relationships and constructive changes in the lives of marginalized society members. But an implicit theme running throughout this book is that many DLs should be designed to make information more readily accessible to a greater variety of people. Agre (chapter 9) speaks of libraries helping nonprofessionals to appropriate professional knowledge.

The Chapters

The first part of the book challenges many assumptions about libraries, digital and traditional, and the documents and collections of which they are comprised. Levy (chapter 2) opens the discussion by saying that if we are to talk about how libraries might evolve in a digital age, we need first ask about the nature of the materials that make up library collections. He takes a social perspective on documents, asking what they are, how they work, how they relate to speech, and how emerging digital materials differ from earlier media. He describes documents as "talking things," as "representational artifacts ... made to carry very particular kinds of messages and in very particular ways." He concludes that we are still working out how to "throw our voices into silicon ... to delegate responsibility" to the digital. His chapter asks us not only to rethink documents and DLs in terms of the work that they do and the slipperiness of materiality but to address our own relationship to change and stability.

While Levy (chapter 2) is concerned with documents, Marshall (chapter 3) is interested in libraries and collections. She questions the concept of the "library without walls" and the popular assumption that digitization will lead to seamlessness. Her focus is on boundaries—technical and social, intentional and unintentional, visible and invisible, actual and interpreted. She investigates three cases: system codevelopment at an image collection in a university library; an ethnographic study and prototype document repository for a work group; and the design of a prototype digital library reading appliance. She concludes with three morals about crossing boundaries and, where possible, blurring them. She says that we should plan our encounters with intentional boundaries, engender realistic expectations about the complexity of unintentional boundaries, and design creatively to navigate around interpreted boundaries.

O'Day and Nardi (chapter 4) propose that we consider settings of technology use as information ecologies and apply the ecological metaphor to physical and digital libraries. This metaphor lends focus to key characteristics of diversity, of a sense of locality, of the presence of keystone species, and of the evolution of different elements over time. Their point is that the ecological perspective raises questions that might otherwise be ignored and, in particular, "highlights important linkages and dependencies" that must be considered in design and evaluation. Their chapter raises new questions about DLs and illustrates not only the uses of an ecological metaphor but the way that new metaphors for DLs can help us think about DLs differently and to see "the interconnections that shape successful practice."

The second part of the book emphasizes the design and evaluation of digital libraries. Borgman (chapter 5) adopts a broad definition of digital libraries that includes the full life cycle of information creation, retrieval, and use. She explores the connections between usability and utility and warns that applying these criteria to design is neither simple nor straightforward. For the coming generation of DLs to be able to serve "every citizen," she says that we will have to know more about information-related behavior. She describes information search as a form of problem-solving behavior, and draws on research on problem solving to describe the search process and the skills needed, which she then relates to DL usability. Her work is grounded in case studies of three very different groups—energy researchers and professionals, undergraduate geography students, and elementary school science students.

Marchionini, Plaisant, and Komlodi (chapter 6) illustrate longitudinal and multifaceted needs assessment for prototype design in three cases—needs assessment for prototype design for the Library of Congress, a design for a system for a community of teachers, and long-term evaluation of Perseus, a system serving teaching and research on ancient Greece. Human-centered design, they argue, must be based on "assessing human information needs and the tasks that arise from those needs and evaluating how the digital library affects subsequent human information behaviors." They conclude with principles that "resonate across the cases": both designers and evaluators must know the users; design and evaluation methods must be concurrent, ongoing, and embedded into DL management; and design and evaluation require multiple views. Their chapter illustrates the way that juxtaposing several unrelated projects can result in useful insights beyond the individual studies.

Bishop, Mehra, Bazzell, and Smith (chapter 7) present a case study of the Afya project, a digital library of health information for African American women designed with heavy participation from users to ensure its appropriateness and usefulness. They used participatory action research, which is particularly appropriate in a study aimed at empowering marginalized members of society. The study is especially relevant to the creation of online collections and services for underserved user groups in the era of the "digital divide." More generally, it addresses the changing role of information users as DLs are customized to specific user communities.

Lynch (chapter 8) notes that much of the research on DLs has focused on research prototypes, which he fears is misleading. Real-world DLs are different. He focuses on three major areas of tension that exist in traditional libraries but are "amplified" in DLs: control and governance, economics and sustainability, and audience. He reflects on the emergence of commercially based digital libraries, which, unlike traditional libraries, operate in the marketplace. Lynch describes how the development of DL services in traditional libraries has caused some "strange and unexpected, and occasionally wonderful, things to happen" but has been limited by institutional factors and by their governance by librarians, intermediaries with a commitment to preexisting models and practices. And he looks at what is happening as traditional libraries either compete or contract with commercial DLs. This chapter is not only an insightful reflection on the evolving world of DLs and traditional libraries and the relationship between them but a critical assessment of the possibilities for socially grounded design and evaluation in such an environment.

The last set of chapters is about DLs as they relate to the practices of knowledge creation and use in a variety of communities. Agre (chapter 9) warns that "society will evaluate digital libraries in terms of the ways that they fit, or fail to fit, into the institutional world around them" and considers how to conceptualize and evaluate this fit. He focuses on the boundary between technology (including DL technology) and institutions (including libraries, which, he says, articulate with other institutional fields in stable and structured ways). Agre considers two cases—the construction of healthy scholarly communities and the processes of collective cognition in a democratic society. He sketches how each works internally and links to the rest of the world and how technology and DLs may contribute to each. Agre concludes that the library, even when it is digital, plays a critical role as the place where a community "can conduct its collective cognition with a reasonable degree of autonomy." Implicit in his discussion is the need to ensure that DLs continue to fulfill this role. This chapter makes an important contribution to DL research by illustrating the uses of social theory; Agre argues that the sociological conceptualization of user communities and institutions is logically prior to the design and evaluation of technical systems.

The chapter by Star, Bowker, and Neumann (chapter 10) is concerned with how, in the design of digital libraries that serve large numbers of people, concepts traditionally seen as individual or psychological scale up in practice. Specifically, they ask how scaling up affects transparency, which they argue is achieved at larger levels of scale through "the convergence of knowledge and resources across groups of users." They explore convergence in three examples at individual, community, and infrastructure scales. The first case examines how becoming a member of a profession makes acquiring information easy. The second shows a professional community, nursing, aligning its codification and accounting procedures with those of other strategically important groups. Finally, they describe a large-scale information infrastructure serving heterogeneous communities, the *International Classification of Diseases*. With increasing scale, they find that transparency "becomes more subject to contention arising from the heterogeneity of the participating social worlds" but that once achieved it acquires coercive power.

Van House (chapter 11) is concerned with understanding the situated, distributed, and social processes of knowledge work. She describes DLs as supporting users' knowledge work and as being the loci of knowledge work. She claims that "the DL challenges existing practices of knowledge work, the boundaries of knowledge communities, and the practices of trust and credibility, all of which are central to the creation and use of knowledge." Drawing on several areas of social theory and an empirical study of data sharing in two environmental science fields, she argues that the ease with which DLs cross the boundaries of knowledge communities "highlights critical issues of trust and credibility in the networked world."

Spasser (chapter 12) makes a strong argument for applying social realist theory as an evaluation framework for DLs. He examines the assembly and vetting of DL content in the Flora of North America project—which he calls "one of the country's largest scientific collaborations"—in the context of complex organizational issues. He utilizes his social realist framework to identify several sets of contradictions within the project.

To Whom Does This Matter?

We envision this book as speaking to three broad and overlapping groups—people whose major concern is with information technology (including but not limited to the design, management, and evaluation of digital libraries); those concerned with information artifacts and infrastructure (documents, classification systems, collections, and libraries, traditional and digital); and those whose primary interest is the creation and use of information and knowledge.

For digital library researchers, implementers, managers, evaluators, or funders, our goal is not to provide recipes for building and evaluating DLs but to challenge readers to broaden their understanding of DLs and question their assumptions about the relationships among technology, information, practices, and people. These chapters do not merely report on how others have built and evaluated DLs. They ask questions that have no simple answers but that must be an ongoing part of the process of creating and assessing DLs.

This book is aimed also at people who are concerned with trends in digital information. Currently there is much discussion about new digital genres and the future of traditional media, including books, newspapers, and scholarly journals. New economic and technical models are being discussed and tested, including models that capitalize on the ability to track, control, charge for, and limit uses of digital materials that were uncontrollable in a paper environment. Computing and telecommunications make it possible to share, modify, and reuse information that was local in the nondigital world. Much of what we have to say about digital libraries also applies to such areas as knowledge management, as well as less readily classified emerging applications areas. We cannot fully anticipate the effects of current choices. Our message is one of respect for the power and effectiveness of existing networks of people, technology, and practices and curiosity about them and their possible successors.

We hope that this book is also of interest to people concerned with broader issues of information, knowledge, work, and social practice. It is our contention that changes in information technology highlight taken-for-granted practices and understandings, making the invisible visible, denaturalizing what has been naturalized, threatening to undermine what have been long-standing relationships. Information technology offers new ways of doing what we have been doing and also offers possibilities for new activity and understanding.

Conclusion

The authors of these chapters began with two questions: How do we evaluate DLs? How do we understand them so that we can build better DLs? But our shared orientation goes far beyond this and asks, How do DLs make a difference in people's lives? How do DLs support (or undermine) our collective efforts to record, know, understand, and order our world and experiences? What can studying DLs tell us about information and knowledge and about processes of cognitive and social order?

Although many of these chapters end with guidelines or suggestions of some sort, perhaps their greatest contribution is to raise questions and concerns made visible through the lens of social practice. As a group, the authors and editors challenge readers to think differently about DLs and about information technology more generally and to engage in conversations that include users and designers, social scientists, and technologists.

One implication of the discussion in this book is that because the relationships among knowledge, technology, and people are dynamic, needs assessment, design, and evaluation must be equally dynamic. Agre (chapter 9) says: "Experience with these appropriations [of technology by users] helps to shape new generations of technology, which are appropriated in turn. These appropriations are famously unpredictable." DL design must be open to continual uncertainty and change. Dynamic evaluation is, as Marchionini, Plaisant, and Komlodi (chapter 6) remind us, "process-oriented and iterative rather than product-oriented and summative."

We also hope that these chapters demonstrate the usefulness of seeing DLs as sociotechnical systems, of considering the mutual constitution of work, technology, communities, and identity. We hope they endorse the value of such concepts as transparency, scale, place, and boundaries in understanding DLs. And we intend to demonstrate the utility of looking to a broad range of social theory for analytical bases for understandings DLs.

Another implication is the need for multiple methods for needs assessment and evaluation, with particular attention to naturalistic inquiry. The sociotechnical system that we call a DL is complex, situated, and unique and needs a variety of methods to produce a diversity of evaluative information.

Most of all, we hope that this book prompts discussions among social scientists, technologists, librarians, users, researchers, and professionals engaged with DLs. Those who come to this book looking for axioms, heuristics, or principles of DL design will be disappointed. The reasons we don't include such guidelines are pragmatic. Digital libraries and socially informed DL research are relatively new. To our knowledge, the research represented in this book presents a significant slice of what has been done. Second, and on a practical level, sweeping pronouncements are not justified. Axioms may be useful under certain limited circumstances, but they are generally incompatible with a situated approach to social research.

This brings us back once again to digital libraries as sociotechnical systems, as much more than technology, contents, and functionality. As Haraway (1997, p. 126) says: "The computer is a trope, a part-for-whole figure, for a world of actors and actants, and not a Thing Acting Alone. 'Computers' cause nothing, but the human and non-human hybrids troped by the figure of the information machine remake the world." So, too, the digital library causes nothing—but stands for a network of people, practices, artifacts, information, and technology that may remake at least parts of our world.

Acknowledgments

The work represented in this book has been supported in a variety of ways by the sponsors and participants in the digital library projects that are identified in individual chapters. The preparation of this book was encouraged and supported financially by Bruce Schatz as part of the Illinois Digital Libraries Initiative Project funded by the National Science Foundation, the National Aeronautics and Space Administration, and the Defense Advanced Research and Projects Agency. This book originated in discussions among the editors, many of the authors, and others as part of the Digital Libraries Initiative. We also wish to acknowledge the emerging intellectual community of people concerned with socially informed DL research, many, but not all, of whom are represented in this volume. We're grateful for the unflagging support of Doug Sery at MIT Press. Finally, we want to thank the authors for their dedication and their patience.

Notes

1. For the purposes of this book, we have deliberately avoided defining the phrase *digital library*. These chapters demonstrate the great variability of systems and applications included under this phrase. Christine L. Borgman (chapter 5) and Clifford Lynch (chapter 8) (despite his protests to the contrary) take on the task of definition.

References

Badre, A., 2002. *Shaping Web Usability: Interaction Design in Context*. Boston: Addison-Wesley.

Beyer, H., and K. Holtzblatt. 1998. *Contextual Design: Defining Customer-Centered Systems*. San Francisco: Morgan Kaufmann.

Bishop, A. P., and S. L. Star. 1996. Social Informatics for Digital Library Use and Infrastructure. In M. E. Williams, ed., *Annual Review of Information Science and Technology* (vol. 31, pp. 301–401). Medford, NJ: Information Today.

Blomberg, J., J. Giacomi, A. Mosher, and P. Swenton-Wall. 1993. Ethnographic Field Methods and their Relation to Design. In D. Schuler and A. Namioka, eds., *Participatory Design: Principles and Practices* (pp. 123–155). Hillsdale, NJ: Erlbaum.

Bowker, G. C., and S. L. Star. 1999. *Sorting Things Out: Classification and Its Consequences*. Cambridge, MA: MIT Press.

Bowker, G. C., S. L. Star, W. Turner, and L. Gasser, eds. 1997. *Social Science, Technical Systems, and Cooperative Work: Beyond the Great Divide*. Mahwah, NJ: Erlbaum.

Brown, J. S., and P. Duguid. 2000. *The Social Life of Information*. Boston: Harvard Business School Press.

Buttenfield, B. P. 1996. GIS and Digital Libraries: Issues of Size and Scalability. In L. C. Smith and M. Gluck, eds., *GIS and Libraries: Patrons, Maps and Spatial Data* (pp. 69–80). Champaign–Urbana: University of Illinois Press.

Coyne, R. 1995. *Designing Information Technology in the Postmodern Age: From Method to Metaphor*. Cambridge, MA: MIT Press.

Dervin, B., and M. Nilan. 1986. Information Needs and Uses. In M. E. Williams, ed., *Annual Review of Information Science and Technology* (vol. 21, pp. 3–33). White Plains, NY: Knowledge Industry.

Dillon, A. 1994. *Designing Usable Electronic Text: Ergonomic Aspects of Human Information Usage*. London: Taylor & Francis.

Hackos, J. T., and J. C. Redish. 1998. *User and Task Analysis for Interface Design*. New York: Wiley.

Hakken, D. 1999. *An Ethnography Looks to the Future*. New York: Routledge.

Haraway, D. 1997. *Modest_Witness@Second_Millenium. FemaleMan©_Meets_Oncomouse™: Feminism and Technoscience*. New York: Routledge.

Joint Conference on Digital Libraries. 2001. *Proceeding of the First ACM/IEEE-CS Joint Conference on Digital Libraries (Roanoke, VA, USA, June 24–28, 2001)*. New York: ACM Press.

Levy, D. M. 2001. *Scrolling Forward: Making Sense of Documents in the Digital Age*. New York: Arcade.

Lyman, P., and N. Wakeford, eds. 1999. Introduction: Going into the (Virtual) Field. *American Behavioral Scientist*, 43(3), 359–369.

Marchionini, G. 1995. *Information Seeking in Electronic Environments*. Cambridge: Cambridge University Press.

Nardi, B. A., and V. L. O'Day. 1999. *Information Ecologies: Using Technology with Heart*. Cambridge, MA: MIT Press.

Paepcke, A. 1996. Digital Libraries: Searching Is Not Enough: What We Learned On-Site. *D-Lib Magazine*, 2(5). ⟨http://www.dlib.org/dlib/may96/stanford/05paepcke.html⟩.

Paisley, W. 1968. Information Needs and Uses. In C. A. Cuadra, ed., *Annual Review of Information Science and Technology* (vol. 3, pp. 1–30). Chicago: Benton.

Pawson, R., and N. Tilley. 1997. *Realistic Evaluation*. London: Sage.

Rubin, J. 1994. *Handbook of Usability Testing: How to Plan, Design, and Conduct Effective Tests*. New York: Wiley.

Savolainen, R. 1998. Use Studies of Electronic Networks: A Review of Empirical Research Approaches and Challenges for Their Development. *Journal of Documentation*, 54, 332–351.

Shapin, S. 1994. *A Social History of Truth: Civility and Science in Seventeenth-Century England*. Chicago: University of Chicago Press.

Suchman, L., J. Blomberg, J. Orr, and R. Trigg. 1999. Reconstructing Technologies as Social Practice. *American Behavioral Scientist*, 43(3), 392–408.

Van House, N. A., B. T. Weil, and C. R. McClure. 1990. *Measuring Academic Library Performance: A Practical Approach*. Chicago: American Library Association.

Van House, N. A., M. J. Lynch, C. R. McClure, D. L. Zweizig, and E. J. Rodger. 1987. *Output Measures for Public Libraries* (2d ed.). Chicago: American Library Association.

Vredenburg, K., S. Isense, and C. Righi. 2001. *User-Centered Design: An Integrated Approach*. Upper Saddle River, NJ: Pearson Education.

Wilson, P. 1996. The Future of Research in Our Field. In J. Olaisen, E. Munch-Petersen, and P. Wilson, eds., *Information Science: From the Development of the Discipline to Social Interaction* (pp. 319–323). Boston: Scandinavian University Press.

Part I

2

Documents and Libraries: A Sociotechnical Perspective

David M. Levy

Introduction

What are libraries in a digital age? What function do they serve at a time when whole new classes of materials—digital materials—appear to be taking center stage? What are digital libraries (DLs), and how do they differ (or how will they differ) from traditional libraries? In these early days of the digital revolution, no firm and final answers to such questions have emerged. But we might observe that whatever libraries are, have been, or will be is necessarily tied to collections of materials. One way or another, all libraries do some combination of housing, organizing, and providing access to collections. It follows that the character of particular libraries is closely bound to, if not defined by, the collections those libraries oversee.

But if there are no firm answers at this time about how libraries, digital or otherwise, will evolve, perhaps we can inquire into the nature of the materials that make up their collections. We might then ask, What are documents? How do they work? To what extent (or in what ways) do newly emerging digital materials differ from earlier forms realized on paper or in other media?

My intention in this chapter is to approach such questions from a social perspective. We live in an era when the term *revolution* is invoked often and easily. Although the word used to refer to social and political movements, it is now often used to refer to changes in technology and their imagined effects on society. An enormous investment in labor and capital is being devoted to questions of bits and bytes. And while discussions of the social aren't entirely absent from either public or private discourse, they tend to be dwarfed by the seemingly insatiable demands of the technology. The political scientist Langdon Winner (1986) has observed that discussions of values, of social concerns, tend to arise as an afterthought to discussions that are primarily about making mechanisms work.[1]

But in taking a social perspective on documents, I do not wish to ignore or discount the technological. If anything in our world has the mark of the technological, surely it is our written forms. However we might wish to define (and dispute) what technology is, can there be any question that paper, ink, printing presses, computers, and display screens are technologies or the products of technology? It might be better to say that I will be taking a *sociotechnical* perspective. In so doing, I align myself with recent work in science and technology studies that aims "to find multidisciplinary ways of talking about heterogeneity: of talking, at the same time, of social and technical relations even-handedly without putting one or the other in a black box whose contents we agree not to explore" (Bijker and Law 1992, p. 5).

Documents and Documentation

When we think of libraries, we of course think of books. The library has a place in the Western mind as the home, or perhaps even the temple, of the book (and not just because the word *library* is from the Latin *liber*, meaning "book"). But we also know that libraries have held and cared for many other types of materials. Serials (journals, newspapers, and magazines), audiovisual materials, organizational records, and personal papers all have their place in modern libraries, archives, and special collections. And long before the introduction of digital technologies, libraries had to contend with a range of media and technologies well beyond the catalogs, stacks, lamps, and desks that support the use of books. Film stock, audio and video tape, microfiche, and the technologies needed to display these have had a home in libraries for decades. Indeed, Klaus Musmann (1993) argues that libraries have been leaders and even innovators in the use of cutting-edge technology.

But given such a range of forms, media, and technologies, what ties these materials together, other than the (surely nonnegligible) observation that they are all cared for or managed by libraries? In what sense, if any, are they *of a kind*? Intuitively, they are all written forms, provided that pictures and other nontextual representations are granted the status of writing. But is it possible to be more specific or clear?

In an article entitled "What Is a 'Document'?" Michael Buckland (1997) has compiled various attempts to clarify the nature and scope of documents in the early days of information science, mainly through the efforts of some of its European pioneers. What is now called information science was first known as *documentation*, and its practitioners were called *documentalists*. Running through their thinking in the first part of the twentieth century were questions about the scope of the field and the breadth of its objects of concern. As Buckland (1997, p. 805) explains:

> Documentation was a set of techniques developed to manage significant (or potentially significant) documents, meaning, in practice, printed texts. But there was (and is) no theoretical reason why documentation should be limited to *texts*, let alone *printed* texts. There are many other kinds of signifying objects in addition to printed texts. And if documentation can deal with texts that are not printed, could it not also deal with documents that are *not* texts at all? How extensively could documentation be applied? Stated differently, if the term "document" were used in a specialized meaning as the technical term to denote the objects to which the techniques of documentation could be applied, how far could the scope of documentation be extended? What could (or could not) be a document?

For Paul Otlet, one of the seminal figures in the documentation movement, writing in 1934, "Graphic and written records are representations of ideas or of objects,... but the *objects themselves* can be regarded as 'documents' if you are informed by observation of them" (Buckland 1997, p. 805). Examples of nontextual objects that could be documents included "natural objects, artifacts, objects bearing traces of human activity (such as archaeological finds), explanatory models, educational games, and works of art" (Buckland 1997, p. 805).

Suzanne Briet, the French librarian and documentalist, declared in 1951 that "A document is evidence in support of a fact" and is "any physical or symbolic sign, preserved or recorded, intended to represent, to reconstruct, or to demonstrate a physical or conceptual phenomenon" (Buckland 1997, p. 806). Buckland (1997, p. 806) adds: "The implication is that documentation should not be viewed as being concerned with texts but with access to evidence." Briet's most unusual example was of an antelope. In the wild, she claimed, it isn't a document, but once captured and placed in a zoo it becomes evidence and is thus transformed into a document.

The Indian theorist Shiyali Ramamrita Ranganathan considered a document to be an "embodied micro thought" on paper "or other material, fit for physical handling, transport across space, and preservation through time" (Buckland 1997, p. 807). He refused to include audiovisual materials: "But they are not documents; because they are not records on materials fit for handling or preservation. Statues, pieces of china, and the material exhibits in a museum were mentioned because they convey thought expressed in some way. But none of these is a document, since it is not a record on a more or less flat surface."

Buckland cites other definitions, but these three should suffice to show something of the range of attempts made to pin down the notion. The contrasts are striking, both in terms of the contradictions among them as well as in the differences of emphasis. While Otlet and Briet embrace three-dimensional objects, Ranganathan insists that only flat surfaces are admissible. For Otlet, a document is something that informs; for Briet, it is evidence; while for Ranganathan, it is a "micro thought." Ranganathan is explicit about the importance of preservation in time. For the other two, this criterion is at best implicit.

None of these attempts to describe a document is meant to function as a dictionary definition. They are not meant to define the word *document* so much as to identify and bound a set of phenomena. These thinkers were answering the question, "What are the proper objects of study within our discipline?" Consequently, there is no objective standard (such as common word usage) against which to assess their accuracy, no answer to the question, "Who is right?" But one thing is clear: all were a response to and an attempt to come to terms with the proliferation of media and formats at the time.

"We Can't Even Say What a Document Is Anymore"

If the question of boundaries (of what is inside and what is out) was of concern fifty or more years ago, how much more urgent is it today when we have not only film and microfiche to contend with but all manner of digital materials? In August 1996, a short article appeared in *Wired* magazine under the heading "What's a Document?" Here is the text in its entirety (Weinberger 1996; p. 112):

Have you noticed that the word *document* doesn't mean much these days? It covers everything from a text-only word processing file to a spreadsheet to a Java-soaked interactive Web page.

It didn't used to be like this. A document was a piece of paper—such as a will or passport—with an official role in our legal system.

But when the makers of word processors looked for something to call their special kind of files, they imported *document*. As multimedia entered what used to be text-only files, the word stretched to the point of meaninglessness. Just try to make sense of the file types Windows 95 puts into the Document menu entry.

The fact that we can't even say what a document is anymore indicates the profundity of the change we are undergoing in how we interact with information and, ultimately, our world.

Several things are noteworthy about this article. First is simply the fact that the question "What is a document?" is being posed in a mainstream publication rather than in a professional journal. The question has moved beyond the narrowly defined realm of documentalists and information scientists. (The writer, David Weinberger, identifies himself as vice president of strategic marketing at OpenText.)

Clearly, changes in information technology are now sufficiently widespread to lead to reflection outside the bounds of academic and other professional communities. Second, the writer has a clear sense of what a document is or at least what it once was, and he feels that it has been violated by the introduction of digital forms. Documents are legal missives realized on paper. The newer digital forms are nothing of the sort. As if that isn't bad enough, a once coherent notion has been violated and reduced to meaninglessness, and we can't even say what a document is any more. Finally, the writer ties the loss of this once coherent notion to profound societal changes now taking place. Like so many people, he seems to decry the pace of change and the disappearance of familiar categories.

Indeed, Weinberger is correct, more or less, in what he says about the word *document*. *The Random House Dictionary* defines a document as "A written or printed paper furnishing information or evidence, as a passport, deed, bill of sale, bill of lading, etc.; a legal or official paper." *The American Heritage Dictionary* calls a document "A written or printed paper that bears the original, official, or legal form of something and can be used to furnish decisive evidence or information." The word *document*, it seems, has something to do with writing, paper, and evidence. Digital materials clearly don't qualify because they're not on paper.

Of course, dictionaries don't prescribe usage; they describe it. People *are* increasingly using the word *document* to name digital materials. (These days, says Weinberger, the word "covers everything from a text-only word processing file to a spreadsheet to a Java-soaked interactive Web page.") The day can't be far off when dictionary definitions will cover digital materials too. This suggests that Weinberger's concern isn't only with the word and its uses but with the coherence of a cultural category. Word processing files don't belong in the same category as wills, nor do Web pages belong in the same category as passports. We can't even say what a document is anymore, he is arguing, because there is no logic by which wills, passports, Web pages, and spreadsheets would be properly grouped.

Talking Things

Narrowly speaking, of course, Weinberger is correct. If we insist on seeing documents as written legal materials, as "official papers," then digital materials must necessarily be excluded. But suppose, taking a leaf from the pioneering documentalists, we aim for a more encompassing view. Is it possible to come up with a coherent story that accounts for both paper and digital forms?

My proposal is that documents are talking things. They are bits of the material world (clay, stone, animal skin, plant fiber, sand) that we've imbued with the ability to speak. One of the earliest characterizations of documents comes from the Book of Genesis, and, curiously, it is a description of human beings, not of written forms: "God formed Adam from the dust of the earth and blew into his nostrils the breath of life, and Adam became a living soul." The parallel between this mythic event and the creation of actual documents is strikingly close, for indeed, when we make documents, we take the dust of the earth and breathe our breath, our voice, into it.

While this doesn't literally bring the inert material to life, it is nonetheless an extraordinary act of ventriloquism. Ventriloquism, according to the dictionary, is "The art of speaking ... in such a manner that the voice does not appear to come from the speaker but from another source" (*Random House*, 2d ed.) When we write, when we make documents, we "throw our voice" into the materials. In some cases (when writing a letter to a loved one, for example), it is clear where the voice comes from. In other cases (such as maps), the source of the voice may be carefully disguised, and the document may appear to speak objectively and to speak for itself (Wood 1992). The ventriloquist may be a person or an anonymous organization.[2]

Thinking of documents in this way is hardly new. Indeed, an awareness of written forms as talking things has ancient roots. It is made explicitly in Plato's *Phaedrus*, which has an extended reflection on the nature of writing. Written forms may speak, Socrates observes, but they are dumb. Indeed, writing has some of the same disadvantages as painting. Painting may produce realistic images, but they are representations and not the real things. If you query one of them, it is unable to respond. "The same holds true of written words: you might suppose that they understand what they are saying, but if you ask them what they mean by anything, they simply return the same answer over and over again" (1973, p. 97).

This is a crucial point. For Plato, it shows up the limits of writing. Written forms are a pale shadow of their human counterparts. They are incapable of dialogue, the Socratic path to wisdom. While this observation is true enough, it fails to grasp what is truly powerful about documents. For it is precisely in their ability to "return the same answer over and over again" that the utility of documents is made manifest. The brilliance of writing is the discovery of a way to make artifacts talk, coupled with the ability to hold that talk fixed, so it can be repeated again and again at different points in space and time. It is something that documents do well and that people by and large do not. It is not that we are incapable of performing in such a manner—a messenger can deliver a singing telegram to multiple hotel rooms—but it is not of our essence to do so. Yet it is exactly of the essence of documents.

Framing documents in this way sets up a strong parallel between documents and people. Each in its own way is a talking thing. This is hardly an accidental parallel. Documents are exactly those things we create to speak *for us*, on our behalf and often in our absence. It is common enough for one person to stand in for and speak for another. Lawyers and agents (press agents, literary agents) do this for a living. Lovers in some of our most famous love stories send their confidants to do the talking for them. Documents do this too, and like their human counterparts, they speak in order to act for us, to take on jobs or roles. Each kind of document, each *genre*, has a specialized form of work to do.

Indeed, we can think of document genres as being akin to the differentiation of human (social) roles. Mail carriers, flight attendants, police officers, and chefs, for example, all have particular kinds of work to do. They talk and act in ways appropriate to their roles and even wear uniforms that signal their social identity. Much the same is true of document genres. Greeting cards, novels, and cash register receipts (as well as wills and passports) all have distinctive looks (recognizable "uniforms"). They speak in ways and carry content appropriate to the social practices in which they are embedded.

Delegating to Humans and Nonhumans

Documents, then, are things we create to take responsibility for some of our concerns. Bruno Latour (1992) calls the process by which people assign tasks to others *delegation*, and he applies the term equally to the assignment of tasks to humans and to nonhumans. As an example of delegation to objects, he cites the door (Latour 1992, p. 228):

So architects invented this hybrid: a wall hole, often called a *door*, which although common enough has always struck me as a miracle of technology. The cleverness of the invention hinges upon the hinge-pin: instead of driving a hole through walls with a sledgehammer or a pick, you must simply gently push the door …; furthermore—and here is the real trick— once you have passed through the door, you do not have to find trowel and cement to rebuild the wall you have just destroyed: you simply push the door gently back.

The door therefore takes on much of the work of keeping the entryway either open or closed. It relieves people of the need to break a hole through the wall only to repair it moments later. The door thus assumes certain responsibility, but in return it asks something of the people who use it; there is a reciprocal relationship. People must know how to open and close the door. They must not only have such competencies (including the skill and the strength to perform these acts), but they must exercise them. Of course, not everyone closes the door after passing through it, and at such times the door's function (to keep the cold out and the heat in or vice versa) is defeated. This job too can be delegated—to a person (a "doorman") or to another artifact, to a spring or some other mechanism, that closes the door automatically (which Latour calls a "groom").

Through examples of this kind, Latour paints a picture of a world realized through the ongoing interactions of human and nonhuman actors. We humans may make artifacts and delegate to them, but they in turn shape our behavior and in effect delegate to us. Ascribing agency to inanimate things may strike some as inappropriately anthropomorphic. To this charge Latour (1992, p. 235) responds: "It is well-known that the French like etymology; well, here is another one: *anthropos* and *morphos* together mean either that which *has* shape or that which *gives shape* to humans. The groom is indeed anthropomorphic in three senses: first, it has been made by humans; second, it substitutes for the actions of people and is a delegate that permanently occupies the position of a human; and third, it shapes human action by prescribing back what sort of people should pass through the door."

To read with any clarity, one generally has to know the writer's intended audience. In this case, Latour (1992, p. 227) is speaking to (or against) his fellow sociologists and their attempts to understand and explain human culture purely in social terms:

[Sociologists] are constantly looking, somewhat desperately, for social links sturdy enough to tie all of us together or for moral laws that would be inflexible enough to make us behave properly. When adding up social ties, all does not balance. Soft humans and weak moralities are all sociologists get. The society they try to recompose with bodies and norms constantly crumbles. Something is missing, something that should be strongly social and moral. Where can they find it? Everywhere, but they too often refuse to see it in spite of much new work in the sociology of artifacts.

But Latour is arguing equally against those who attempt to understand the world (and in particular our technological society) purely in technological terms. According to this view, technological determinism (which has a near strangle hold on our culture and is held by many of today's technological visionaries), technology development obeys its own inner logic and drives and determines the shape of society.

In contrast to these two views, Latour and his colleagues in social studies of science and technology have been working for more than a decade to formulate a view of human culture that privileges neither the social nor the technological and in which neither is reducible to the other. As Wiebe E. Bijker (1995, p. 273) puts it: "The relations I have analyzed ... have been simultaneously social and technical. Purely social relations are to be found only in the imaginations of sociologists or among baboons, and purely technical relations are to be found only in the wilder reaches of science fiction. The technical is socially constructed, and the social is technically constructed."

According to this view, the central element of study is neither the social actor nor the technological artifact but networks of mutually stabilizing activities and objects (variously called *actor networks* or *sociotechnical ensembles*). In Latour's example above, for instance, neither the door, the hinge, nor the automatic door closer can be understood in isolation. Each is what it is only relative to the other two elements but also only relative to any number of other factors, including the way buildings are designed, how humans are socialized to enter and leave them, and so on.

More Than Text

From this perspective, then, *all* artifacts are fundamentally social. All human-made things (the word *artifact* literally means "made with skill") have a "social life" (Brown and Duguid 1996). But as should now be clear, the term *social* (as I am using it here) is not opposed to or distinct from the technical. Artifacts are bits of the material world that have been molded and shaped to participate with *us*, in our world.

Documents, of course, are artifacts, and in this sense they are fundamentally social too. But they are a particular class of artifacts—those capable of speech—and this property makes them even more evidently and intensively social. We may delegate social roles to all manner of artifacts, but it is only to documents that we have delegated one of our most distinctive and highly celebrated of human capabilities. (Indeed, what I am suggesting is that documents are, by definition, exactly those artifacts to which we have delegated the human gift of speech.)

There is a sense in which all artifacts may be said to speak. Each particular thing we encounter has its own story to tell. It was created at a certain time and place. It comes out of a certain community, exemplifies a certain style and design aesthetic. It has been subject to wear and tear, has perhaps been broken and repaired, and bears evidence of the passage of time. A Shaker cabinet has much to tell about Shaker life and living. A house—any house—speaks not only of the era in which it was built but of the lives that have been lived in it. Cars have much to tell us about their owners, about how they see themselves, and the kind of image they wish to present to the world.

Documents certainly speak in these ways, too, but this form of talk is not what they are fundamentally about. Unlike cabinets, houses, and cars, documents are *representational* artifacts; they are made to carry very particular kinds of messages and in very particular ways. Perhaps the simplest instance of such a representational stance is given by Michael T. Clanchy (1993) in describing the transition in thirteenth-century England to an increasingly literate society. Up until the thirteenth century, Clancy reports, people were required to witness and thereby validate financial and legal transactions. To transfer a piece of property (real estate) from one person to another, the donor would speak his intentions aloud in the presence of witnesses. At the same time, the donor would hand over a symbolic object—for example, a knife or a small piece of earth from the land being transferred. Should there be a dispute, the witnesses to the event would be required to testify. By the end of the thirteenth century, however, a written document could serve as both the statement of intent and as witness to the facts of the matter.

In this instance, the knife or piece of turf clearly had a symbolic or representational function. It was meant to stand for, to bear witness to, and to call to mind the declaration through which property was transferred from one owner to another. Needless to say, this is an extremely limited case of the kind of talk that documents exhibit. The knife as an undifferentiated whole was meant to stand for the fact of property transfer. It was only in the context of particular witnesses and their memories that one could have any idea of what the knife was saying or that it had anything to say at all.[3] Documents talk much more powerfully and broadly when their representational abilities are more fully articulated.

The most obvious case of such further articulation is written language—the use of alphabetic or other symbolic marks to create composites that correspond more or less directly to human utterances. But it is not only through alphabetic and textual representations that documents can be said to speak. There are many other "written" forms, including maps, diagrams, pictures, photographs, and all manner of other conventional and well-articulated nonverbal representation schemes. It might seem odd to say that a Mathew Brady photograph of Lincoln or a map of the New York subway system "speaks," but certainly these forms stand in for us too, they tell stories, they represent.

Traditionally, this parallel has been conceived of as a distinction in language, between spoken and written language or between orality and literacy. There is a long history of speculation about the relationship between these two modes of communication. For much of this century, following Saussure's lead, linguists have taken writing to be a straightforward derivative of speech and undeserving of sustained attention. As recently as 1964, the linguist Otto Jespersen (1969, p. 17) declared that "language is primarily speech" and the written word only a poor substitute for the spoken word. Earlier in the century, another linguist was even more blunt: "Language is basically speech, and writing is of no particular theoretical interest" (Bloomfield 1933, quoted in Sampson 1985, p. 11). As a result (McKenzie 1999, p. 34),

Other formalized languages, or more properly perhaps, dialects of written language— graphic, algebraic, hieroglyphic and ... typographic—have suffered an exclusion from critical debate about the interpretation of texts because they are not speech-related. They are instrumental of course to writing and printing, but given the close interdependence of linguistics, structuralism, and hermeneutics, and the intellectual dominance of those disciplines in recent years, it is not surprising perhaps that the history of non-verbal sign systems, including even punctuation, is still in its infancy.

Within recent linguistics, there has been an increased interest in exploring the properties of written language in their own right—and not as a mere transcription or derivative of speech (Sampson 1985). In other disciplines, such as philosophy, some have been bold enough to move beyond the text. Goodman (1976), for example, explored the nature of various nontextual representation schemes or, as he called them, "notations," including diagrams, paintings, and musical notation. Recalling attempts made by pioneering documentalists (although apparently unaware of them), D. F. McKenzie (1986) recommended that the field of bibliography[4] expand beyond its narrow concern with printed alphabetic materials to encompass even landscape if it serves a narrative function within a community. For the Australian aboriginal Arunta tribe, he claims there is a strong link between talk and landscape (McKenzie 1986, pp. 32–33):

These visual, physical features form the ingredients of what is in fact a verbal text, for each one is embedded in a story, has a specific narrative function, and supports in detail the characterization, descriptive content, physical action, and the symbolic import of a narration....

The argument that a rock in Arunta country is a text subject to bibliographical exposition is absurd only if one thinks of arranging such rocks on a shelf and giving them classmarks. It is the importation into Arunta land of a single-minded obsession with book-forms, in the highly relative context of the last few hundred years of European history, which is the real absurdity.

In McKenzie's understanding of the role of landscape in Arunta culture, rocks and other natural features have come to function as documents, to assume some of the responsibility for telling culturally significant narratives.

Going Digital

This last example most dramatically illustrates how focusing on the social role of documents allows us to move not only beyond print, text, and alphabetic materials but also beyond an exclusive concern with paper. Indeed, any material that can be made to talk repeatedly gives us the possibility of creating documents.

For many centuries, documents have been realized primarily by fixing marks (symbols) in a two-dimensional substrate. Just in the last 100 years, we have established how to record activity (sounds and images) via film, audio tape, and video tape. In such instances, holding marks fixed on a surface won't work since activity, by definition, involves change over time. What these newer technologies do is to allow us to replay—to repeat—patterns of sound and image. Sameness of *performance* is substituted for sameness of static image. These new communicative forms, to paraphrase Plato's words, "go on telling you the same thing again and again." A different technological means is being used to achieve the same end.

Clearly, then, what we are witnessing today is a working out of the means—both social and technical—to make talking things out of digital materials. Over the last few decades, the technical means to produce, modify, distribute, and consume digital materials have been steadily refined. Thanks to the proliferation of personal computers and the increasingly broad availability of the Internet and the World Wide Web, the beginnings of a global infrastructure are now in place. The result is what has been called a *digital convergence*. A single medium or representational format (ones and zeros) is now capable of representing all the forms of talk we have so far managed to create: text and graphics, voice, and moving images. And a single device is capable of making all these forms manifest. Talking things that were once realized in the form of bound books, serials, audio tapes, and video tapes can now be produced and consumed on a personal computer.

But are these digitally induced forms of talk repeatable? A strong rhetorical current runs through discussions of the new media, suggesting that fixity and repeatability are properties of earlier materials (most notably paper) but are not inherent in digital forms. According to this view, the fixity of print, the stability and permanence of the bound book, will ultimately give way to the fluidity, to the instability and impermanence, of digital materials. This strain of thinking has been most clearly and forcefully articulated within the hypertext community. Jay David Bolter (1991, p. 31) states it this way:

Electronic text is the first text in which the elements of meaning, of structure, and of visual display are fundamentally unstable. Unlike the printing press or the medieval codex, the computer does not require that any aspect of writing be determined in advance for the whole life of the text. This restlessness is inherent in a technology that records information by collecting for fractions of a second evanescent electrons at tiny junctions of silicon and metal. All information, all data, in the computer world is a kind of controlled movement, and so the natural inclination of computer writing is to change, to grow, and finally to disappear.

To be sure, there is something quite appealing about this argument for the "death of fixity." It speaks to the longing to be freed from past constraints, which Paul Duguid (1996) has called "liberation technology." Unfortunately, the attempt to paint paper documents as fixed and digital varieties as fluid is simplistic and wrong (Levy 1994). Paper documents have never been fixed completely. They too admit change. Paper documents of all kinds are regularly annotated (see, for example, Marshall 1997), and even books, our symbols of long-term stability, undergo change. Individual copies are annotated, and new editions supplant earlier ones. Paper documents, in other words, exhibit a kind of controlled movement.

As for digital materials, they are no more fully fluid than paper documents are completely fixed. What, after all, is the point of having a Save function in a text or graphics editor if not to be able to fix a document in a particular state of completion? And current interest in figuring out how to preserve digital materials (Garrett and Waters 1996; Levy 1998) is a clear indication of the desire to guarantee some measure of stability—for some of our digital materials, at least.

Rather than think of paper as fixed and digital as fluid, we would do better to realize that all documents are fixed *and* fluid. Indeed, each genre has its own pattern or rhythm of fixity and fluidity, of stasis and change. This is only natural since each genre is a kind of talking thing that has been tailored to do a certain kind of work. Based on the nature of the work it does, each genre will needs to hold talk fixed for certain periods but be open to modification at others. A typical shopping list, for example, has a short lifetime—often a matter of hours—and is frequently modified during that time, but a book, a particular physical volume, may undergo little change over the course of decades. It is possible that digital documents (certain genres, at any rate) will undergo a faster rate of change, and in this sense, they may be more fluid (more rapidly changing) than some, or even most, genres of paper documents. This, however, should not be taken to mean that all digital talk will simply be unfixable or unrepeatable.

Communicative Stability and Social Order

What the "death of fixity" argument misses most importantly is the profound significance of communicative stability. The ability to keep talk fixed, to guarantee its repeatability, has become an essential cornerstone of human social organization, and we humans are not about to forego it any more than we are about to forego our ability to make fire, to build shelter, or to fashion clothing. The root of the word *communicate* (which is also the root of the words *community* and *communion*) is the Latin *communis*, meaning "common." Putting talk into stable external forms allows it to be shared, to be held in common. Just as the stability of the earth allows us to build stable structures on top of it, so written forms provide stable reference points that help to orient us in social space.

Indeed, communicative stability is at work in all our major social institutions. Science, law and government, religion, education, the arts, commerce, and administration all rely on the stabilizing power of documents to accomplish their ends. In the form of books and journal articles, documents are carriers of scientific knowledge. As sacred scripture, they are the central artifacts around which religious traditions have been organized. As written statutes, charters, and contracts, they play a crucial role in constructing and regulating lawful behavior. As works of literature, paintings, and drawings, they are the tangible products of artistic practice. As textbooks and student notes, they are crucial instruments around which learning practices are organized. As receipts and accounts, memos, and forms, they are critical ingredients in the organization of commerce and indeed all bureaucratic conduct. In each of these cases, the ability to hold talk fixed—to provide communicative stability—is crucial.

These institutions are essentially the cultural mechanism by which we create and maintain a meaningful and orderly social world. Science and religion are quests for meaning, order, and intelligibility. Media, the arts, and entertainment are means by which we tell ourselves (and continually reinforce) stories about who we are and why we are here. Education is concerned with socializing our young—bringing them into the social order we've constructed and training them to carry the meaning-making and order-making project forward. Government, law, commerce, and administration are all concerned with regulating human conduct—the exchange of goods and services, the orderly procession of human affairs.

Through their extensive participation in all these institutions, documents therefore play a crucial role in supporting—in making and maintaining—the social order. But documents not only support social order; they are part of it. They themselves need to be tended and taken care of, just like everything else in our world. Without physical maintenance, documents will decay. Without constant organizing, they will become inaccessible. Without continual updating, without ongoing interpretation and reinterpretation, they will lose their currency and their ability to speak effectively in their situations of use.

This means that ongoing human practices are required to stabilize documents so that they in turn can stabilize us, our practices, and our institutions. It is a process of mutual stabilization. The library, of course, is one of the institutions whose mission has been to stabilize documents—through cataloging, reference, conservation services, and the like. But it is worth noting that libraries, like all modern bureaucratic institutions today, also *use* documents to stabilize their own internal practices. Libraries thus use documents to maintain documents (Buckland 1992).

A Changing Order

These are anxious times. Nearly all our social institutions are in the process of rethinking themselves. Some of the turmoil is the result of the destabilizing effects of the introduction of digital document technologies. It is hardly surprising that libraries would be particularly vulnerable. As Francis Miksa (1996) points out,[5] libraries have lived through a number of different "eras." The most recent of these, which he calls the "modern library era," may be coming to an end. We cannot yet know what new institutions will replace the modern library or even whether the change will be great enough to merit a new name. But one thing is certain: not only will we continue to collect documents, but these collections will need to be stabilized through organized human intervention.

Much of today's rhetoric is about what is new, about what is revolutionary. Without question, digital technologies have novel features. Without question, they provide many powerful opportunities for communication quickly and at a distance. But we will fail to see the current transformation correctly unless we also see the ways in which current developments are deeply continuous with the past. In this chapter, I have offered a view of documents as talking things. While new forms are emerging out of new technological ensembles, the social impulse to make such surrogates or "second selves" is quite old.

In these early days of the digital transition, we are working out how to throw our voices into silicon; we are working out (in Latour's language) how to delegate responsibility to ensembles of ones and zeros. But the ventriloquist's dummy cannot talk without the collusion of both a ventriloquist and an audience who are willing to pretend that the voice is coming from a block of wood. The door can't do its part without the support of travelers. So digital materials cannot speak their mind and do their work without the collaboration of a vast infrastructure of people, institutions, and other artifacts. This is the effort we are all engaged in, and the (re)fashioning of libraries and library practices is an essential part of this socio-technical work.

Notes

1. The notion of values, Winner (1986, p. 180) suggests, "plays an important role in the bureaucratic and technological vocabulary. Its definition in that lexicon is something like 'the residual concerns one needs to ponder after the real, practical business of society has been taken care of.'"

2. Making a document is thus the separating out of a part of ourselves to do the talking *for us*. The actress Candice Bergen has written about her father, Edgar Bergen, the famous ventriloquist, and his relationship with his dummy, Charlie McCarthy. Charlie was a kind of alter-ego for Edgar, behaving in ways that he himself could not (Bergen 1984, p. 23): "My father—while not *himself* the perfect hero—had by now created someone who was. Awkward, silent, socially unsuccessful, Edgar created someone who caught people's fancies when he, most often, could not. Gradually he began leaving things to this dummy—so saucy, witty, self-assured—and learned to let *him* take over while behind his left shoulder, bashful, sort of beautiful, stood Edgar, as if by accident, listening in amusement while Charlie just wowed 'em."

3. This story also nicely illustrates the process of delegation. Initially, the work of witnessing was distributed among human and nonhuman actors (human witnesses plus the knife or piece of turf). Gradually, though, as the nonhuman came to be more fully trusted, more of the task could be delegated to the inanimate material, to the document.

4. Bibliography is "the discipline that studies texts as recorded forms, and the processes of their transmission, including their production and reception" (McKenzie 1986, p. 4).

5. See, for example, Fran Miksa's (1996) argument that the era of the government-funded public library conceived of as a social agency is coming to an end.

References

Bergen, C. 1984. *Knock Wood*. New York: Linden Press/Simon and Schuster.

Bijker, W. E. 1995. *Of Bicycles, Bakelites, and Bulbs: Toward a Theory of Sociotechnical Change*. Cambridge, MA: MIT Press.

Bijker, W. E., and J. Law. 1992. General Introduction. In W. E. Bijker and J. Law, eds., *Shaping Technology/Building Society: Studies in Technological Change* (Vols. 1–14). Cambridge, MA: MIT Press.

Bloomfield, L. 1933. *Language*. London: George Allen and Unwin.

Bolter, J. D. 1991. *Writing Space: The Computer, Hypertext, and the History of Writing*. Hillsdale, NJ: Erlbaum.

Brown, J. S., and P. Duguid. 1996. The Social Life of Documents. *First Monday*, 1(1). ⟨http://www.firstmonday.dk/issues/issue1/documents/⟩.

Buckland, M. 1992. *Redesigning Library Services*. Chicago: American Library Association.

Buckland, M. K. 1997. What Is a "Document"? *Journal of the American Society for Information Science*, 48(9), 804–809.

Clanchy, M. T. 1993. *From Memory to Written Record: England 1066–1307* (2nd ed.). Oxford: Blackwell.

Duguid, P. 1996. Material Matters: The Past and Futurology of the Book. In G. Nunberg, ed., *The Future of the Book* (pp. 63–102). Berkeley: University of California Press.

Garrett, J., and D. Waters. 1996. *Preserving Digital Information: Report of the Task Force on Archiving of Digital Information*. Washington, DC: Commission on Preservation and Access and The Research Libraries Group.

Goodman, N. 1976. *Languages of Art*. Indianapolis: Hackett.

Jespersen, O. 1969. *Essentials of English Grammar*. University: University of Alabama Press.

Latour, B. 1992. Where Are the Missing Masses? The Sociology of a Few Mundane Artifacts. In W. E. Bijker and J. Law, eds., *Shaping Technology/Building Society: Studies in Technological Change* (pp. 225–258). Cambridge, MA: MIT Press.

Levy, D. M. 1994. Fixed or Fluid? Document Stability and New Media. In *Proceedings of the European Conference on Hypertext Technology '94, Edinburgh, Scotland* (pp. 24–31). New York: ACM.

Levy, D. M. 1998. Heroic Measures: Reflections on the Possibility and Purpose of Digital Preservation. In *Digital Libraries '98: The Third ACM Conference on Digital Libraries, Pittsburgh, PA, June 23–26* (pp. 152–161). New York: ACM.

Marshall, C. C. 1997. Annotation: From Paper Books to the Digital Library. In *Digital Libraries '97: Proceedings of the Second ACM International Conference on Digital Libraries (Philadelphia, PA, July 23–26, 1997)* (pp. 131–140). New York: ACM Press.

McKenzie, D. F. 1999. *Bibliography and the Sociology of Texts*. Cambridge: Cambridge University Press.

Miksa, F. 1996. "The Cultural Legacy of the 'Modern Library' for the Future." *Journal of Education for Library and Information Science*, 37(2), 100–119.

Musmann, K. 1993. *Technological Innovations in Libraries, 1860–1960*. Westport, CT: Greenwood Press.

Otlet, P. 1990. *International Organization and Dissemination of Knowledge: Selected Essays*. Amsterdam: Elsevier.

Plato. 1973. *The Phaedrus*. Trans. W. Hamilton. New York: Penguin.

Sampson, G. 1985. *Writing Systems: A Linguistic Introduction*. Stanford: Stanford University Press.

Weinberger, D. 1996. What's a Document? *Wired*, 4(8), 112.

Winner, L. 1986. *The Whale and the Reactor: A Search for Limits in an Age of High Technology*. Chicago: Chicago University Press.

Wood, D. 1992. *The Power of Maps*. New York: Guilford Press.

3

Finding the Boundaries of the Library without Walls

Catherine C. Marshall

Introduction

In the first years of digital library efforts, there was much talk of the "library without walls." The introduction of electronic resources into the traditional library showed promise for bringing about seamless "anytime, anywhere" access. How has this seamless access played out as digital collections grow and as an increasing number of institutions and individuals have come to maintain and use them?

Contrary to this picture of seamlessness, the practical everyday reality of workplaces and public institutions like libraries is rife with *boundaries* that shape human interaction and any associated engagement with technology and documents. What do I mean by boundaries? Instead of relying on a hard-and-fast definition, I appeal to the reader's intuition: a boundary is something that tends to separate, to interpose; a boundary is a perceptible seam in the social fabric, the technological infrastructure, or a physical setting or may span all three.[1] Physical and social factors—such as work setting, organizational structure, and the introduction of noninteroperable technologies—produce some of the most noticeable boundaries.

For example, studies have demonstrated that people tend to work with those colleagues who are close by (Kraut and Egido 1988). Tightly coupled work performed over a distance (using technology like videoconferencing) may be reorganized so that it is performed by colocated team members (Olsen and Teasley 1996). Distance, then, imposes a perceptible boundary that is not necessarily bridged by videoconferencing technology. On the other hand, complex work situations might have boundaries because, for example, various technologies don't interoperate, or a physical setting has particular limitations. But boundaries can't be predicted by looking at these factors in isolation; detailed ethnographies of work make it apparent that human actions in such places can actually be well coordinated (Suchman 1998).

How does this notion of boundaries play out in "the library without walls"? In this chapter, I take a look at the less tangible but still crucial boundaries that are introduced, amplified, or even overcome by technology design and the mix of physical and digital document technologies that are common in many workplaces and today's heterogeneous libraries.

These technologically derived boundaries may not be accidental. Intentional boundaries arise through explicit decision. For example, consider the firewalls that are used to separate institutional information resources or other intellectual assets from the outside world or the authentication services that help protect copyrighted or for-pay materials.

Technological boundaries may also come about through more implicit limitations and contingencies. For example, the remote users of digital materials that are part of a larger, mixed physical-digital collection may be able to access the Online Public Access Catalog (OPAC) records for the entire collection but not the physical materials that correspond to some of the records. Oversimplification or simply an over-optimistic assessment of the digitization process tends to amplify this existing boundary, for it seems straightforward at the outset of digital library projects simply to scan materials and put them online. If such an assumption were true, the boundary between physical and digital materials would be far less conspicuous than it is.

Other boundaries are interpreted or constructed through design. For example, digital collections are often, in fact, distributed, and the perception of the collection as a single entity arises primarily through the interface that has been developed to access it. The simplest example of this kind of interpreted boundary is found on library Web pages that are themselves lists of uniform resource locators (URLs) that point to a set of related resources. What appears to the patron, at first glance, to be a seamless collection is actually a distributed set of resources that may be maintained by entirely separate organizations and institutions. An analogous kind of boundary occurs in the construction of individual entities. Where are the edges of the documents? In most cases, this seems either self-evident (of course, a bound book is logically an entity) or covered by existing cataloging rules and conventions (for graphical materials that are part of a series, for example). However, it is easy to find circumstances in which these document boundaries are more difficult to interpret, such as archival personal correspondence in which a book is sent with a letter.

Finally, some boundaries arise through interpretation but not through explicit decision. For example, uneven, inconsistently coded, or incomplete metadata may create such a boundary. Certain portions of a collection will never appear as

Table 3.1
Types and Examples of Boundaries within the Library without Walls

	Actual	Interpreted
Intentional	Firewall Authentication service	Collection boundaries Document boundaries
Not intentional	Digital/paper divide	Incomplete or uneven metadata

the result of a search that uses such a metadata element, in spite of an apparently effective *search strategy* (a strategy that retrieves other relevant documents). Much like digitization, the difficulty in creating good metadata, whether by human intervention or automated processing, is frequently underappreciated (Marshall 1998). Table 3.1 summarizes this continuum of boundary types and examples.

Such boundaries become particularly important when we talk about the design of digital library technologies because notions of access, media, community, and use are all painted with an exceptionally broad brush. In typical digital library design scenarios, the users may be remote, difficult to observe, and possibly unknown to the designers; the access platforms may include anything from obsolete personal computers to high-performance workstations; and the collection itself may include or grow over time to include media types of great diversity.

This chapter examines three very different digital library use and maintenance situations. Following some background about each situation, I explore the different types of boundaries that I have encountered in field work and technology design. The consequences and implications of these boundaries conclude the discussion.

A Space of Digital Library Use and Maintenance Situations

I use three sources of examples and experiences in this chapter. The three projects and field studies are by no means methodologically equivalent, but all are compelling for different reasons and have contributed to my understanding of digital library boundaries in fundamental ways. Two projects involve sophisticated field methods and a long-term relationship with the field sites, and the third and most recent stems from a need to put an innovative digital library technology into use.

I decided not to draw all the examples from a single in-depth ethnography in a library setting because the breadth of sources allows me to see things from a variety of use and maintenance perspectives. The three situations I describe represent heterogeneous library collections, locally maintained workgroup collections, and personal collections. Furthermore, all of the sites and communities that I discuss are undergoing transition. They are coping with analogous problems of digitization, complex mixes of old and new media and technologies, the need to create metadata, and the changes that new digital collections are bringing to a well-developed set of practices.

The first case arises from recent field work and systems codevelopment efforts at the Educational Technology Center, a part of the university library at a large institution. A field work and cooperative case-based prototyping effort at a district office of a large state government agency is the source of the second set of examples. Finally, I use a technology deployment effort that places a digital library reading appliance (an e-book) into users' hands as a third illustrative situation. I describe each effort, along with relevant aspects of its associated collection and documents.

An Image Collection That Supports Instruction
The ethnographic study I draw on here took place in a university library's Educational Technology Center, a service organization that supplies the faculty with visual resources for use in the classroom. The university library serves a teaching-oriented institution of about 8,000 students. The study, part of a larger participatory effort to develop tools for the collection's users, involved diverse participants from each site. Anthropologists, designers, and computer scientists from Xerox[2] collaborated with catalogers, archivists, software specialists, the library's directors, the media production staff, and faculty members over the course of the project. The image collection itself has its roots in what started as a circulating set of about 50,000 to 75,000 35 millimeter slides used by faculty members in the course of instruction. Because the university has strong programs in art, architecture, design, and other visually oriented disciplines, the slide collection is an important resource.

At the point we began field work at the university library, the collection and ancillary resources (such as a one-of-a-kind poster collection) were in the process of being digitized and cataloged as digital images. This transition was also viewed as an opportunity to revisit and cull the physical slide collection, since many of the slides had deteriorated with age, and the newer slides had been cataloged and accounted for using a different scheme than the one used for the older ones.

During this early period of digitization, the library and educational technology center's staff decided to store records of the new images in the online catalog, so the transition represented a full shift in both medium (physical to digital) and record keeping (local to centralized). The image collection itself, which was mostly Web-based, remained distributed over different file servers.

The physical slide collection was originally used as a faculty resource. Faculty members would request that slides be produced from existing materials (anything from advertising brochures to plates in art books) so that they could be projected in the classroom. The faculty members who use these slides generally have ready-to-hand ways of putting together presentations (for example, using light tables and light boxes) and had personal slide collections that could be interwoven with the library materials. Because of a growing emphasis on distance learning and the availability of high-quality digital image-projection facilities, faculty members were beginning to develop a practice around the use of digital images in place of 35 mm slides. These digital images would then be available for student use, too. Support for a broader circle of users, including users outside the immediate university community, was under discussion as the digital images were being prepared;[3] copyright restrictions on some of the materials made this a thorny issue indeed.

For the purpose of this discussion, I concentrate on the site's problems with digitizing the images, controversies surrounding the possibility of user-supplied or decentralized metadata, the desire to enforce copyright restrictions, and the need to interpret collection boundaries. All of these factors contributed to construction of boundaries. For brevity, I refer to the staff members at the Educational Technology Center and the library staff members (such as the cataloger and the archivist) as *the ETC*.

A Workgroup Document Repository

The second source of examples is derived from a long-term field work and co-operative prototyping engagement at a Xerox customer site, a district office of CALTRANS, the State of California's transportation agency.[4] The part of this engagement that I discuss revolves around setting up a self-sustaining repository of shared working documents for a bridge replacement project. This effort includes setting up a means to scan in documents (both as they arrive and from a collected backlog of documents that are pertinent to the project) and developing a means to access and maintain them. The documents are highly varied in genre, including correspondence, spreadsheets, maps, plans, internal memos, and penciled calculation. The documents additionally are highly varied in physical dimensions.

Although not a principal participant in this project, I accompanied the four main project members to the district office field site on numerous occasions, during which time a new personal computer and scanner were installed and on-site scanning began. I also sat in on the group's design meetings, which were centered on the workgroup document repository and its metadata. I have played a role in shepherding scanned documents through the processing required to put them in the repository. Thus I present a narrowed focus on the construction of the group's working document collection as a basis for illustrating points about boundaries.

The repository itself is Web-based (and transitional, as newer document management substrates become available) and includes searchable metadata. Documents are represented in the repository as images and as recognized text. Document metadata include a variety of tailored hand-assigned attribute-value pairs such as dates, document type (such as letter, memo, or map), and classification numbers. CALTRANS uses an institutional filing scheme, which has been modified by the engineers working at the field site to be more in line with local needs. A computed index has been created to support full-text searching as well.

Because the collection includes potentially sensitive documents and portions of the processing technology itself are proprietary, the digitization and repository building process is necessarily complex. My description, in fact, will not do it justice; much of its richness is eliminated for brevity's sake. The documents are first scanned using a Xerox workgroup scanner and commercial off-the-shelf software for processing the document image and saving it in the appropriate format. A form-based interface allows a collection maintainer to assign document metadata (possibly very sketchy) and upload the tagged image file format (TIFF) file that represents the document image. This package—the document image and its metadata—is composed into an e-mail message. As e-mail, the package crosses the firewall and, through a partially automated process, put through a series of steps to add it to the corpus. The new results are then pushed back over the firewall, where it is available to engineers at the field site using authentication-based access (again, through a Web interface).

For this discussion of boundaries, I focus on the efforts necessary to interpret document boundaries, scan in documents, assign metadata to describe them, and send them across the firewall and process them. In this discussion, I refer to the effort as the *CALTRANS project*.

XLibris: A Personal Digital Library Reading Appliance
Early on, the need for a specialized reading appliance that allows library patrons to work with digital library materials was recognized. Indeed, the Bibliothèque de

France included a project to develop prototype reading stations, Computer-Assisted Reading Environments, for the library. These reading machines would "allow the reader to work on a corpus of digitized documents culled from the library's immense reserves. Among the diverse possibilities presented by this tool, the functions of comparison, annotation, and indexing are doubtlessly the most attractive" (Bazin 1996, p. 157). The third source of examples comes from our experiences putting an innovative digital library reading appliance (an e-book) into use.

The prototype e-book software, XLibris, runs on a family of portrait mode displays and devices (Schilit, Golovchinsky, and Price 1998). Some are tethered digitizing pads (they serve as reading displays on an ordinary PC). Readers may use them as they would use a document alongside their normal computer display. Others are pen computers, making the reading device self-contained and significantly mobile.

As a specialized reading device, XLibris promotes unself-conscious engagement with documents. Readers can mark directly on the pages in different colors of ink and highlighter with a stylus as if they were annotating paper, an important requirement revealed by studies of practice (Marshall 1998). They can also move the device to change the reading angle and position as if they were working with a paper document, another important requirement (Gujar, Harrison, and Fishkin 1998). In fact, the prototype developers refer to XLibris as being based on a paper document metaphor (Schilit, Price, and Golovchinsky 1998).

The project has thus far had a dual focus on usability, including critical factors such as legibility, form factor, and basic navigation functions like page turning and utility. These facilities make the e-book useful within the context of a particular activity.

How do documents come to be on an e-book? In theory, they are seamlessly downloaded from the digital libraries of the reader's choice. In practice, however, not all documents are digital, nor are all digital documents in a convenient format. Nor are the "originals" (the electronic or paper documents the readers have in hand) uniformly legible. Some are taken from online scanned page image files whose quality is beyond our (or the readers') control. Other significant documents arrive via fax (there are many fax-based information services). Still others are poor-quality reproductions in which the figures and pictures are dark and unintelligible.

In this case, I am interested in the transitions—the boundary crossings—that a document must make to reside on the device and not those used in other kinds of activities. Besides the physical/digital divide, document and collection boundaries are subject to interpretation. Thumbnails (the reduced document representation that the prototype currently uses to present multiple documents or multiple pages of a single document) constitute the metadata of interest for this discussion. In this effort to put a personal reading device into use, issues associated with real intentional boundaries have not yet arisen, although they doubtless will as more copyrighted material comes into play and as we connect the device to more remote information resources.

An Examination of Boundaries

Now that the stage is set, I can take a closer look at each type of boundary and how it was manifested in the three efforts. How did these seams come about? Certainly, they have made our lives as technology developers (or collection maintainers and users) more difficult. Moreover, they are often paradoxical, arising though competing, equally valid, interests; in these circumstances, it is instructive to look at the side-effects and consequences attendant to the boundary.

Intentional, Actual Boundaries: Firewalls, Authentication, and Domain Detection

As the use of the Internet grows, both public and private institutions feel an increasing need to protect their intellectual assets and to more carefully observe copyright restrictions. Many intranets now sit behind firewalls, intermediary computers that secure network data from Internet traffic. Of course, firewalls are semipermeable boundaries. Depending on the implemented services, they still allow computers working behind them to access data outside, and they allow limited access from outside to inside. This restriction forms the basis for one type of problematic (although necessary) boundary that arises in two of our example situations.

The cooperative technology development efforts described in each of the first two cases required a significant negotiation of boundaries. The first case developed an end-user tool to work with the image collection. The second designed a workgroup tool to retrieve documents and maintain the shared repository. Every organization maintained either a firewall or a scheme of authentication- (or domain-) based access. Boundary crossing, then, can require both getting in (gaining access to the remote site) and getting out (pushing processed documents or protocols beyond one's own protective infrastructure).

Getting In   Paradoxically, the real intentional boundaries in each of these cases work to narrow the use of the collection. They implement hard-and-fast rules about who's an insider and who's an outsider, where in practical circumstances there are many exceptions.

In the CALTRANS case, the public might benefit from access to parts of the collection. As it stands, they may even be able to order these documents through other channels. An environmental impact statement or minutes from public meetings, for example, may be of great interest to the people living near the construction project.[5] The nearby residents are, in fact, insiders for the purpose of a portion of the collection. On the other hand, some working documents are not public and are indeed in the province of the project's engineers. For these documents, the local residents remain outsiders. For a more extensive discussion of a collection's multiple constituencies, see Lisa R. Schiff, Nancy A. Van House, and Mark H. Butler's (1997) description of the social aspect of the University of California at Berkeley's Digital Library Initiative project.

In the case of the Educational Technology Center's image collection, scholars external to the university may benefit from access to unique parts of the collection for which the university holds copyrights. In fact, the library has gone to some effort to make those portions of the collections public. However, other portions of the collection are for members of the local university community. They have been made available as educational resources and, because of copyright restrictions, cannot be made generally accessible over the Internet. The ETC and the library have explored different ways of constructing these boundaries so that domain-based restrictions do not put restricted collection elements beyond the reach of a legitimate distance learning constituency. The intention is that materials that should be accessible to a broader scholarly community remain that way.

Getting Out   Cooperative prototyping can easily involve more than getting into the site that holds and uses the collection. It may also exhibit the reciprocal problem of getting out of a site where the development or processing is taking place. This may be as simple as the developers' need to do the bulk of the work on their home ground, where they have the tools and the time to write and debug prototype software. But often it's more complicated. The technological innovations that come about through the prototyping effort may constitute intellectual assets for the development organization; hence, it must be protected just as the collection is protected. The first two projects I described both involve temporary holes in the firewall. The firewall holes made possible the development of a Z39.50-based client or permitted "pushing files over the firewall"—that is, returning documents processed by protected software to become available once again back at the field site.

Is this a real issue for our hypothetical library without walls? Or is it a red herring, an awkwardness brought about by the fact that we're all new at this? There is an initial desire to dismiss these intentional boundaries: we've put them up; surely we can take them down when the need arises. This is not, strictly speaking, true. Many of these boundaries represent the compromise wrought by competing interests within the same organization.

Unintentional Actual Boundaries: The Physical/Digital Divide and Media-Based Boundaries

Digital libraries are never, in fact, wholly digital; digital collections and collections in other nondigital media coexist and are brought together through use (Levy and Marshall 1995). But certainly a large part of the movement to digital libraries involves making physical things digital and then rerendering them in a physical form as need be. The digitizing and printing functions appear at the outset to be very straightforward and well worked out. After all, scanners that turn physical documents into digital documents now have automatic page feeders, and inexpensive printers produce paper documents from digital ones quickly at high resolution. Indeed, it would seem like the transitions between digital and physical document forms have never been easier. The physical/digital divide must be becoming more and more imperceptible. But observing our three use situations, the boundary is still very much in evidence.

Digitizing   In the early days of the project at the Educational Technology Center, we made schedules based on an optimistic assessment of how long it would take to digitize existing resources. But the digitization activity was less straightforward than anyone originally imagined. One ETC project (an effort to create high-quality digital images from a poster collection) required many transitions, from physical artifact to film, photo CD, local storage, and finally cleaned-up published image. Time-consuming problems always seem to arise, and hard decisions must be made in the translations between forms. An ETC staff member talked through a portion of one of these transitions this way: "the images are shifted and twisted, and the color is really bad a lot of times. So you have to go in there and match them back to the slide, which you hope is close to the original."

Digitization was also an important part of the CALTRANS project. In this case, digitization seemed plausible through a more straightforward route—"ordinary" scanning. To get the practice started, Xerox project members digitized the materials themselves. In fact, sometimes the CALTRANS workers at the field site were busy with other matters while we scanned materials. It is from this direct involvement with the scanning process that I began to see some of the difficulties inherent in the task. Each document proved an exception in some way.

For example, if we were scanning in a skewed copy of an original, did we want to deskew it? The skew made the reproduction look visually distinct from the original and from the other documents in the collection. The user interface to the document repository presented search results as thumbnails. It would be easy to distinguish this particular document by its skew.[6] Some documents included (or consisted of) oversized pages. How did we want to handle those: Reduce them and then scan them? Scan them in parts? Omit the margins that were seemingly not important? The overhead of making these decisions on so many "exceptional" documents increased an apparently straightforward digitizing task to formidable proportions.

It is interesting to note that eventually a CALTRANS intern was charged with the scanning task. This partitioning of the work is not unexpected. The engineers attend to other more pressing matters (for example, bridge design). Will each document get the time and attention it did when members of Xerox group are scanning? Probably not. The intern may well feel more pressure to work quickly and get through the overwhelming numbers of documents designated as scanning fodder.

The XLibris digital library document appliance also turned up a set of paper/ digital boundary problems. In several of our early efforts to put the device into use, the materials that users wanted to read arrived on paper. In one case in particular, at first glance, the solution looked unproblematic. The documents were technical papers, printed out at different sites on a regular laser printer, and appeared to make perfect candidates for an automatic feeder. Or were they? This is an account an XLibris user gave of the scanning experience:

They [the documents] started out a lot in A4 format. And so we tried to fix that up in the scanning process, and whatever we did was pretty much ineffectual. What we had originally tried to do is use image shift to automatically center it. And that didn't work. So then Gene put stuff on the platen, in an attempt to push it up but in order to not confuse it about what the size of the paper was. He felt pretty constrained that he had to place it so that it would end up at the 8 by 11 corner. And that meant that he couldn't shift up as much as he wanted to. Probably over half of the time that I ended up with stuff down at the bottom, maybe even three-quarters of the time, I could still make out what it was.

To increase the stakes introduced by the vagaries of digitization, it is important to note that each of the three use situations (an institutional collection, a workgroup collection, and a personal collection) suffered from the same bootstrapping problem. The repository is far less useful when it's incomplete for whatever task the user has in mind.

Getting Documents Back Out   Much digital library material ends up back in print, if just for many of the factors that Bill N. Schilit, Morgan N. Price, and Gene Golovchinsky (1998) cite. Ignoring the environmental consequences of this transition back to paper (or to paper, in the case of documents created electronically), haven't high-speed printers or inexpensive, high-resolution local color printers, for that matter, solved this problem? It is instructive to look again at circumstances of two projects.

At the Educational Technology Center, recall that faculty members use the image collection in classroom teaching. Many of them rely on 35 mm slides in personal collections or on 35 mm slides they can request from the ETC. For the purpose of assembling and projecting a lecture, it is inconvenient to mix physical and digital images. So if the digital collection is to be useful, a faculty member needs to find the entirety of materials for a lecture in that form. Nor is printing to a 35 mm slide as easy as printing to paper.

In the case of XLibris, the same sort of question arises: What is the role of the read and marked-up document in the larger scope of the reader's activities? Will the document be printed out on paper again? Printing is not as difficult as producing 35 mm slides, but the technology's users still have questions about whether and when to print and exactly what makes it back across the digital/physical divide. For example, they wonder if they should print the document to make it more portable or to archive it.[7] They wonder what will be rendered on the printed page. Will a page on a piece of paper display exactly the same as on the device? One Xlibris user said: "I really want this to be easily shiftable to home or wherever. I don't want to have to go to the device, and I didn't know really what printing capabilities I had at first."

If we think of library, workgroup, and personal collections as continuing to be largely heterogeneous, and look at the practice surrounding digital/physical or physical/digital transitions, there is no reason to believe they won't be. The unintentional but very real media-related boundaries will continue to be a factor in designing digital library technologies.

Intentional Interpreted Boundaries: Collection and Document Boundaries
Interpreted boundaries usually arise through decision. Where are the edges of the collection? What will we maintain, curate, and digitize? In what order should we perform the expensive and time-consuming task of digitization? Who is the collection's constituency? Where are the edges of individual documents? Inherent in each decision is the identification of a new boundary.

Collection Boundaries   The most interesting example of the interpretation of collection boundaries that I encountered comes from the Educational Technology Center's image collection. In discussions at the site, we never arrived at a completely consensual description of what was in the image collection and what was not. This is a common state of affairs when distributed digital resources come into the picture. Does a link to a departmental Web server constitute inclusion of that set of images in the collection? It's hard to say. The interpretation of the collection's boundaries depended crucially on perspective.

Because the collection began as a media-based 35 mm slide collection, it is tempting to carry over the media type as the distinguishing feature of the resource. The slide collection has been transformed, through the magic of digitization, into the digital image collection. But then are the 35 mm slides still available? What about the case of digital videos?

Another interpretation is rightfully derived from the collection's use: these images are used in teaching. Much of the original collection has been built up from faculty requests. But what about the unique archival images that are being added to the collection? They may be of more value to scholars than in classroom instruction. Do they now fall outside the boundaries of the collection?

Storage and maintenance forms yet a third basis for discrimination. The collection's images are stored and are maintained by the library and the Educational Technology Center. Are the images and the 35 mm slides kept on a departmental server or in a faculty office part of the collection? Since in many cases it's likely that the Educational Technology Center has produced all of them. Production is part of its mission. They could just as well be stored in centralized slide cabinets, notebooks, and servers.

In a very conventional way, the metadata that is stored in the library's OPAC may (and by some interpretations, should) circumscribe what's in and what's out of the collection. A collection element should have a metadata component in the catalog. Ideally, the library, the Educational Technology Center, and, equally important, the faculty and students would know exactly what is in the collection. It's easy to observe that this interpretation of collection boundaries (a definition that uses centralized metadata to describe a distributed set of image resources) may not fit the way that the digital portion of the collection is growing.

Finally, image genre comes into play as a boundary definer. The original collection contained slides of works of art, architecture, and design but not of biological specimens or microbiological images. Some library staff members explicitly included the possible expansion to these other genres as within the scope of the collection. Others conceived of the collection boundaries as conforming to the original image genres.

Document Boundaries   Talk about digital libraries often makes it seem as if document boundaries no longer matter. Geoff Nunberg (1993, p. 22) makes this observation: "Reading what people have had to say about the future of knowledge in an electronic world, you sometimes have the picture of somebody holding all the books in the library by their spines and shaking them until the sentences fall out loose in space."

Yet document boundaries have mattered in each of the three cases I use here. There has always been the opportunity to ask, usually in the face of a particularly problematic example, "Is this a document?" The question usually arises when handling something that could be construed as a document part on its own (a digital video clip, for example) or a composite document (papers that are intentionally paper-clipped together). The answer becomes even more elusive in the case of the workgroup collection. Is an envelope that contained the letter part of the composite document? What about the routing slip? Neither is meaningful without the rest of the composite, but it's not even clear when and if they're meaningful as they stand. In the case of the routing slip, a physical fastener—a staple—has introduced an apparent document boundary.

Why not, in this case, just "define" the smallest possible component as the document boundary and allow for the possibility of composite documents? Small elements of the working document collection would then be available to the CAL-TRANS group, and the person responsible for digitization might not need to make so many judgment calls about document boundaries. We must be careful, however, when we assume that finer-grained access and the simplification of document boundaries are wholly desirable outcomes. Document disaggregation, as Ann Peterson Bishop (1999) explores in some depth, has an effect on a user's ability to interpret the content. A document is more than the sum of its parts. Suggesting a component-based solution to the interpretation of document boundaries seems, on the whole, unwise.

In the third case, XLibris document and collection boundaries are based on personal use. This constraint, personal use, should radically simplify the interpretation of these fluid boundaries. Does it? Not entirely. Our experiences with long documents—for example, books or long reports—suggest that they might be better represented and manipulated as individual chapters. As in the case of the documents at CALTRANS, document boundaries may be circumstantial—the technical paper that's broken into two electronic files, the paper's text, and the figures. Collection boundaries are even dicier. Do all the documents a person uses at a given time constitute a workspace, given a document appliance metaphor? Or does the notion of workspace suggest a grouping according to activity?

Answering questions about how to construct intentional interpreted boundaries is especially difficult. Often the construction of these boundaries imposes an extra burden on someone in the picture. Decisions about what's in and what's out of a collection and what's part of and what's not part of a document often begin in the name of policy but end up as a catalog of exceptions.

Unintentional Interpreted Boundaries: Metadata-Based Boundaries

Metadata schemes, in theory, are bridging techniques. By describing documents according to a uniform framework, they bring coherence and accessibility to a collection. They are, then, surprising places to find boundaries. Interest in universal metadata schemes, long in use in libraries and other stores of institutional documents, has grown considerably with the availability of Internet resources. Human-created and automatically generated metadata, then, have every good intention of promoting intelligible resource description and uniform seamless access across many different kinds of collections. Why then do I use them as an example of the opposite?

First, metadata are idealized solutions to a very thorny problem. Even the most thorough, dedicated, and well-trained cataloger, using the most unambiguous of schemes (one tailored for the sort of resource the cataloger is describing, much as the Simons-Tansey code is used for slide collections) and the best of authority lists must engage in some amount of interpretation. She must speculate about the needs of the collection's constituency and the specific application of the coding scheme.

Uneven, Incomplete, and Legacy Metadata   At the Educational Technology Center, early on, the Simons-Tansey code had been applied, in fact, to organize the slide collection. The staff still had the book describing the code, with an earlier slide librarian's notes about decisions that she made as she cataloged difficult slides. In spite of these notes and the apparent appropriateness of the Simons-Tansey scheme when it was originally selected, it had fallen out of use.[8]

There are two things to notice about this brief account. First is the amount of interpretation that is done to bring a general (if complete) scheme into a local setting. Second is that we are left with a legacy metadata scheme that's still descriptive but is no longer being actively used in the organization of the collection. I'll come back to the issue of legacy metadata. First I illustrate interpretation in action as one of the catalogers working with the ETC codes a digital image that represents a William Wegman photograph:

I have a picture of William Wegman's dogs. Say I want a picture of a dog, and I don't know what we have. You do a keyword search with *dog*. But I just want to get slides. So I can type "dog and [slide identifier]," which would be a unique qualifier. If I had two slides with a dog in it, I would ask myself, "Should I search on the artist's name also?"

It is readily apparent from this scenario that the cataloger is trying to anticipate how students will be looking for this image as they search the collection. She is, in fact, foreseeing a far different use, well beyond the impetus for assembling the collection in the first place. In fact, filling in the two other work settings also tends to be an exercise in putting oneself in a future circumstance, one in which the document will actually be sought and be used.

If we consider the professionally cataloged ETC image collection, it's easy to see that there is a natural (and desirable) unevenness to metadata. Some of the images from the original physical 35 mm slide collection have a metadata element that the ETC plans to preserve in a new OPAC-based coding scheme (the legacy Simons-Tansey classification). The new cataloging scheme provides for descriptive keywords (*dog* and *Wegman*, for example). But these are interpreted according to the cataloger's perception of future searches, a perception that is bound to differ among different catalogers and is bound to change over time as the collection grows.[9] From observation, it's also easy to see that most metadata schemes are partial. Hard-to-code or inappropriate elements are at least deferred, if not omitted.

Why does this natural unevenness introduce boundaries? Partial description will interact with retrieval techniques to reduce the recall rate. There is usually no indication that the search covers only part of the collection. In effect, an invisible boundary has been introduced between the metadata haves and have-nots. The user no longer knows the shape of the collection she is searching.

The second facet of this problem arises from the less carefully controlled realm of user-created metadata. In theory, user-created metadata will align the description of collection elements more closely with the collection's various uses. Likewise, the drawback of an uncontrolled vocabulary (or inexact use of a controlled vocabulary) is introduced. At the ETC, the cataloger expressed exactly these concerns:

> Whoever does it [assigns keywords], you have to be familiar with whatever thesaurus or source that we decide to use. [But you] still need to [have] a little know-how or learn how to make the best use of it. If you think of damage control,… even if you do not know what you're doing, [and] you're using a term that someone else may say, "no way!"… at least you'll have a *variety* of wrong terms. [laughter]

Designers of the metadata facilities for the CALTRANS working document collection voiced similar concerns: Would there be a proliferation of keywords? Would the existing terms be applied in a uniform way? Would the person coding the documents be sufficiently familiar with the institution's filing scheme?

In practice, these concerns may well be overwhelmed by the simple shortfall of metadata. Like digitization, metadata creation is often more time-consuming than we expect. The payoff doesn't tend to come until later, when someone needs to retrieve a document. And the benefit may not even accrue to the person who does the work of assigning the terms or properties. It is easy to become overly optimistic about the quantity of metadata people will assign.

Consequences for Technology Design

Is there a reason to be mindful of these boundaries, for surely they will always arise, whether by circumstance or by design? Ultimately, it seems that understanding the boundaries that we construct, and the ones that are already firmly in place, will contribute to our ability to develop useful digital library technologies.

Usability tends to be a real focus of attention in human-computer interaction. And this is rightly so. As Don Norman (1988) has pointed out, it is all too easy to ignore basic design principles and render an artifact unusable, whether a simple door or a sophisticated telephone system. However, it takes a deep understanding of work to make an artifact useful; an elegant design is no guarantee of utility.

Richard Harper (1997, p. 66) points out in discussing his own use of ethnographic methods to study the IMF that when we design information technology, we come to the table with an agenda. I do not point out these boundaries as a sly way of asserting "This digitizing stuff will never work" or "See what these people are going through" but rather as a way of suggesting a concrete set of places to explore as potential sites for new kinds of sociotechnical intervention.

Surely some of these technologies and practices will continue to interject boundaries into libraries and workplaces. Firewalls, for example, have been put up for a well-motivated reason, not to thwart our own implementation efforts. Ascertaining the edges of documents, collections, workspaces, resources, and metadata will continue to be a challenge. Catalogers and archivists have long grappled with such questions. The digital/physical boundary may become increasingly permeable as technologies to cross it grow in sophistication and reach. Three-dimensional scanners are available today, for example. But of course, boundaries will continue to exist. Documents and other entities in computers will continue to be representations of physical artifacts, and physical renderings will continue to be representations of what's inside the computer.

Why then have I focused on boundaries if they are inevitable? The stories I have told suggest at least three different morals about crossing—and where possible blurring—boundaries.

First, we should plan our encounters with real intentional boundaries. As developers, we should share strategies for ways around firewalls. For example, the CALTRANS project took a particularly ingenious e-mail route to avoid the firewall in its initial document image processing work-around. Furthermore, we should build our use cases and scenarios with awareness of the effects of boundaries like authentication. For a good example of this level of awareness, see Bishop's (1998) account of the effects of login in the use of DeLIver.

Second, we should engender realistic expectations about the complexity of unintentional boundaries, in particular (from my own observations) those crossed through digitization and metadata creation. This observation seems to have been borne out in many earlier digital library projects, especially if one tracks discussion participants' questions and accounts on listservs like DIGLIB and IMAGELIB.

Finally, we should design creatively to get around the interpretive boundaries like document, collection, and metadata boundaries. Cooperative prototypes should enfranchise as many constituencies as possible to articulate what the appropriate boundaries should be. This is actually already true of the cases I describe. Existing digital library facilities should then take advantage of opportunities for exposing the invisible boundaries, especially metadata boundaries, and making the effects of these boundaries clear.

Notes

1. I use the word *boundary* in a slightly different way than Susan Leigh Star and James Griesemer (1989) do when they talk about boundary objects. While boundary objects are "objects which are both plastic enough to adapt to local needs and the constraints of the several parties employing them, yet robust enough to maintain a common identity across sites" (Star and Griesemer 1989, p. 393), my emphasis is on a looser sense of sociotechnical boundaries and specifically on the technologies that (intentionally or not) impose them.

2. My Xerox collaborators in this effort were Susan Anderson, Françoise Brun-Cottan, Brinda Dalal, Dave Lindahl, Nelson Mejias-Dias, Andrea Mosher, Susan Stewart, and Pat Wall. I should clarify at the outset that I am one of the computer scientists.

3. As it turned out, the images were both prepared through local efforts and acquired though membership in a prominent consortium. The latter route, sharing common digitized resources through a consortium, seems very promising in the light of the sum of my experiences with digitization. See the discussion of the unintentional—but very real—boundaries we encountered crossing the physical/digital divide.

4. The principle body of research and cooperative prototyping has been performed by members of the Work Practice and Technology group at Xerox PARC: Lucy Suchman, Jeanette Blomberg, Randy Trigg, and David Levy.

5. This example of insider and outsider for the CALTRANS collection is hypothetical and as such presents a much less complicated situation than the one that actually exists. The hypothetical is based on my sense of what is in the repository and who might be interested in it, based on a conversation Julian Orr and I had with a CALTRANS archivist.

6. The importance of visual recognition of desired papers in a working document collection is discussed in the context of a related case-based prototyping project carried out by my colleagues Jeanette Blomberg, Lucy Suchman, and Randy Trigg (1996).

7. Oddly enough, it seems relatively common for individuals to archive digital documents on paper. See D. M. Levy (1998) for an insightful discussion of why archiving digital materials is so hard.

8. This example deserves a much longer story, but for the sake of brevity, I am omitting the details of the change in organization schemes. A longer account is provided in Marshall (1998).

9. We can also see that automatic indexing techniques won't precisely address this unevenness either, since image description goes well beyond what is actually depicted by the image.

References

Bazin, P. 1996. Toward Metareading. In G. Nunberg, ed., *The Future of the Book* (pp. 153–168). Berkeley: University of California Press.

Bishop, A. P. 1998. Logins and Bailouts: Measuring Access, Use, and Success in Digital Libraries. *Journal of Electronic Publishing*, 4(2). ⟨http://www.press.umich.edu/jep/04-02/bishop.html⟩.

Bishop, A. P. 1999. Document Structure and Digital Libraries: How Researchers Mobilize Information in Journal Articles. *Information Processing and Management*, 35(3), 255–279.

Blomberg, J., L. Suchman, and R. Trigg. 1996. Reflections on a Work-Oriented Design Project. *Human-Computer Interaction*, 11(3), 237–265.

Gujar, A., B. L. Harrison, and K. P. Fishkin. 1998. A Comparative Empirical Evaluation of Display Technologies for Reading. In *Proceedings of Human Factors and Ergonomics Society Forty-second Annual Meeting (Chicago, IL, October 5–9, 1998)* (vol. 1, pp. 527–531). Santa Monica, CA: Human Factors and Ergonomics Society.

Harper, R. 1997. *Inside the IMF: An Ethnography of Documents, Technology, and Organizational Action.* San Diego: Academic Press.

Kraut, R., and C. Egido. 1988. Patterns of Contact and Communication in Scientific Research Collaboration. In *Proceedings of the Conference on Computer-Supported Cooperative Work* (pp. 1–12). New York: ACM Press.

Levy, D. M. 1998. Heroic Measures: Reflections on the Possibility and Purpose of Digital Preservation. In *Digital Libraries '98: Proceedings of the Third ACM Conference on Digital Libraries (Pittsburgh, PA, June 23–26, 1998)* (pp. 152–161). New York: ACM Press.

Levy, D. M., and C. C. Marshall. 1995. Going Digital: A Look at Assumptions Underlying Digital Libraries. *Communications of the ACM*, 38(4), 77–84.

Marshall, C. C. 1998. Making Metadata: A Study of Metadata Creation for a Mixed Physical-Digital Collection. In *Digital Libraries '98: Proceedings of the Third ACM Conference on Digital Libraries (Pittsburgh, PA, June 23–26, 1998)* (pp. 162–171). New York: ACM Press.

Norman, D. 1990. *The Design of Everyday Things.* New York: Doubleday.

Nunberg, G. 1993. The Places of Books in the Age of Electronic Reproduction. *Representations*, 42, 13–37.

Olson, J. S., and S. Teasley 1996. Groupware in the Wild: Lessons Learned from a Year of Virtual Collocation. In *Proceedings of the Conference on Computer Supported Cooperative Work* (pp. 419–427). New York: ACM Press.

Schiff, L. R., N. A. Van House, and M. H. Butler. 1997. Understanding Complex Information Environments: A Social Analysis of Watershed Planning. In *Digital Libraries '97: Proceedings of the Second ACM International Digital Libraries (Philadelphia, PA, July 23–26, 1997)* (pp. 161–168). New York: ACM Press.

Schilit, B. N., G. Golovchinsky, and M. N. Price. 1998. Beyond Paper: Supporting Active Reading with Free Form Digital Ink Annotations. In *Human Factors in Computing: Conference Proceedings CHI 98 (Los Angeles, CA, April 18–23, 1998)* (pp. 249–256). New York: ACM Press.

Schilit, B. N., N. M. Price, and G. Golovchinsky. 1998. Digital Library Information Appliances. In *Digital Libraries '98: Proceedings of the Third ACM Conference on Digital Libraries (Pittsburgh, PA, June 23–26, 1998)* (pp. 217–226). New York: ACM Press.

Simons, W., and L. Tansey. 1970. *A Slide Classification System for the Organization and Automatic Indexing of Interdisciplinary Collections of Slides and Pictures.* Santa Cruz: University of California.

Star, S. L., and J. R. Griesemer. 1989. Institutional Ecology, "Translations" and Boundary Objects: Amateurs and Professionals in Berkeley's Museum of Vertebrate Zoology, 1907–39. *Social Studies of Science*, 19(3), 387–420.

Suchman, L. 1998. Centers of Coordination: A Case and Some Themes. In L. B. Resnick, R. Saljo, C. Pontecorvo, and B. Burge, eds., *Discourse, Tools and Reasoning: Situated Cognition and Technologically Supported Environments.* Heidelberg, Germany: Springer-Verlag.

4

An Ecological Perspective on Digital Libraries

Vicki L. O'Day and Bonnie A. Nardi

The Internet and other online venues have transformed the possibilities of what a library can be. In the process, the boundaries among sites, tools, collections, and services for information gathering have become blurred. How should both users and designers of these hybrids evaluate, invent, and integrate new information technologies and practices? When should new services replace or augment older ones, and how can this be accomplished in a way that adds value to the whole information environment? To approach these questions, we need to look carefully at more than the technologies themselves. Our view must encompass the contexts in which they are used.

There is always a gap in understanding between technology development and use. Technology developers tend to focus on the particular tools they are designing, and typically, they do not know in advance the myriad ways their tools will be used. Users, on the other hand, know their own settings but are unlikely to understand the capabilities of new tools as thoroughly as designers do. To bridge this gap, a variety of user-centered design methodologies have been successfully employed. These include design scenarios (Carroll 1995), user studies (Winograd, Bennett, Young, Gordon, and Hartfield 1995), participatory design workshops and long-term collaborations (Schuler and Namioka 1993), contextual inquiry (Beyer and Holtzblatt 1997), and usability testing (Nielsen 1994). Each practice allows technology developers and users to communicate with each other and inform the process of design. Each method gives developers insight about situations of use and also can give users insight about, and thus influence with, new technologies.

In this discussion, we propose a way of thinking about technology use that complements these design methodologies. We consider settings as *information ecologies*, which we define as systems of people, technologies, practices, and values (Nardi and O'Day 1999). The ecology metaphor provides one particular way of looking at the usage of digital libraries (or other new technologies) that allows, and in fact prompts, certain kinds of important questions and discussions. This is not new methodology but rather a particular way of developing a certain consciousness about technology, its meanings, and its impacts.

To see what is different about ecological thinking, it is helpful to examine some of the other metaphors people commonly use for technology. We use the word *metaphor* very loosely to refer to choices of language and images. Each metaphor we implicitly or explicitly adopt guides us to think along certain pathways and steers us away from others. In the next section, we discuss some common metaphors and see where they lead. We follow with an outline of our own ecological metaphor. We describe how a physical library can be understood as an information ecology. To illustrate ecological thinking, we raise a variety of strategic questions about the technologies and practices associated with digital libraries. These questions might be framed by users or designers. In either case, finding their answers should be a creative and worthwhile process that generates new possibilities for digital libraries of the future. We close the chapter with reflections on why ecological thinking is useful for users and developers of digital library tools and services.

Metaphors for Technology

It is reasonable to assume that language shapes our thinking in some way, even if we do not know exactly how it works. The effort in the past few decades to adopt gender-inclusive terms is a reminder of the perceived power of language to influence behavior. Similarly, we believe that the language we use for technologies influences what we do with them and the kinds of conversations we have about them. All of the current metaphors for technology have something to recommend them. Each captures something important about technology's role in our lives. But each metaphor also leaves something out, and we would like to examine familiar metaphors to understand what actions and ideas they capture and what they omit.

Tool Metaphor
First, let us look at the notion of technology as a tool. This is one of the most common ways of understanding what technology is about. We don't mean to suggest that technologies are *not* tools but rather that this language is a way of casting technology into a particular role. A tool is used to accomplish a task, usually

involving some kind of work. Talking about a technology as a tool suggests that it is an object and that it is under the control of its user. Tools have appropriate places, and they are matched to tasks and situations. We have tools for cooking, painting, car repair, surgery, housework, writing, and most other human occupations. We are always inventing new tools, which both emerge from and enable new activities.

Some tools are best for experts, and some are fine for novices. Tools may require training, or they may be easy to use the first time. When tools are well designed, people are able to step through the necessary actions, whether the task is answering a telephone or adding a column to a spreadsheet. Tool language leads us to think about how people, tasks, and technologies fit together.

Because tools are encountered most often in the context of work, key concerns that arise when we evaluate technologies as tools are productivity, utility, usability, skill, and learning. The questions we ask about technological tools include "What does the tool do? Will it make me more efficient? and Can it be used easily?"

While these questions are important, other good questions typically do not come up. In general, tool language leads people to focus on the interactions between individuals and their tools. Some tools (such as telephones) are intended for use by more than one person. Tool-centered thinking tends to focus rather narrowly on the actions of people while they are engaged with technologies rather than on broader social, organizational, or political contexts underlying their use. When we talk about gadgets (tools that are especially small, clever, or fun), we are even less likely to think about social contexts than when we consider more utilitarian tools.

Text Metaphor

An entirely different metaphor is one of technology as a kind of text with meanings that can be read by users. This is a rather uncommon way of looking at technology, encountered more in social science literature than in everyday talk (Latour 1995). But it is an interesting perspective, and it offers different handles on the problem of understanding and evaluating technology than does the tool metaphor.

For example, consider a signal light at an intersection (borrowing from Latour 1995). From a functional perspective, the signal light flashes different colors to show drivers when to stop and go. But in addition, the signal can be read as a physical representation of part of the traffic code.

If we choose to think about technologies as texts, we are led to think about what they communicate and prescribe to their users. The key concerns here are about intention and meaning rather than toollike functionality. We might ask questions such as "What is the message of this technology? Who is producing this message? and What does the technology implicitly tell me to do?" These are particularly interesting questions to ask of new Internet services and tools if we want to understand how consumers and producers are expected to benefit from using them.

The text metaphor adds a different dimension to thinking about technology, and we believe that the lines of inquiry it suggests are useful. However, it puts users into a fairly passive role. As users of technology, we would like to do more than read its meanings, though such reflection is a good thing. Users also should be making active choices about which technologies they want to use and how they want to use them. Thinking about tools has some leverage for this kind of evaluation process, but the criteria suggested by tool-centered thinking mostly revolves around productivity, as we have discussed. What about other criteria, such as how well a technology reflects and supports values? Neither the tool nor the text metaphor steers us to think about technology in this way.

Assistant Metaphor

A third metaphor often referenced (especially in advertisements) is the idea of technology as assistant or agent. In this perspective, the technology does not exactly extend human capabilities as most tools do (think about eyeglasses, hammers, potholders, and calculators). Instead, the technology acts on our behalf while we are busy with more important activities. It takes over the drudgery of routine tasks.

An obvious example is an information search agent, an increasingly common feature of online information services. In general, a search agent is activated with search criteria (such as particular keywords of interest) and information sources (such as the daily newspaper) and alerts the user whenever new information that matches the criteria becomes available. This metaphor develops parallels between technology and human assistants, such as librarians. It emphasizes certain key ideas, including expertise and personalized service. If we wish to evaluate a technology that evokes this metaphor, we are led to ask questions such as "Will it do what I want? Is it reliable? and What are the lines of accountability when it fails?"

The language of technological assistants is more expansive than tool language in some ways because it extends beyond functionality to something like the overall effects of using a tool. However, the emphasis on personalized service keeps us focused on the individual user to the exclusion of others who might be located in the same physical or online environment.

System Metaphor

Sometimes people talk about technology as a system, a complex arrangement of social and technical forces, rather than as objects with particular features and capabilities. We have encountered powerful analyses of technological systems in the writings of Jacques Ellul (1967), Langdon Winner (1977), Neil Postman (1993), and others. Ellul's book *The Technological Society* was first published in 1954, long before computer chips became pervasive in offices, homes, libraries, cars, toasters, telephones, and everywhere else. But Ellul's arguments are independent of particular technologies. He writes about the complete intertwining of our social and technical arrangements for living that develop in close relationship no matter what our intentions may be. For Ellul and others, our involvement with technology is pervasive and inescapable.

The systemic view tends to be pessimistic, yet it is easy to find examples that illustrate the widening impact of technological changes. Consider how the availability of cars and inexpensive fuel affects shopping patterns, which in turn affect the composition of neighborhoods and the social experience of people who live in them. When we enlarge our perspective beyond particular artifacts to include all the relations they participate in, the complexity can become overwhelming. Can we grasp the ways tools and practices fit together for large diverse groups of people? How can we imagine the implications of adopting a new technology? Can both designers and users effectively shape how technology is used? Is technology neutral in fact, as many developers see it, or does it have its own agenda that sweeps us along despite any contrary intentions we may have? These questions sound as if people who are very much against technology might ask them. But the issues they raise are real and important and are not so extreme as they may seem at first. Unintended consequences are common when new technologies are invented. Many of these can be considered useful and valuable, but some are, on balance, more negative than positive. Sometimes unintended consequences lead to the withdrawal of a product, such as certain pesticides or even the database of consumer information that Lotus published on a CD-ROM and then decided to recall. In other cases, such as the amazing success story of the World Wide Web, unintended consequences take a positive turn.

Tool language leads us to think about technologies as being under people's control (that is, smaller than human scale), and system language leads us to think about technologies as being out of people's control (that is, larger than human scale). From a broad perspective that encompasses large social movements and historical change, the systemic view is probably right. However, we find the large-scale systemic perspective adopted by Ellul, Winner, and other sociologists to be pessimistic. Adopting such a broad perspective feels unsatisfying and incomplete because it leaves out the presence of individuals with choices and ideas. We need a stronger sense of the local and particular.

Ecology Metaphor

We arrive at the metaphor of an information ecology because we want a way of thinking about technology that acknowledges complex interdependencies between tools and practices and also admits the possibility of diverse local variations. We define information ecologies as systems of people, technologies, practices, and values. Let's consider a few examples to give an idea of what we include in the idea of an information ecology.

A doctor's office may be considered as an information ecology. A variety of people can be found there, including doctors, nurses, physician assistants, receptionists, and patients. Most doctor's offices are filled with both medical and accounting tools. The practices include examinations, treatments, scheduling, and tracking insurance forms. The values are to provide high-quality care to improve people's health through good teamwork.

A small business office provides another example of an information ecology. Its staff might include a manager, secretary, sales people, and people who produce goods or services. Computers are used in most offices for a variety of purposes, including keeping track of budgets, creating and filing documents, and exchanging e-mail. Phone messaging systems are a common office technology, and there are also low-tech tools such as paper calendars and pencils to complement the computers. The values of the office include customer satisfaction and worker safety.

We explicitly include local values in the definition of an information ecology because discussions of values can be fruitful in the creation and evaluation of new technologies. For example, privacy issues motivate both users and developers. These discussions do not take place if we focus only on the productivity and efficiency afforded by new technologies. They are important to the success of a new technology implementation.

Some of the major technology failures came about because too little time was spent clarifying motivations and ensuring that the policies around technology use were consistent with local values. It is easy to skip over discussions of local values and principles, especially when people are enthusiastic about new technologies. There are many practical problems in adopting new technology, and it is tempting to leap ahead and start deciding what to buy, where to put it, and how to get it networked. But before moving to know-how questions, it is important to consider know-why questions.

Our aim in proposing the ecology metaphor is to prompt conversations that might not otherwise take place. For users, reflecting on local values can be helpful in choosing new technologies and deciding how they will be used. For designers, talking with users about the guiding principles behind users' enterprises can highlight new and interesting design problems.

Drawing on the information ecology metaphor leads us to focus on certain key characteristics, such as diversity, a sense of locality, the presence of keystone species, and the coevolution of different elements over time. These characteristics are associated with biological ecologies, and they are also found in information ecologies if we look for them. To return to a point we made earlier, the idea of an information ecology is good to think with. It is a conceptual tool for directing our attention to aspects of technology that might otherwise be missed.

Each of the metaphors for technology that we have discussed so far (tool, text, assistant, and system) provides a particular kind of leverage for thinking about design and use. We do not want to leave these other metaphors behind. Tool-centered thinking, for example, will always be helpful in considering the quality of fit between a person, tool, and tasks to be done. But we find that the ecology metaphor is especially useful when we want to look at dependencies between technological tools and social practices without getting overwhelmed. We believe that this metaphor, along with the others we have mentioned, will challenge users and designers and clarify some of their common concerns. Together, these metaphors make up a powerful repertoire for evaluating technology and fostering conversations among users and designers.

In the remainder of this discussion, we concentrate on how ecological thinking can inform the development and use of digital libraries. To ground this discussion, we first describe how physical libraries work. Many of the people, tools, practices, and values of physical libraries remain important in the development of digital libraries.

Library as Information Ecology

A library is a busy and complex environment. In this section we provide only a brief summary of the aspects we find most relevant to a consideration of digital libraries. This summary is based on ethnographic studies that O'Day conducted at the Hewlett-Packard corporate library and Nardi conducted at the Apple corporate library. More complete descriptions of this work appear elsewhere (Nardi and O'Day 1996, 1999; O'Day and Jeffries 1993a, 1993b). The specific features of library activity that we wish to highlight here are the reference interview, the search expertise of librarians, and library clients' search patterns. We focus on these three aspects of library experience because they are less visible and commonly known than many other library practices.

At the libraries we studied, clients could search for materials themselves or ask a librarian for help. Often they did some of both, though some clients always helped themselves and others always asked for help. The libraries had a range of physical and online information sources and materials, including books, journals, newspapers, Web access, specialized databases on CD-ROMs, and online subscription services to many other databases. Each library had a staff of several librarians.

When a library client chose to ask a librarian for help in finding information, the librarian asked questions to clarify what the client was looking for. This is called a reference interview, and it is part of librarians' professional practice. From a client's perspective, this feels like a brief conversation rather than an official interview. But a few key questions usually appear. One is how recent the information should be to be useful to the client. Another is how many books or articles the client is looking for and what form of information the client is interested in (such as news articles, journal articles, corporate reports, government publications, or books). The search results would be more scattershot without this kind of information. When the librarian knows what the client is looking for, the search results can be tailored more closely to the client's needs and interests. When a client's information need is especially vague, the interview can help to narrow the criteria for accomplishing the first step.

For librarians, reference interviews are valuable beginnings to any search. Clients, however, usually don't know that anything important is going on in this conversation. More than one library client we interviewed reported that the interaction consisted of passing on keywords to the librarian, who then simply applied these keywords to different databases. The reference interview is a transparent event, even to some of the people who have experienced it many times. One of the values of the library ecology is service, and this includes making the client's experience as easy as possible.

Librarians' search strategies are also transparent to users, since most searching in corporate libraries takes place behind the scenes, between the time of a client's request and the delivery of results. It was clear from our studies that far more takes place than a simple application of user keywords to different databases. Hundreds of databases and other information sources are available, and an early part of the search strategy is making an informed decision about where to look. Some search problems involve looking in paper-based sources (for example, to clarify terminology) before going online. In other searches, librarians make creative leaps from one online source to another, carrying new ideas for terms and phrases to try. As they search, librarians are using their knowledge of the clients' interests, preferences, and even their work activities and current knowledge about a topic to direct the progress of the search. Searching is a complex process, though clients usually do not see its complexity revealed in their interactions with librarians.

A third aspect of library experience that is not apparent on casual observation is the interconnected nature of many searches. Library clients often bring information needs to the library that are related to the needs they brought last time and the ones they will bring next time. Some searches are repeated regularly, such as monitoring the financial profiles of major business competitors every quarter. Other searches are open-ended and exploratory, as people digest the results of one search and then use them to discover where to look next. People learn more about what they want to know as they go along. They also learn about terms, subtopics, and good information sources. They can't always express an information need clearly and succinctly until they get started. In these circumstances, a librarian is especially helpful in framing the opening search question so that people are not inundated with material. Often, the librarians we interviewed kept notes on their clients' searches when they had an intuition that the client would be back for further information.

In general, we found the libraries we studied to be complex places that offered clients a range of experiences, from self-help browsing to customized search services. There was diversity of materials, tools, and styles of support.

Physical libraries located in corporations, universities, or towns continue to integrate new information tools and services as they become available. Along with service, broad access to information is a value that guides library practice. As new online sources come along, they make their way into libraries. But what about the reverse situation, when people access digital library collections from desktops? Which features of the physical library experience carry over to the new locales? Which features would we like to carry over, and which should be different to take advantage of new technological advances? How might these features depend on the environment in which digital library collections are used—that is, the local information ecology, whatever that might be? These are questions of interest to both users and designers of digital library technologies.

We can sharpen these questions by considering digital libraries ecologically. In the following sections, we use four characteristics of information ecologies (diversity, locality, keystone species, and coevolution) to identify interesting issues raised by digital libraries. Our goal is to raise useful questions for communities of developers and users. For developers, the questions may suggest new design possibilities. For users, the questions may point to areas that each local information ecology must address in light of its own needs, practices, and values.

Diversity

In a physical library, people occupy different ecological niches. We have already noted diversity among library clients who bring different information-finding styles to the library. In addition, librarians have their own styles and preferences. In the Hewlett-Packard library, for example, one librarian specialized in business searching, another in technology and chemistry, and another particularly enjoyed tracking citations and searching for patents on any topic. Libraries also have a variety of tools, from paper notebooks to high-tech databases.

What would it mean for digital libraries to be marked by a similar level of diversity? Here are some of the questions we might consider.

Who makes digital libraries work? Who shapes content and access?   The first and most obvious question about diversity has to do with people. Who is involved? Are there niches for new kinds of contributions? One of the distinguishing features of the Web is that anyone can be a publisher of information, even of large collections. Noncommercial providers are more numerous and visible in this setting, and the possibilities and needs for expert mediation grow along with the breadth and variety of information available. Who can help users locate appropriate sources, compare and evaluate sources, translate terminology, or learn effective search strategies? Are lightweight consulting services available? How could consulting services be configured to reach as many clients as possible? How might a consulting practice be developed as an extension of a digital library collection, either sponsored by the collection's publishers or independently organized? What additional features should user interfaces to digital libraries offer to support interactions between consultants and clients?

What paths exist to a digital library collection? Just as online content is produced more broadly than it used to be, the paths that users can follow to reach online content have been multiplied. The Web is a densely connected network of linked information, so one may arrive at a particular destination through any number of paths. This fact offers interesting possibilities for supporting diverse user needs. Examples of novel paths include a hospice Web page that offers a link to an online bookstore that has a special "shelf" of books about hospice volunteer work and that gives the hospice a percentage of sale proceeds from the "shelf." In general, how can information collections, or parts of them, be made available in a variety of contexts? And when this is done, how can users know the underlying sources of the information they find so they can apply their own evaluation criteria?

What are the complementary low-tech tools? In physical libraries, small slips of paper and pencils are often placed next to catalogs. These low-tech tools have survived the shift from paper catalogs to online catalogs in every library we have visited. In the development of digital libraries, we tend to focus on the high-tech part, such as different media representations or powerful search engines. But what low-tech tools are needed to help these high-tech tools work well? This question must be answered in part by the inhabitants of each local information ecology from which online information collections are accessed, such as schools, offices, homes, or public libraries. If people need slips of paper next to their computers, they should be provided in each setting. But there are additional design possibilities for both tool developers and users. The general idea is to expand beyond thinking specifically about online collections and services to thinking about a whole suite of complementary technologies and trying to imagine the ways users will move fluidly from one tool to another.

What are the bridges between the physical and digital worlds? This question may be one of the most fruitful unexplored areas for designers of digital library services. Paper is highly portable and easily markable, among other things. Online representations can be searched easily, can be flexibly organized, and rapidly communicated. How can digital and paper combinations be developed to take advantage of the advantages of each? The availability of fast high-quality scanners and printers invites us to think about bridges between the digital and paper worlds. Each information ecology may make different choices about what is accessed on paper and what is accessed online (and what is replicated in both worlds), but in each ecology there is a need for bridges between these domains.

Locality

An information ecology is an environment with fuzzy boundaries, just as is a biological ecology, but it is not unbounded. We might consider an engineering department to be an information ecology, but we would not consider a company with thousands of people to be one. A large company probably includes many different information ecologies. An information ecology should be understood as an environment in which people know each other and carry out interrelated activities. The environment does not need to correspond to a physical location. A network community can be an information ecology if it has stable participation and practices. A strong sense of locality is important in bringing intelligibility and coherence to the activities and values of the ecology. As with the idea of diversity, the ecology metaphor prompts questions of how digital libraries might strengthen their support for locality.

How could a digital library collection be tailored for the different information ecologies in which it is used?   Some (perhaps most) online information collections have a multitude of potential uses. Consider Medline, for example, which is an extensive collection of medical research literature that is now available on the Web. It is easy to see how Medline could be used in the context of a school, doctor's office, law office, senior center, business, or home. Seniors might be particularly interested in medical issues related to aging. Students may want to learn basics about health. Patients may gather recent research results related to their own illnesses, lawyers may trace trends of medical accountability, and business owners may study the latest research on common work-related injuries. Although several different Web sites provide Medline searching, the sites we have encountered are all intended for a general user population. Yet in each of the examples we have suggested, people might need particular filtering, vocabulary translation, or other framing of this huge collection to make it more useful for their own situations. How can publishers and tool providers support different forms of locality? How can users be involved in framing digital library collections to meet their needs?

What opportunities can digital library collections offer for different online organizations and presentations of content?   Some physical libraries have a little bit of everything, while others develop strengths in certain topical areas. Physical libraries can specialize in how they arrange their spaces to support different activities, such as children's story time or quiet study. What might the corresponding specializations look like in the online world? Many digital library collections include content and access mechanisms bundled together. For example, the only way to reach the Association for Computing Machinery's collection of computer science materials is to visit the ACM Web site. There can be compelling economic and technical rationales for integrating content and access, but interesting design and use possibilities may be raised by decoupling them. It might be interesting, for example, to develop a site that emphasizes casual browsing over searching by cleverly filtering and composing selected materials. What might an online version of a reading room with only current journals look like? Or a digital library collection might be partitioned into subtopics, with each one available through specialized access paths.

Keystone Species

One thing biologists look for in ecologies is the presence of *keystone species*, species whose contributions are so central to the dynamics of the ecology that without them the ecology would disintegrate: "The loss of a keystone species is like a drill accidentally striking a powerline. It causes lights to go out all over" (Wilson 1992, pp. 347–348).

We believe that information ecologies have keystone species too, although their contributions may be invisible at first glance. We see keystone species in information ecologies as those people whose special contributions stitch together people, tools, and practices, filling gaps and helping the whole enterprise to run well. In general, we find that keystone species are often people who translate, localize, and otherwise create necessary bridges. Often their contributions are unofficial, though this is not always the case.

In physical libraries, librarians are a keystone species because without their efforts in assembling well-rounded collections, arranging for convenient access, and providing helpful, unobtrusive assistance, libraries simply would not be able to offer adequate information services. In schools that are getting wired for Internet access, competent teachers who are enthusiastic about experimenting with the Internet and developing curriculum using it are a keystone species because they bridge the significant gap between technology and classroom teaching. The idea of keystone species suggests several questions for digital libraries.

What roles can librarians play for digital library collections and services?   We have dwelt on the importance of what librarians do in physical libraries because many users of library services and developers of information access software are not aware of it. Once the value of librarians' contributions in the library is acknowledged, we should ask how online settings can benefit from their expertise. It does not make sense to adopt a new technology that has strengths in some areas but diminishes the overall quality of information-finding experiences. The personalized consulting service available in physical libraries is not practical for digital libraries, unless they have a strong sense of place and offer a variety of communication modalities. Indeed, one of the affordances of digital library tools is powerful do-it-yourself searching for casual users. But users who are stuck or users who want to expand expertise in the finer techniques of information finding would benefit from the assistance of librarians.

What needs exist for mediators of different kinds?   When we talked about diversity, we raised the question of who might be involved in digital library access. The idea of mediation is one way to sharpen this question. There are certain to be gaps between digital library collections and services and the particular needs of user communities. One avenue to consider might be translation from one domain to another, not just translation of language but also translation of ideas and conceptual frameworks. In schools, teachers who are early adopters of the Internet help their colleagues translate the concepts of the online world into ideas that make sense for the classroom. Localization of any kind suggests the need and opportunity for mediation.

Coevolution
Biological ecologies change all the time, and so do information ecologies. There are mutual adaptations between tools and social practices and between tools and other tools. What opportunities do these adaptive processes bring up for designers and users?

What happens to the reference interview?   In the library, the reference interview accomplishes several important things. It narrows the search to specific time periods, specific kinds of information, and the specific amount of information that the user is looking for. Many casual searchers do not understand the impact that these few factors can make in trimming search results to a manageable and useful set, and they don't volunteer this information unless a librarian asks for it. We have noticed that some search interfaces for large collections prompt users for these basic criteria, which appears to be a successful adaptation of online interfaces based on face-to-face reference interview techniques.

Another feature of the reference interview is that it reveals how open-ended or specific a user's information-finding problem is. Many people follow an exploratory search process, beginning with a broad information need, such as, "I want to know about plastics" or "I want to know about midsized chip manufacturers." Their search progresses to specific areas of interest as they learn more. The reference interview *reveals* the exploratory nature of an information problem, but it may not *change* it. Sometimes the librarian can narrow the search considerably by finding out more about the context of a client's interest, why they want the information, and what they will do with it. Sometimes the information request is open and vague because that's just the kind of information needed right now. The client just needs something to get started and will be able to fine-tune the request later on. Most current search interfaces don't cope well with broad exploratory information requests, although these requests are among the most common. Many search interfaces do not help people narrow their searches by allowing them to provide contextual information, and when users' requests lead to an avalanche of information, that is unfortunately what they get. Many opportunities can be found for designers to respond to the diversity of information-finding problems and styles that people bring to a digital library collection.

What about the converse issue? How does a reference interview (or any consulting interaction) change when people seek help after they have already done some poking around online on their own? People should not have to start over when they move to a different search modality. Instead, they should be able to carry their current status with them. Information gathering may start with online interactions, move to physical settings, and move back online. What would it mean to support these transitions with online tools? How can information consultants in different information ecologies take advantage of other people's previous work?

How will digital library services adapt to the common pattern of extended interconnected searches? In our summary of how libraries work, we indicated that people often carry out multistep searches, where successive steps may take place a few days or even a few months apart. Between forays into information collections, people digest what they have found so far and decide where they want to go next.

Librarians adapt to this behavior by keeping records on their interactions with clients so they can pick up where they left off. How can digital library services adapt as well? Conversely, will people's search patterns change when they are using online collections more frequently, and how might this change the effectiveness of the ways they use the information they gather?

What happens to paper? How do new and old tools interoperate?   In the Hewlett-Packard library study, we interviewed a number of library clients. We looked at examples of past searches people had done and talked about how they had incorporated their search results into their work. We were able to look at previous searches because nearly every client we interviewed had printed them out, marked them up, and saved them. For the most part, these large collections of abstracts and articles had been delivered through e-mail. What people would have liked to do (but couldn't) was transfer their annotations back to the digital world so that the annotations could be shared with coworkers or used as pointers for further information gathering.

Similarly, people wanted to transfer search results easily into spreadsheets (to do exploratory analyses on quarterly financial report data, for example) or into other tools, but this could be accomplished only with considerable effort. The development of digital library collections and tools has been impressive, but integration with existing media and tools remains minimal.

How do digital libraries coexist with physical libraries?   One of the authors (O'Day) spent several years working with an elementary school in Phoenix, Arizona, that was integrating a virtual world (a multiuser domain, or MUD, called Pueblo) into classroom activities. To help the younger students learn about the geography of the virtual world and the syntax of navigation commands, one of the teachers attached large cardboard labels to doors, stairways, and other areas of the school. These labels were similar to the text labels of the virtual world (such as "Stairway … ⟨up⟩" or "Library … ⟨west⟩"). Within the wide-ranging geography of the virtual world with its thousands of rooms was an area that modeled the elementary school, where students could locate virtual classrooms created by their own teachers. This is a lighthearted example of an interesting bidirectional movement between physical and online settings. Digital library services and physical library settings do not directly correspond, by any means, but they have clients whose interests may encompass both physical and digital resources. How might these resources point to one another in useful ways? How can such references be kept up-to-date to maintain their integrity over time?

The aim of this discussion has been to use the information ecology metaphor to generate ideas and questions that are relevant to the development and use of digital library services. We conclude with some reflections on why users and designers should bring ecological thinking to bear in their own activities.

Ecological Thinking

We have used the ecology metaphor to generate a lengthy list of questions that could be asked of digital libraries. We'd like to step back and ask, "So what?" Would these questions have been asked anyway? Would all of them have been asked? Do all of them need to be asked?

Some of the questions we present here have been raised in the extensive ongoing research on digital library topics. The problems of how to bridge paper and electronic worlds or how to find information through multiple search paths have been pondered in many research labs. However, we have seen little discussion of many other questions we raise. We question the role of human librarians in making digital libraries work, the uses of complementary low-tech tools, tailoring digital libraries to meet the needs of specific local settings, localized mediation, and changes to the reference interview. These questions are stimulated by attention to the four basic features of information ecologies presented here.

We believe that looking at the broad picture is more important than focusing only on the details of particular technology innovations. Even when a new technology is meant to serve a general purpose, exposure to the richness of users' environments is a valuable resource for design insight and creativity. Design problems get harder—and more realistic—as more interconnections among people, tools, and practices are revealed. A technological innovation may look good when considered in isolation and yet turn out to be problematic or incomplete in actual settings of use. Ecological thinking highlights important linkages and dependencies that developers need to know about.

Users also need to approach digital libraries from a more comprehensive, less tool-centered perspective. As digital library technologies emerge into widespread use, they transform the way we find and work with information. When people look only at technical features when they make decisions about how to apply new technologies, they are likely to miss some of the interconnections that shape successful practice. Organizations can be redesigned in naive ways that end up compromising service, such as reducing the presence of librarians in the belief that digital library tools can replicate their work. The successful adaptation of digital libraries calls for sustained, thoughtful conversations among developers and among participants in each information ecology where they are used.

References

Beyer, H., and K. Holtzblatt. 1997. *Contextual Design: A Customer-Centered Approach to Systems Designs*. San Mateo, CA: Morgan Kaufman.

Carroll, J. M., ed. 1995. *Scenario-Based Design*. New York: Wiley.

Ellul, J. 1967. *The Technological Society*. New York: Vintage Books. (Original work published 1954)

Latour, B. 1995. Mixing Humans and Nonhumans Together: The Sociology of a Door-Closer. In S. L. Star, ed., *Ecologies of Knowledge*. New York: State University of New York Press.

Nardi, B. A., and V. L. O'Day. 1999. *Information Ecologies: Using Technology with Heart*. Cambridge, MA: MIT Press.

Nardi, B. A., and V. L. O'Day. 1996. Intelligent Agents: What We Learned at the Library. *Libri*, 46(2), 59–88.

Nielsen, J. 1994. *Usability Engineering*. Cambridge, MA: AP Professional.

O'Day, V. L., and R. Jeffries. 1993a. Information Artisans: Patterns of Result Sharing by Information Searchers. In *COCS '93: Proceedings of the Conference on Organizational Computer Systems (Milpitas, CA, November 1–4, 1993)* (pp. 98–107). New York: ACM Press.

O'Day, V. L., and R. Jeffries. 1993b. Orienteering in an Information Landscape: How Information Seekers Get from Here to There. In *Conference Proceedings on Human Factors in Computing Systems (CHI '93, Amsterdam, The Netherlands, April 24–29, 1993)* (pp. 445–483). New York: ACM Press.

Postman, N. 1993. *Technopoly: The Surrender of Culture to Technology*. New York: Vintage Books.

Schuler, D., and A. Namioka. 1993. *Participatory Design: Principles and Practices*. Hillsdale, NJ: Erlbaum.

Wilson, E. O. 1992. *The Diversity of Life*. New York: Norton.

Winner, L. 1977. *Autonomous Technology: Technics-out-of-Control as a Theme in Political Thought*. Cambridge, MA: MIT Press.

Winograd, T., J. Bennett, L. D. Young, P. S. Gordon, and B. Hartfield, eds. 1995. *Bringing Design to Software*. Reading, MA: Addison-Wesley.

Part II

5

Designing Digital Libraries for Usability

Christine L. Borgman

Introduction

If national and global information infrastructures are to serve "every citizen" (Europe and the Global Information Society 1994; Computer Science and Telecommunications Board 1997), then digital libraries should be reasonably easy to understand and to use. But how easy can we make them? While some suggest that information systems should be as easy to use as automatic teller machines (ATMs), the comparison is unfair. ATMs support only a few procedures for withdrawing or depositing funds. Other widely adopted information technologies such as radios, televisions, and telephones support only a small set of actions, but even these technologies are becoming more complex and harder to use. Turning on a television, changing channels, and adjusting the volume are easy, yet programming a video cassette recorder to schedule the recording of a television program is notoriously difficult. Similarly, most people are capable of making and receiving telephone calls but find that advanced telephone features such as call forwarding, call waiting, or three-way calling can be prone to error.

One reason that technologies become more complex as features are added is that the relationship between task and tool becomes less visible. An ATM has few enough features that each one can correspond to a single key or menu choice. Tape-based telephone answering machines have a direct mapping between task and action: press Play to hear messages, press Delete to erase messages. Voice message systems have these functions and many more. The result is that the mapping becomes abstract: for example, log in to the system with a user identification number, give a password for new messages to play automatically, and delete a message by pressing 76.

Desktop computers are especially abstract in their relationship between form and function. A computer cannot be inspected to identify its functions in the ways that a telephone, television, or other information technology designed for a single application can be inspected. Keyboards, pointing devices, function keys, and screen displays can be programmed to support almost any imaginable application. Because of this generality, the same physical actions may produce different results in each software application.

In the abstract world of computing, the real-world clues are gone, replaced by pull-down or pop-up menus, screen displays, searching tools, and lists to browse. Usability depends heavily on users' abilities to map their goals onto a system's capabilities. Also missing in automated systems is the safety net of human assistance. Instead of the store clerk, librarian, or other intermediary who listens carefully to an ambiguous question and responds with an interpretation (such as "I see. What you may be looking for is ..."), the options for assistance may be an automated help system, an e-mail query, or a telephone help line. These are not acceptable alternatives in most cases. Information systems will achieve wide acceptance only if they are easy to learn and use relative to perceived benefits.

This chapter explores behaviors involved in understanding and using digital libraries. First, the term *digital libraries* is defined, setting a context for the discussion. The second section of the chapter examines usability issues in the design of information systems, and the third section looks at the knowledge and skills involved in searching for information. My focus is on the individual user and on searching as a form of problem solving. Other chapters in this book set these issues in a broader context of groups and organizations and address additional uses of digital libraries, including the creation and use of digital documents.

What Are Digital Libraries?

Despite its popularity, *digital library* remains a problematic term. Clifford Lynch (1993) was prescient in noting that the term obscures the complex relationship between electronic information collections and libraries as institutions. Douglas Greenberg (1998, p. 106) proposes that "the term 'digital library' may even be an oxymoron: that is, if a library is a library, it is not digital; if a library is digital, it is not a library." Patricia Battin (1998, pp. 276–277) rejects the use of the term *digital library* on the grounds that it is "dangerously misleading." Indeed, a review of definitions reveals that *digital library* describes a variety of entities and concepts (Bishop and Star 1996; Fox 1993; Fox, Akscyn, Furuta, and Leggett 1995; Greenberg 1998; Lesk 1997; Levy and Marshall 1995; Lucier 1995; Lyman 1996; Lynch and Garcia-Molina 1995; Schauble and Smeaton 1998; Waters 1998; Zhao and Ramsden 1995).

Of these many definitions, the most succinct one arising from within the computer and information science research community originated in a research workshop on scaling and interoperability of digital libraries (Lynch and Garcia-Molina 1995): "A digital library is a system that provides a community of users with coherent access to a large, organized repository of information and knowledge."

In contrast, the most succinct definition arising from the community of library practice is that set forth by the Digital Library Federation (DLF): "Digital Libraries are organizations that provide the resources, including the specialized staff, to select, structure, offer intellectual access to, interpret, distribute, preserve the integrity of, and ensure the persistence over time of collections of digital works so that they are readily and economically available for use by a defined community or set of communities" (Waters 1998).

As discussed in more depth elsewhere (Borgman 1999, 2000), researchers are focusing on digital libraries as networked information systems and as content collected on behalf of user communities, while librarians are focusing more on digital libraries as institutions or services. These communities are not mutually exclusive, of course, and most large digital library research projects involve librarians as well as scholars from information and computer science.

Both of these notions and more are encompassed in the two-part definition that arose from the Social Aspects of Digital Libraries research workshop (Borgman et al. 1996) in which several of the contributors to this book participated:

1. Digital libraries are a set of electronic resources and associated technical capabilities for creating, searching, and using information. In this sense, they are an extension and enhancement of information storage and retrieval systems that manipulate digital data in any medium (text, images, sounds; static or dynamic images) and exist in distributed networks. The content of digital libraries includes data, metadata that describe representation, creator, owner, reproduction rights, and metadata that consist of links or relationships to other data or metadata, whether internal or external to the digital library.

2. Digital libraries are constructed—collected and organized—by [and for] a community of users, and their functional capabilities support the information needs and uses of that community. They are a component of communities in which individuals and groups interact with each other, using data, information, and knowledge resources and systems. In this sense they are an extension, enhancement, and integration of a variety of information institutions as

physical places where resources are selected, collected, organized, preserved, and accessed in support of a user community. These information institutions include, among others, libraries, museums, archives, and schools, but digital libraries also extend and serve other community settings, including classrooms, offices, laboratories, homes, and public spaces.

The above definition extends the scope of digital libraries in several directions, reflecting the contributions of scholars from a dozen disciplines. It moves beyond information retrieval to include the full life cycle of creating, searching, and using information. Rather than simply collecting content on behalf of user communities, it embeds digital libraries in the activities of those communities, and it encompasses information-related activities of multiple information institutions. The above broad definition of the term *digital libraries* is assumed in this chapter.

Uses, Users, and Usability of Digital Libraries

Usability issues in digital libraries and other forms of information systems persist, despite the technological advances of the last two decades. Many of the challenges identified early in the 1980s have yet to be resolved (Borgman 1984, pp. 33–34):

The change in the use of computing technology is a fundamental one. Once the computer began to be used by people who were not experts, the access requirements changed dramatically. The technology-oriented expert who uses a system daily can learn (eventually) to use almost any mechanism, no matter how poorly designed. The situation is different with the new community of users. Most of them lack both a technological orientation and the motivation to invest in extensive training. The new class of users sees a computer as a tool to accomplish some other task; for them, the computer is not an end in itself. This new generation of users is much less tolerant of "unfriendly" and poorly designed systems. They have come to expect better systems and rightly so.

The technology has moved much more rapidly than has our understanding of the nature of the tasks for which we use it or our understanding of the human ability to adapt. Indeed, an important issue is whether the user should adapt to the computer or the computer adapt to the user. Computers have turned out to be much harder to use than we had expected, and design and training problems have resulted. We have had many calls for more "user friendly" systems, but we don't understand human-computer compatibility well enough even to agree on what "user friendly" means. Thus we are left with several distinct challenges: 1) we need to determine what factors make computers difficult to learn and use; 2) we need to define a set of characteristics for "user friendly" systems; and 3) we need to apply the research to design.

Although these same three challenges remain, a larger array of design factors now are recognized. Research in human-computer interaction in the 1980s was just that—the relationship between an individual user and the computer in direct interaction. "User friendly" design addressed screen displays and functional capabilities but did not delve deeply into task motivation, much less into the relationship between a computer user and the work, educational, or leisure context from which the task arose. People were expected to adapt to systems, and considerable effort was devoted to user training. Today people have higher expectations of information systems. Digital libraries should be easy to learn, to use, and to relearn. They should be flexible in adapting to a more diverse user population.

And yet, as noted in the introduction to this chapter, digital libraries will never be as easy to use as automatic teller machines or other single-purpose technologies. People must make some investment in learning to use them effectively. We focus first on issues of making digital libraries easier to use and then on the knowledge and skill requirements for using them.

Usability Criteria

Perspectives on usability have shifted substantially over the course of this century. The initial purposes of ergonomics were to place people into the technological order. Human skills were measured relentlessly so that people could be matched with the machine task to which they were best suited and machines could be operated by those with the requisite capabilities (Edwards 1995; Gilbreth 1921). By the early 1980s, the focus of ergonomics (also known as *human factors*) had shifted toward shaping technology to human capabilities and needs. This period also marked the transition from mainframe computing systems operated by skilled professionals to desktop computing for end users.

This transition includes several landmarks. The first conference held by the Association for Computing Machinery (ACM) Special Interest Group on Computer-Human Interaction (SIGCHI) was held in 1982 and has now become a major annual international conference. The transition is marked by publication of the first edition of Ben Shneiderman's (1987) textbook, *Designing the User Interface: Strategies for Effective Human-Computer Interaction* (1987) and of Donald Norman's popular book, *The Psychology of Everyday Things* (1988). Conferences held during the period helped to disseminate shifts in thinking—for example, the Scandinavian movement toward the work-oriented design of computer artifacts (Ehn 1988). The first participatory design conferences in the United States (Namioka and Schuler 1990) were held during this time. University courses in human-computer interaction and user interface design, first offered in departments of computer science and information studies, later spread to the social sciences, the humanities, and other fields. A large body of research on human-computer interaction now exists, which in turn has led to general principles and guidelines for the design of information technologies.

Despite these advances, establishing generalizable benchmarks for usability remains problematic due to the variety of applications and the diversity of user communities served. Many criteria and guidelines for usability have been derived from the findings of research in human-computer interaction. Perhaps the most general are the requirements for "being fluent with information technology" (FIT) (National Research Council 1999). "FITness" skills, according to the report, include (1) contemporary skills in using today's information technology, such as practical experience on which to build new competence; (2) foundational concepts, such as basic principles and ideas of computers, networks, and information that are suffi-cient to understand information technology opportunities and limitations; and (3) intellectual capabilities, including the abilities to apply information technology in complex and sustained situations, to manipulate information technology advanta-geously, and to handle unintended and unexpected problems as they arise.

If we are to achieve a goal of having "every citizen interface ... [with] the nation's information infrastructure" (Computer Science and Telecommunications Board 1997, p. 45), then information technologies—and particularly digital libraries—need to meet certain criteria. Systems should be easy to understand. They should be easy to learn, error tolerant, flexible, and adaptable. They should be appropriate and effective for the task. They should be powerful and efficient, inexpensive, por-table, compatible, and intelligent. They should support social and group inter-actions. They should be trustworthy (secure, private, safe, and reliable), information centered, and pleasant to use.

Other applicable criteria are the user interface design rules established by Ben Shneiderman (1992, 1998) and adapted to information retrieval (Shneiderman, Byrd, and Croft 1997). By these criteria, the systems should strive for consistency, provide shortcuts for skilled users, offer informative feedback, design for closure, offer simple error handling, permit easy reversal of actions, support user control, and reduce short-term memory load. Jakob Nielsen (1993) identifies five usability attributes for information systems as well as other applications: learnability, effi-ciency, memorability, errors, and satisfaction.

These principles offer general guidance for design but are far from a cookbook for constructing an individual information system. Principles such as "easy to learn" must be applied relative to the application and the user community. A system that supplies daily weather reports to the public must be much easier to learn than one that supplies geophysical data to researchers, for example. Determining appropriate benchmarks for any given system involves evaluation with members of the target audience and comparisons to similar applications. Issues of evaluation are set in an organizational context in other chapters in this book by Agre; Levy; Lynch; Marchionini, Plaisant, and Komlodi; Marshall; O'Day and Nardi; and Star, Bowker, and Neumann.

While the value of making systems easier to use may be self-evident to users, it is not always self-evident to software vendors, programmers, or even the managers who acquire software on behalf of end users. The literature on human-computer interaction abounds with studies indicating that companies release software without basic human factors testing. Due to the belief that market timing, number of features, price, and other factors are more important to business success than is usability, usability testing itself is often seen as too expensive or as ineffective (Computer Science and Telecommunications Board 1997; Landauer 1995; Nielsen 1993; Sawyer, Flanders, and Wixon 1996; Shneiderman 1998). Studies to determine the veracity of such beliefs reveal hard evidence that improving usability is cost-effective, both for software producers and for the organizations that implement software (Computer Science and Telecommunications Board 1997; Nielsen 1993). Almost half the code in contemporary software is devoted to the user interface (Myers and Rosson 1992). The greatest source of cost overruns in software development lies in correcting usability problems (Nielsen 1993). Even a small amount of usability evaluation in the development process can pay for itself several times over in cost savings from lost productivity (Computer Science and Telecommunications Board 1997; Landauer 1995; Nielsen 1993; Sawyer, Flanders, and Wixon 1996).

Content and Community

Design guidelines and evaluation criteria can be employed to build more usable systems but only to the extent that design goals are appropriate for the application. At the core of effective digital library design is the relationship between the content to be provided and the user community to be served. Design goals can originate from either perspective.

Design often originates with an existing collection and a goal of making the content available in a digital library. An organization may own (or hold the rights to) one or more collections such as photographic images of animals, maps of a region, historic literary texts, or instructional materials. Any of these collections could be used in a variety of ways to serve a variety of purposes. In deciding which features of a digital library will support these materials, the next step should be to determine who would use the content, how, and why.

For example, a set of animal images could be valuable in biology classrooms at the elementary, secondary, and college levels. In a digital library to serve any of these applications, each image could be described by common and biological names. The search capabilities could be simple and learnable by students in a few minutes. The display capabilities could support one or a few images at a time on basic desktop machines available in classrooms. If the same set of animal images were to serve an audience of biological researchers, a more elaborate taxonomic description would be required, as well as requiring more extensive searching and display capabilities. If the animal images were intended for advertising purposes, then the images could be described for the emotional impact sought, such as "peaceful," "pastoral," "aggressive," "leadership," or "tension." Colors, image size, granularity, and the cost of using the image in different media would be essential descriptive elements. Search capabilities would need to be simple enough for nontechnical users and yet support browsing through various combinations of elements. High-quality displays on larger screens would be necessary as well.

Alternatively, digital library design can begin with the audience to be served. Law firms, for example, serve the information needs of their attorneys with multiple digital libraries. Attorneys need resources on statutory and case law that apply to their current cases. They often need related technical, social, or policy materials as well. Librarians, paralegals, or attorneys, all of whom are familiar with legal terminology and resources, may do the actual searching of digital libraries. Many information needs can be satisfied with commercially available digital libraries of statute and case law. The content of these digital libraries is collected and organized for the information needs and work practices of the legal profession. Accordingly, these systems provide sophisticated searching features that assume legal expertise. Some initial training and continuing education are required, which is acceptable because these systems are used frequently. Similarly, work product, litigation support, and other databases of materials internal to law firms are designed for the information needs and work practices of the firm.

These same digital libraries of legal resources contain materials of considerable value to members of the lay public who may need legal information for contracts, wills, real estate transactions, or landlord-tenant disputes. However, the systems that are commercially available to the legal community are rarely usable by a lay audience. It is unclear whether this is due to the technical expertise required to use the systems, the time investment necessary to learn the systems, or their cost. This does not mean the systems are poorly designed, however. Usability is relative and must be judged based on intended goals. The lay audience is served by a complementary set of resources tailored to the information needs, technical skills, and financial resources of nonlegal professionals.

The tradeoffs involved in designing digital libraries for single or multiple communities are exemplified by the seemingly simple query, "Why is the sky blue?" This question can be answered at many levels. Most children are curious about this topic by the age of eight years, yet it is of interest to astrophysicists as well. Even if phrased in similar terms, the child and the astrophysicist intend different questions and expect different answers. The child is happy with a simple answer that a few sentences, a picture book, or a multimedia science game might provide. In contrast, the scientist probably wants recent journal articles, a data set of observations from a satellite, or maybe an experimental kit to explain concepts such as light, color, and atmosphere to an introductory college class.

While this is a reasonable question to pose to a global digital library, it contains few clues as to the results desired. In what form or in what medium (text, images, sounds) should the digital library produce an answer? Should the answer be delivered only in the language in which the question was asked or in other languages if appropriate content exists? For what kind of computer and operating system should the results be formatted? What text, image, video, audio, and statistical software does the user possess to manage the results?

More generally, how much diversity in user populations or in content can a given system support? When should design be based on providing one community with access to multiple collections? When should design be based on providing one collection to multiple communities? What baseline capabilities are needed to provide access to multiple communities? When should a community's needs be supported by a single collection, and when by independent access to multiple collections? When should access to multiple collections be aggregated in a single system? To answer the "Why is the sky blue?" question, for example, scientists may need sophisticated data analysis facilities, while children may need rich but simple-to-learn browsing capabilities. The scientists' user interface may require high-end hardware and software, advanced computing skills, and extensive domain expertise and thus be usable only by that small and specialized user community. Conversely, the children's user interface to those same data may run on low-end hardware platforms, require minimal computing skills and domain expertise, and be usable by a broad audience. The ability to make a set of documents useful to multiple communities for multiple purposes is an important focus of current research in digital libraries (Phelps and Wilensky 2000; Wilensky 2000).

Knowing the Users

The degree that a digital library design will be tailored to particular user communities will depend on the goals of the application, the profile of the community, the amount of user participation in design, and the characteristics of the application. If the scope of the user community is well defined, such as employees of a company or students in a university, then a representative sample can be studied, and design participation can be solicited or appointed by management.

If the scope of the user community is less well defined, such as the prospective users of a new product or service, designers still can sample from the target audience. Marketing studies may provide baseline data. Research on primary and perhaps secondary target audiences may identify common elements and requirements as well as the degree to which their needs and interests vary. Such studies provide a starting point for design. Prototypes can be tested on samples of the target audience and the design refined. However, digital libraries on a global information infrastructure will serve larger, more diverse, and more geographically distributed audiences than will most systems of today. Scaling methods of design and evaluation to this complex environment is one of the greatest challenges of constructing a global information infrastructure (President's Information Technology Advisory Committee 1999).

Individual Differences

Another consideration in designing digital libraries is the range of skills, abilities, cognitive styles, and personality characteristics that are found within a given user community and that may affect usability. Collectively, these factors are known as *individual differences*. Studies of human-computer interaction with information-retrieval systems, word processing software, and computer programming reveal a variety of individual differences that influence human performance (Egan 1988). Population characteristics known to influence usability of digital libraries include computer skills, domain knowledge, and familiarity with the system. Other influences include technical aptitudes such as reading and spatial abilities, age, and personality characteristics such as those measured by the Myers-Briggs tests (Borgman 1989; Egan 1988). Social and cultural factors are thought to influence usability but are even harder to isolate and study (Computer Science and Telecommunications Board 1997; Leventhal et al. 1994; Shneiderman 1998). Even when selecting content and organizing collections for well defined communities, one must accommodate considerable differences within groups.

Three brief case studies from my research (with collaborators as indicated) illustrate the process of designing digital library applications for specific user communities.

Case Study: Energy Researchers and Professionals

This set of studies had two goals. One was to identify behavioral characteristics of energy researchers and professionals that influenced usability of an operational system. The second was to apply the results of the behavioral study to make the system easier to use. The system had a terse Boolean interface typical of its day and a large database of bibliographic records. Neither the content nor the organization of the database could be changed, but usability could be improved by constructing a front-end client to the system and by developing a simple instructional module. The extant body of research on the information-related behavior of scientists provided a baseline for designing the study (e.g., Meadows 1974; progress in this area of research was later summarized in Meadows 1997 and Borgman 2000). In the first phase of the study, we interviewed a sample of the scientists currently using the existing system so that we could identify their information needs and uses (Case, Borgman, and Meadow 1986). The interviews revealed considerable individual differences in information-related behavior within the community on factors such as frequency of use, skills, habits, and purposes.

In the second phase of the project, we designed the client. Most of the interviewed scientists and professionals used only basic system features, so we focused on simplifying those features rather than on developing specialized techniques. Most respondents were intermittent users, so design also focused on reducing the time to learn and relearn the system. During the third phase of the study, we evaluated the client and the instructional module in an experimental setting, with subjects drawn from the user community. Results indicated that the client-user interface provided significant usability gains over the native system (Borgman, Case, and Meadow 1989; Meadow, Cerny, Borgman, and Case 1989).

Case Study: Elementary School Students Studying Science

The Science Library Catalog Project grew out of a project based at the California Institute of Technology whose goal was to improve instruction in elementary school science, specifically biology and physics (Borgman, Gallagher, Hirsh, and Walter

1995). An identified weakness of science instruction was that students were not learning how to search for new information beyond what they learned in the classroom. Our goal was to supplement "hands-on science" with "hands-on information searching." The audience included school children ages eight to twelve.

The project constructed online catalogs of science and technology materials, with longer-term goals of extending the system to include full text, images, students' reports, and other materials. The Science Library Catalog was developed and tested iteratively with multicultural populations in southern California over a five-year period. Research was conducted in public and private schools and in a major public library. In most of the experiments, the catalog data were those of the schools or public libraries studied.

At the time of the initial study, little prior research existed on children's information-related behavior. Few had studied how children search for information either in paper-based or electronic environments. Lacking a baseline specific to information-related behavior, we started by identifying what was known about the cognitive development and technical skills of children in the target age group. Research in education and in psychology revealed that children ages eight to twelve typically lacked basic skills requisite for the online catalogs of the day, such as typing, spelling, alphabetizing, and science vocabulary. However, children have other skills that could enable using alternative designs. These skills include the ability to use a pointing device, to browse, and to recognize terms and concepts that they may not be able to recall from memory.

The design was radically different from online catalogs, information-retrieval systems, or other digital libraries available at the time. Science and technology topics were displayed on cascading bookshelves, and only a mouse was needed to navigate. Topics were presented in a subject relationship (based on the Dewey Decimal Classification system, although the numbers were not displayed), providing context that is not evident in most online or card catalogs even today. Catalog records were reformatted to display as pages of the book. Basic catalog data were displayed as a title page in the familiar form they would appear in a children's book. Page corners were dog-eared so they could be turned to reveal more information where available. A map of the library was tucked into the book pocket. When the map was clicked, footprints traced a path from the location of the computer to the location of the bookshelf where the item was held.

Children found the metaphor familiar and appealing. Most could find books of interest in a 1,500-record database in a minute or so. We refined the user interface in a series of experiments, improving screen displays and navigation features. Ultimately, we developed a system that was easy to learn and highly effective for this user community, and results were consistent across the schools and public library studied.

We studied the system in enough different situations to believe that the results could be generalized to other elementary school-age children. In the hope of achieving generalizable results, we relied on widely available hardware, software, and content. The Science Library Catalog was developed in HyperCard on Macintosh computers, which was a common platform in elementary schools at the time. Input consisted of catalog records in the MAchine Readable Cataloging (MARC) format, following international standards.

However, we encountered problems of scaling and of migration to subsequent generations of technology. The largest database studied was about 8,000 records. That size strained the usability of the hierarchical browsing structure. Schools did not yet have internal networks, so the catalog could be used only from computers in the library. The searching metaphor was tied to a physical location so that a path could be traced from the computer to the physical location of the items described in the catalog. Considerable redesign, based on additional studies of user behavior, would be required to maintain the same level of tailoring while adapting the system to operate with larger databases, in networked environments, or with other systems in real time.

Case Study: Undergraduate Students Studying Geography
Our current research on the Alexandria Digital Earth ProtoType (ADEPT) addresses the design, development, deployment, and evaluation of a geographic digital library ("geolibrary") in undergraduate education (Borgman et al. 2000; Leazer, Gilliland-Swetland, and Borgman in press; Leazer, Gilliland-Swetland, Borgman, and Mayer 2000). Our thesis is that digital library services that provide instructors and students with a means to discover, manipulate, and display dynamic processes will contribute positively to undergraduate instruction and to student learning of scientific processes. Our research design involves a variety of qualitative and quantitative methods and is part of a five-year project (1999–2004) funded by the U.S. Digital Libraries Initiative, Phase Two (National Science Foundation 1999).

The Alexandria Digital Library (ADL) was developed at the University of California, Santa Barbara (UCSB) under the first Digital Libraries Initiative (1994–1998). ADL is an operational digital library that allows users scattered across the

Internet to access collections of maps, images, and other georeferenced materials from a 1.5 terabyte (and growing) collection of materials from UCSB's Map and Imagery Laboratory (⟨http://www.alexandria.ucsb.edu⟩). The operational version of ADL provides users with access to services that allow them to answer such questions as "What information is available about a given phenomenon at a particular set of places?" ADL also provides new types of library services based on gazetteers and other information access tools. ADL went online in fall 1999 as part of the California Digital Library (⟨http://www.cdlib.edu⟩).

Evaluation of the Alexandria Digital Library is reported in Hill et al. (2000). The ADEPT project extends and enhances the ADL for undergraduate instruction. We are taking a convergence approach to design, with the education and evaluation team focusing on needs assessment, evaluating prototypes in active use, and identifying system requirements. Concurrently, the ADEPT implementation team is focusing on evolving the ADL test-bed architecture and services, such as interface specifications, service prototypes, interoperability, and collection growth and diversity. Ours is an iterative and collaborative approach to development, with evaluation integrally embedded in design. Needs are identified from the user and collections perspective, prototypes are constructed and evaluated, and the results are fed back into the design and development process.

At this writing, we have completed the first year of our project. We now have baseline data, initial protocols and instruments, and basic system architecture. Our initial observations are first, that faculty approaches to teaching the same core course vary widely in intellectual framing of course content, teaching styles, and presentation of topics, which has implications for the design of ADEPT tools and resources. Second, faculty wish to integrate additional materials into ADEPT, which has implications for system design, management of intellectual property, and sharing of resources. Third, even faculty who employ high-end technology in their research tend to rely on chalkboards, overheads, and slide projectors for instruction, which has implications for technology adoption. A fourth observation is that display, layout, and other presentation features are important considerations. Visual context must be provided by clear labeling, zooming, use of recognizable geographic features, and other means. This observation has implications for metadata, for retrieval mechanisms, and for display capabilities. Continuing reports on the project will be provided on ADEPT Web sites at UCLA (⟨http://is.gseis.ucla.edu/adept⟩) and UCSB (⟨http://www.alexandria.ucsb.edu/adept⟩).

In sum, whether digital library design begins with the content, the collections, or the user community to be served, understanding the behavior, context, practices, expertise, and requirements of the prospective users is essential for improving usability.

Search Knowledge and Skills

Information technology applications involve complex cognitive tasks. Workers, learners, and users at all levels need to understand a variety of general computing concepts as well as concepts and skills specific to applications (National Research Council 1999). Viewing digital libraries from a variety of theoretical and practical perspectives sheds light on the knowledge and skills required for effective use.

Information needs, variously defined, are the usual starting point for studying information-seeking behavior. Other approaches consider problem situations, anomalous states of knowledge, or user goals (Belkin, Oddy, and Brooks 1982a, 1982b; Dervin and Nilan 1986; Hert 1996; Ingwersen 1984, 1996). Approaches may focus on the interaction between users and systems in searching for information (Belkin and Vickery 1985). Yet another approach is to view the search for information as a form of problem-solving behavior.

Problem Solving in Digital Libraries

Problem solving has been studied much more comprehensively than has information-seeking behavior. Problem solving offers a model for examining the nature of information-related problems, for studying methods of finding solutions, and for studying expert and novice behavior. An information need is a type of problem, and the solution is the information that fills the need. This section examines problems and solutions from this perspective.

Problems   From a cognitive perspective, all problems have three basic components (Glass, Holyoak, and Santa 1979, p. 392):

1. A set of *given information* or a description of the problem;
2. A set of *operations* or actions that the problem solver can use to achieve a solution;
3. A *goal* or description of what would constitute a solution to the problem.

Multiple types of problems exist, as do multiple types of knowledge that may contribute to solving them. Problems can be classified by the degree to which they are well defined or ill defined (Reitman 1964; Simon 1973). Well-defined problems are those in which the given information, operations, and goal are clearly specified. An example is an elementary algebra problem. Ill-defined problems tend to be open-ended: the given information, operation, or goal is not clearly specified. Information problems usually fall in the latter category.

The lack of definition of some problems is immediately apparent (such as "How can this device be improved?"). Others (such as "How many Japanese cars were manufactured in 1999?") may appear at first to be well defined but, on further exploration, turn out to be ambiguous. In a global economy, design and assembly can be distributed over multiple countries making it difficult to determine national responsibility (Reich 1992). If the query were interpreted as "How many cars were manufactured in Japan in 1999?," then cars manufactured by Japanese companies in plants outside Japan would be excluded. Also ambiguous are the terms *cars* and *manufactured*. Import and export regulations distinguish between passenger vehicles, utility vehicles, trucks, and vans; thus cars could be counted in different ways for different purposes. Similarly, *manufactured* could mean design, production of individual parts, or assembly.

Regardless of how the question is interpreted, the answers could lie in a number of different digital libraries, each with different representations and search capabilities. Documents containing automobile industry statistics are likely to vary in structure and content depending on their origin (for example, U.S. government trade statistics, Japanese government trade statistics, statistics of other governments, the automobile industry, the manufacturing companies, and industry analyses in the popular and trade presses). Statistics might be found in television news broadcasts, company promotional films, Web sites, and many other places. To obtain precise results with the intended meaning, considerable expertise would be required in the organization of content in individual databases in the mechanisms for controlling terminology and in the functional capabilities of each digital library searched. The final answer is of little value unless qualified by an explanation of how the concepts are interpreted and represented.

The Japanese cars problem is ostensibly one with a factual answer. Even more difficult to articulate clearly are questions "about" something. Most concepts can be expressed in multiple ways, and individual terms frequently represent multiple concepts, which vary by context. People typically generate queries from what they know about their problem, rather than what they know about information resources or what they know about the representation of concepts in those resources. Consequently, their initial queries may contain clues that lead them toward their goal or away from it.

The effect of initial queries on the path of a search can be illustrated by the question "Do you have any books about Brazil?" On further elaboration by a human (or automated) intermediary, this might become, "Do you have any books on Brazilian fish, written in English, and published after 1980?" If taken literally, the system would answer with records on books containing the words *fish* and *Brazil* or *Brazilian*, written in English, with a publication date of 1981 or later. Whether the result is relevant to the underlying information need is another matter. Further discussion with a skilled information professional may reveal that "Brazilian fish" is only the entry point for a much different problem, such as the following:

I'm gathering some background information for my neighborhood campaign to prevent a new development project that might pollute our river. I've heard that the Amazon River fish population was severely damaged by development in the 1980s. Maybe some of the data the environmentalists gathered there would be useful for our testimony, but I can't read Portuguese. What I really need is environmental data on local species and local river conditions, and Brazil seemed like a good place to start.

The most useful information in response to this query may have little to do with Brazil. A better result would be environmental studies performed in conditions similar to the local river, experiences of community groups in challenging development projects, and guidance in presenting testimony to government agencies. Furthermore, the most relevant content may exist in papers, records, reports, articles, videos, films, or tapes rather than in books. It is possible (though not evident in the refined query) that Brazil could be relevant for other reasons. Perhaps one of the people hearing testimony has experience there and would find Brazilian examples particularly salient.

Another relevant finding from the problem-solving literature is that the degree to which a problem is well defined or ill defined is partly a function of the knowledge and skills of the problem solver (Glass, Holyoak, and Santa 1979). For example, in the queries above, an expert such as an automotive industry analyst or an environmentalist could articulate the queries more specifically and completely than could a novice to the domain. Similarly, an expert in the use of a particular information system can specify a problem in terms appropriate to that system better than can someone unfamiliar with that system. Human search intermediaries combine their knowledge of a subject domain and of information-seeking behavior with their skills in searching information systems to assist people in articulating their problems. They often ask people about the purposes for which they want information (such as obtaining a job, finding child care, researching a term paper, establishing a business) and elicit additional details concerning the problem.

Solutions   Most of the work on problem solving follows from George Polya's (1957) classic model, which has been applied in contexts ranging from mathematical problems to creative thinking (Glass, Holyoak, and Santa 1979; Koberg and Bagnall 1972). It is particularly useful as a model for solving information problems.

Polya (1957) divides the problem-solving process into four steps: (1) understanding the problem, (2) planning a solution, (3) carrying out the plan, and (4) checking the results. These steps are iterative. Checking the results occurs at multiple points in the process. For example, a plan for a solution may begin with some preliminary searching to explore the problem and then continue with more detailed searching along the most promising paths identified. Interim results are assessed to determine subsequent actions.

The amount and type of planning that goes into solving an information-related problem is a function of several factors. One factor is the degree of problem definition. Well-defined problems such as finding an e-mail address require less planning than ill-defined problems such as finding a birthday gift for a friend. Information problems that appear to be well defined often can turn out to be ambiguous, as illustrated above. In many cases, some initial searching is required to determine the scope of the problem before developing a plan.

A second factor is the amount of expertise the searcher has in the problem domain. Expertise is relative to the problem at hand. Everyone is an expert with regard to some things and a novice with regard to others. Even in an area of expertise, the amount of knowledge about a given problem may vary by stage of search. People gain more knowledge of a problem through exploring it, which influences subsequent steps in the search process.

A third factor that influences planning is knowledge about the resources and operations available to solve the problem. In the case of information problems, searchers need to have knowledge about relevant information sources and strategies to search them. As people become familiar with the range of sources and search capabilities, their planning and searching improve. For example, students in medicine and law gradually become more proficient searchers as they become more knowledgeable about technical terminology, information resources in their fields, metadata available to represent those resources, and the search capabilities for each system and collection.

Search Process

Despite the number of years people spend in formal education, few receive formal instruction either in general problem solving or in searching for information. Information searching is a process that most people learn through experience. Knowledge may be gathered through experiences in libraries, archives, museums, and laboratories and by using information systems. Some people are able to extract general principles, apply them to multiple systems and situations, and become expert searchers. A few become proficient intermittent users, although most remain "permanent novices" with respect to searching digital libraries.

Research on the use of information-retrieval systems reveals great disparities in the use of system features. Novices tend to rely on the most basic features, often engage in short search sessions, and rarely take advantage of sophisticated search refinement capabilities. Intermittent users may use a few more features but rely on a small set of familiar capabilities. Experts are those who use a combination of features, often taking an iterative approach that tests multiple strategies for finding the information sought. Experts are able to combine features in sophisticated ways appropriate to a given problem. Although experts may draw on a common set of known strategies, they tend to develop individualized approaches to searching. Given the same statement of a problem, expert searchers often produce diverse sets of results from the same system due to differences in interpretation of the problem, in choice of terminology, and in choice of features (Borgman 1989).

Studies of expert searchers reveal knowledge and skills that contribute to effective and efficient searching. These techniques can be taught to novices. Some techniques can be incorporated into system features, such as offering users prescribed tactics for broadening or narrowing searches. The requisite knowledge and skills for searching can be categorized in a variety of ways. General knowledge of computing has been divided into syntactic and semantic categories (Shneiderman 1980), object and action dichotomies (Shneiderman 1998), conceptual, semantic, syntactic, and lexical categories (Foley, Van Dam, Feiner, and Hughes 1990), or contemporary skills, foundational concepts, and intellectual capabilities (National Research Council 1999). Combining models for computing, problem solving, and information seeking, the author (Borgman 1986b, 1996) proposed a model of the knowledge and skills required to search for information in digital libraries. The model includes

• Conceptual knowledge of the information retrieval process,

• Semantic and syntactic knowledge of how to implement a query in a given system, and

• Technical skills in executing the query.

These three categories are explored, drawing examples from a broad range of digital libraries: online catalogs, bibliographic and full-text databases, geographic information systems, and World Wide Web–based information systems.

Conceptual Knowledge   The term *conceptual knowledge* refers to the user's model or understanding of a given type of digital libraries, spreadsheets, or word processing system. Users employ their conceptual knowledge of the search process to translate an information need into a plan for executing a search. Experts analyze the problem, determine goals, break the problem into component parts, survey the sources available that may contain relevant information, and make a plan for conducting the search in one or more digital libraries. They carry out their plan, continually checking progress toward their goals, and revise their strategy accordingly.

In the 1970s and 1980s, studies of skilled searchers on bibliographic retrieval systems were distilled into strategies and tactics for information retrieval that could be taught to novices and codified in textbooks for online searching (Bates 1979, 1981, 1984; Borgman 1989; Borgman, Moghdam, and Corbett 1984; Lancaster and Fayen 1973). More recent studies of the World Wide Web and geographic information systems yield similar results about the role of conceptual knowledge in searching. The ability to construct a mental model of an information space continues to be a key predictor of searching success in multiple types of digital libraries (Dillon 2000; Dillon and Gabbard 1998; Priss and Old 1998).

Expert searchers manage searching processes that typically confound or discourage novices. Searchers commonly encounter one or more unsatisfactory situations: search failures (no matches), excess information (too many matches) (Larson 1991b), and irrelevant matches. Studies of searching the World Wide Web reveal similar patterns, with about 30 percent of searches resulting in no matches on some search engines despite the massive amount of content online (Shneiderman, Byrd, and Croft 1997).

When experts encounter no matches, they typically expand the search by framing the topic differently. They refer to term lists, thesauri, or other tools to identify synonyms or more general terms that will improve recall, for example. They may release constraints on the topic, such as date, language, or format. Experts are aware that known facts such as personal names, places, titles, manufacturers, and dates often are incorrect and that searching for them should be generalized. Similarly, when experts encounter too many matches, they may frame the topic in narrower terms, add constraints, or search for a subset of the problem, all of which may improve precision. When experts encounter too many irrelevant matches, they recognize that their choice of terms or parameters may not match those in the digital library adequately. They may reframe the search with other terms and tools in that digital library or look for other collections that may be more suitable.

Vocabulary continues to be the most difficult aspect of searching for any type of information, whether for text, images, numeric data, audio, or any combination of these. Documents, places, ideas, and objects are described differently by those who create them and those who seek them. Metadata play an essential role in access by describing and representing content in consistent ways (Dempsey and Heery 1998; Gilliland-Swetland 1998; Lynch and Preston 1990, 1991; Lynch et al. 1995; Marshall 1998). Even so, mapping from searchers' *entry vocabulary* or starting points to unfamiliar metadata vocabularies remains difficult (Batty 1998; Buckland et al. 1999). People searching for train schedules from Rome to Naples must map their vocabulary to terms such as *rail*, *railway*, or *Eurorail* rather than *train* and to *Roma* and *Napoli* rather than *Rome* and *Naples*, for example. These simple mappings are relatively unambiguous and often automatic. Less obvious, and less familiar to those not speaking the local language, are the equivalence of *Vienna* (English), *Wien* (German), and *Bécs* (Hungarian) for the capital of Austria or of *Prague* (English) and *Praha* (Czech) for the capital of the Czech Republic.

More complex mappings require more conceptual knowledge of how vocabularies are structured. Michael Buckland and his colleagues (1999) offer the example of searching for *rockets* in the Census Bureau U.S. Imports and Exports database, which employs a specialized categorization scheme. The plural term *rockets*, for example, yields only one category: "bearings, transmission, gaskets, misc." while the singular term *rocket* yields three other categories: "photographic or cinematographic goods," "engines, parts, etc.," and "arms and ammunition, parts and accessories thereof." The term *rocket* appears only in a subcategory of the latter term, which is "missile and rocket launchers and similar projectors." Missing altogether from a search on the term *rocket* or *rockets* are general categories that probably are of interest: "guided missiles" and "bombs, grenades, etc."

Specialized vocabulary structures such as these enable subject domain experts to specify precise categories yet also require that searchers explore the structure sufficiently to identify all possibilities. Automatic mapping between terms is difficult because relationships depend on context. Categories that are synonymous for one problem are not for another, unlike the simple mapping between *Vienna* and *Wien* on rail schedules. Even the terms *rail* and *train* are synonymous only in certain contexts.

These are but a few of many examples of strategies and tactics that experts use in planning and executing online searches. General patterns exist, such as "berry picking" relevant results from multiple digital libraries over multiple searches (Bates 1989). Some patterns are indirect, such as "pearl growing," which starts with a core of one or a few known relevant documents and spirals outward for other materials that are related to the starting set (Borgman, Moghdam, and Corbett 1984).

In comparison to expert searchers, studies of novices and intermittent users of online catalogs and other digital libraries reveal little evidence of search planning or search refinement strategies. Nonexpert searchers are more likely to search intuitively than to use advanced features intended to make searches more efficient and effective. For example, novices will use familiar terminology as keywords without verifying that their chosen terms exist in the database. They have particular difficulty recovering from problems involving subject terminology.

In contrast, experts will employ tools such as subject thesauri, classification structures, and name authority files to identify the most promising terminology and appropriate synonyms. When experts retrieve unsatisfactory results, their reflex is to reframe the search. Novices, however, often are unaware of what they are missing and fail to distinguish between poor results due to the contents of the digital library and poor results due to an inadequate strategy. Vocabulary problems arise in all types of digital library searching (Bates 1986, 1989; Berger 1994; Bilal 1998; Blair and Maron 1985; Borgman et al. 1995; Crawford, Thom, and Powles 1993; Efthimiadis 1992, 1993; Hildreth 1993; Hirsh 1998; Lancaster, Connell, Bishop, and McCowan 1991; Larson 1991a, 1991b, 1991c; Leazer 1994; Markey 1984, 1986; Markey and Demeyer 1986; Matthews, Lawrence, and Ferguson 1983; McGarry and Svenonius 1991; Micco 1991; O'Brien 1994; Rosenberg and Borgman 1992; Taylor 1984; Tillett 1991; Walker 1988; Walker and Hancock-Beaulieu 1991). Even in online shopping, inconsistent description of products is emerging as one of the greatest sources of searching difficulties (Lohse and Spiller 1998).

Semantic and Syntactic Knowledge   Conceptual knowledge of the information-retrieval process is used to plan and refine searches. Semantic knowledge is understanding the operations available to execute a plan, such as choosing among types of capabilities that exist for searching. Syntactic knowledge is an understanding of the commands or actions in a specific system (Shneiderman 1992).

Expert searchers' semantic knowledge includes familiarity with capabilities common to most information systems, such as keyword searching, Boolean combinations, browsing thesauri, typical sorting and display options, hypertext features, and so on. These searchers also are knowledgeable about capabilities that may be specific to types of systems (such as text, numeric, image, geographic) and multiple implementations of each. Before searching an unfamiliar digital library, experts usually analyze the documentation and other explanatory materials to determine its features, capabilities, and data representations and then plan their search accordingly. Search capabilities operate somewhat differently on each system, so experts know to examine general retrieval functions and database-specific features and may experiment to determine the interactions between them (Borgman 1996).

In the example of the search for *rockets* (Buckland et al. 1999), novices might do only a keyword search on the term *rockets* and retrieve an incomplete and unrepresentative set of results. Experts, by comparison, usually would employ their semantic knowledge of vocabulary structure to explore the hierarchy of categories. Experts want to know how results are achieved and judge the completeness and accuracy of results accordingly. For example, they need to know whether a term such as *rocket* is being matched in its singular and plural forms and whether the term is a preferred form that is picking up synonyms as well. Novices are more likely to accept the results provided, lacking critical skills to assess how the results are achieved.

Because Boolean operators are implemented in a variety of ways, experts pay considerable attention to Boolean execution algorithms in judging the results from a system. For example, some systems would treat a search for the book title *Usability Engineering* (Nielsen 1993), if entered in that form, as an implicit AND. Others would treat it as an implicit OR or as a "bound phrase" in which both terms, in the specified order, must appear. If treated as AND, only documents containing both terms would be retrieved. This could be a large set if each term could appear anywhere in a full-text document and a much smaller set if the search were restricted to titles, for example. If treated as OR, a massive set could be retrieved, containing all the documents that contain either term. Given the frequency of the term *engineering* in technical databases, this is an unwelcome outcome. If treated as a bound phrase, then only phrases with these two terms in this sequence would be retrieved, which is a desirable result for this particular search. Alternative treatments include retrieving documents with these words in this sequence but allowing a small number of terms to intervene, stemming each term such that matches on words beginning with *usab* or *engineer* are retrieved, or ignoring the second term and retrieving only on *usability*.

The market for information systems has changed profoundly since the early days of information retrieval. When expert intermediaries conducted searching, extensive documentation was provided on search features. Searchers learned how to manipulate systems precisely, based on semantic and syntactic knowledge of specific functions and on knowledge of the comparative features of different systems. When the same database was available on multiple search services, searchers would select which to use based on search engine capabilities as well as on price and other features (Borgman, Moghdam, and Corbett 1984). Most digital libraries now are intended for end users, with minimal documentation or formal training provided. The exceptions are commercially provided scientific, technical, medical, and legal databases that are marketed to experts and professionals and priced accordingly. Specific information on how search functions are executed that was once well documented now may be considered proprietary, especially by Internet search engines. Even expert searchers cannot ascertain how the terms they enter are being treated and thus how to assess the completeness or reliability of results. In the present situation, it is often difficult to obtain adequate semantic or syntactic knowledge about a system to evaluate or compare results.

Technical Skills   A prerequisite to developing syntactic, semantic, and conceptual knowledge about the search process is basic computing skills. Implicit technical skills needed to search any digital library include knowing how to use computer keyboards and pointing devices and familiarity with conventions such as the arrangement of screen displays and pressing Return or Enter after typing a command. Sometimes users must recognize that on-screen buttons such as Enter, Start, Search, or Return are equivalent.

User interfaces have become more consistent in recent years with the adoption of interface design guidelines specific to hardware platforms and operating systems. At a minimum, users can expect consistency in basic operations such as opening and closing windows, pulling down menus, and cutting, copying, and pasting text or objects. Beyond core features, however, each system remains unique, and users still must learn where to point and click and what, where, and when to type.

These technical skills are obvious to proficient computer users but should not be taken for granted in the general population. On first encountering a computer, the uninitiated often start by pointing a mouse directly at the screen like a remote control device, for example. The need to move a mouse on a flat surface perpendicular to the plane in which its reflection occurs is counterintuitive, with few real-world analogies. Telephone help lines are plagued with new users who ask where the Any key is located in their attempts to follow instructions such as "Press any key to begin." This level of naiveté may be declining in the United States and other countries with extensive penetration of personal computers. It continues to exist to varying degrees with new users and will be a factor in the introduction of computers to other parts of the world.

The global information infrastructure is intended for broad penetration in homes, schools, libraries, museums, offices, and other institutions that support diverse communities. The level of technical skills required for searching digital libraries varies widely by application, from the most basic to highly advanced. As digital libraries are designed for more general audiences, a broader range of skill levels will need to be accommodated in many applications. In addition to people who lack general literacy skills, people with physical, sensory, or cognitive disabilities account for 15 to 20 percent of the U.S. population (Computer Science and Telecommunications Board 1997, p. 38). Many disabilities involve reading, vision, manual dexterity, or other factors that limit use of computers. Systems that can accommodate people with disabilities tend to be easier for most people to use (Computer Science and Telecommunications Board 1997).

Summary and Conclusions

This chapter explores usability issues in digital libraries, ranging from ease-of-use criteria to the knowledge and skills necessary for searching. As tasks become more complex and the relationships between task and action become more abstract, technologies become more difficult to use. Real-world analogies disappear, replaced by commands, menus, displays, keyboards, and pointing devices.

Minimum criteria for usability, as derived from research on human-computer interaction, are that systems should be easy to learn, tolerant of errors, flexible, adaptable, and appropriate and effective for the task. While evidence is mounting for the economic value of usability evaluation and iterative design, applying these criteria to design is neither simple nor straightforward. Systems vary widely in audience and application, and the criteria must be applied accordingly. The design of digital libraries can begin with available collections or with a user community to be served. In either case, design must be driven by who will use the content, how, and why.

People make judgments about all aspects of seeking, using, and creating information. They judge the usability of systems, the value of the content retrieved, and its relevance to their problem. Although relevance is subjective, it remains a useful construct for framing searches. Search goals can stress precision, casting the net narrowly to find a few good matches, or can stress recall, casting the net widely to find as many relevant matches as possible.

Searching for information in digital libraries is a form of problem solving. The problem-solving process can be divided into four steps—understanding the problem, planning a solution, carrying out the plan, and checking the results. Several kinds of knowledge are involved in solving information problems. Conceptual knowledge is applied to framing problems and formulating plans for solving them. Semantic knowledge enables searchers to choose among operations for solving problems. Syntactic knowledge is used to execute the plan. Technical skills in the use of computers are needed to employ all of the other knowledge. Experts plan their searches and reformulate them when too many, too few, or the wrong matches are retrieved. Novices, in comparison, often are stymied by unsuccessful searches. They abandon searches rather than reformulate them and show little evidence of planning or strategic actions.

Experts have a variety of strategies and tactics to overcome poor design in digital libraries. Novices do not. Nor will novices tolerate poor design if they have other alternatives. The audience for digital libraries has changed radically since the early days of information retrieval, from expert search intermediaries to every citizen who has access to the network. The next generation of digital libraries must serve a large and diverse community and provide a large and diverse collection of information resources. While we do not yet know how to build such a system, a starting point is to employ what is known about information-related behavior in the systems of today toward building better systems tomorrow.

Acknowledgments

Portions of this chapter are drawn from Christine L. Borgman, *From Gutenberg to the Global Information Infrastructure: Access to Information in the Networked World* (Cambridge, MA: MIT Press, 2000), also in the MIT Press Digital Libraries and Electronic Publishing Series.

References

Bates, M. J. 1979. Information Search Tactics. *Journal of the American Society for Information Science*, 30(4), 205–214.

Bates, M. J. 1981. Search Techniques. In M. E. Williams, ed., *Annual Review of Information Science and Technology* (vol. 16, pp. 139–169). New York: Knowledge Industry for American Society for Information Science.

Bates, M. J. 1984. The Fallacy of the Perfect Thirty-Item Online Search. *RQ*, 24(1), 43–50.

Bates, M. J. 1986. Subject Access in Online Catalogs: A Design Model. *Journal of the American Society for Information Science*, 37(6), 357–376.

Bates, M. J. 1989. The Design of Browsing and Berry-Picking Techniques for the Online Search Interface. *Online Review*, 13(5), 407–424.

Bates, M. J. 1990. Where Should the Person Stop and the Information Search Interface Start? *Information Processing and Management*, 26(5), 575–591.

Battin, P. 1998. Leadership in a Transformational Age. In B. L. Hawkins and P. Battin, eds., *The Mirage of Continuity: Reconfiguring Academic Information Resources for the 21st Century* (pp. 260–270). Washington, DC: Council on Library and Information Resources and the Association of American Universities.

Batty, D. 1998. WWW: Wealth, Weariness, or Waste. *D-Lib Magazine*, 4(11). ⟨http://www.dlib.org/dlib/november98/11batty.html⟩.

Belkin, N. J., R. N. Oddy, and H. M. Brooks. 1982a. ASK for Information Retrieval: Part I. Background and Theory. *Journal of Documentation*, 38(2), 61–71.

Belkin, N. J., R. N. Oddy, and H. M. Brooks. 1982b. ASK for Information Retrieval: Part II. Results of a Design Study. *Journal of Documentation*, 38(3), 145–164.

Belkin, N. J., and A. Vickery. 1985. *Interaction in Information Systems: A Review of Research from Document Retrieval to Knowledge-Based Systems*. Library and Information Research Report No. 35. London: British Library.

Berger, M. G. 1994. Information-Seeking in the Online Bibliographic System: An Exploratory Study. Ph.D. dissertation, University of California, Berkeley, School of Library and information Studies. UMI #AAI9504745.

Bilal, D. 1998. Children's Search Processes in Using World Wide Web Search Engines: An Exploratory Study. In R. Larson, K. Petersen, and C. M. Preston, eds., *ASIS '98: Proceedings of the 61st American Society for Information Science Annual Meeting (Pittsburgh, PA, October 24–29, 1998)* (vol. 35, pp. 45–53). Medford, NJ: Information Today.

Bishop, A. P., and S. L. Star. 1996. Social Informatics for Digital Library Use and Infrastructure. In M. E. Williams, ed., *Annual Review of Information Science and Technology* (vol. 31, pp. 301–401). Medford, NJ: Information Today.

Blair, D. C., and M. E. Maron. 1985. An Evaluaton of Retrieval Effectiveness for a Full-Text Document Retrieval System. *Communications of the ACM*, 28(3), 289–299.

Borgman, C. L. 1984. Psychological Research in Human-Computer Interaction. In M. E. Williams, ed., *Annual Review of Information Science and Technology* (vol. 19, pp. 33–64). White Plains, NY: Knowledge Industry.

Borgman, C. L. 1986a. The User's Mental Model of an Information Retrieval System: An Experiment on a Prototype Online Catalog. *International Journal of Man-Machine Studies*, 24(1), 47–64.

Borgman, C. L. 1986b. Why Are Online Catalogs Hard to Use? Lessons Learned from Information Retrieval Studies. *Journal of the American Society for Information Science*, 37(6), 387–400.

Borgman, C. L. 1989. All Users of Information Systems Are Not Created Equal: An Exploration into Individual Differences. *Information Processing and Management*, 25(3), 237–252.

Borgman, C. L. 1996. Why Are Online Catalogs Still Hard to Use? *Journal of the American Society for Information Science*, 47(7), 493–503.

Borgman, C. L. 1999. What Are Digital Libraries? Competing Visions. *Information Processing and Management*, 35(3) (special issue on Progress toward Digital Libraries), 227–243.

Borgman, C. L. 2000. *From Gutenberg to the Global Information Infrastructure: Access to Information in the Networked World*. Cambridge, MA: MIT Press.

Borgman, C. L., M. J. Bates, M. V. Cloonan, E. N. Efthimiadis, A. Gilliland-Swetland, Y. Kafai, G. L. Leazer, and A. Maddox. 1996. *Social Aspects of Digital Libraries*. Final Report to the National Science Foundation; Computer, Information Science, and Engineering Directorate; Division of Information, Robotics, and Intelligent Systems; Information Technology and Organizations Program. Award 95-28808. ⟨http://www-lis.gseis.ucla.edu/DL/Report.html⟩.

Borgman, C. L., D. O. Case, and C. T. Meadow. 1989. The Design and Evaluation of a Front End User Interface for Energy Researchers. *Journal of the American Society for Information Science*, 40(2), 86–98.

Borgman, C. L., A. L. Gallagher, S. G. Hirsh, and V. A. Walter. 1995. Children's Searching Behavior on Browsing and Keyword Online Catalogs: The Science Library Catalog Project. *Journal of the American Society for Information Science*, 46(9), 663–684.

Borgman, C. L., A. J. Gilliland-Swetland, G. L. Leazer, R. Mayer, D. Gwynn, and R. Gazan. (2000). Evaluating Digital Libraries for Teaching and Learning in Undergraduate Education: A Case Study of the Alexandria Digital Earth Prototype (ADEPT). *Library Trends*, 49(2), 228–250.

Borgman, C. L., S. G. Hirsh, and J. Hiller. 1996. Rethinking Online Monitoring Methods for Information Retrieval Systems: From Search Product to Search Process. *Journal of the American Society for Information Science*, 47(7), 568–583.

Borgman, C. L., D. Moghdam, and P. K. Corbett. 1984. *Effective Online Searching: A Basic Text*. New York: Dekker.

Buckland, M. K., A. Chen, H.-M. Chen, Y. Kim, B. Lam, R. Larson, B. Norgard, and J. Purat. 1999. Mapping Entry Vocabularies to Unfamiliar Metadata Vocabularies. *D-Lib Magazine*, 5(1). ⟨http://www.dlib.org/dlib/january99/buckland/01buckland.html⟩.

Case, D. O., C. L. Borgman, and C. T. Meadow. 1986. End-User Information-Seeking in the Energy Field: Implications for End-User Access to DOE RECON Databases. *Information Processing and Management*, 22(4), 299–308.

Cleverdon, C. W. 1964. *Evaluation of Operational Information Retrieval Systems. Part 1: Identification of Criteria*. Cranfield, UK: Cranfield College of Aeronautics.

Computer Science and Telecommunications Board; Commission on Physical Sciences, Mathematics, and Applications; National Research Council. 1997. *More Than Screen Deep: Toward Every-Citizen Interfaces to the Nation's Information Infrastructure*. Washington, DC: National Academy Press.

Crawford, J. C., L. C. Thom, and J. A. Powles. 1993. A Survey of Subject Access to Academic Library Catalogues in Great Britain. *Journal of Librarianship and Information Science*, 25(2), 85–93.

Dempsey, L., and R. Heery. 1998. Metadata: A Current Review of Practice and Issues. *Journal of Documentation*, 54(2), 145–172.

Dervin, B., and M. Nilan. 1986. Information Needs and Uses. In M. E. Williams, ed., *Annual Review of Information Science and Technology* (vol. 21, pp. 1–25). White Plains, NY: Knowledge Industry.

Dillon, A. 2000. Spatial-Semantics: How Individual Users Perceive Shape in Information Space. *Journal of the American Society for Information Science*, 51(6), 521–528.

Dillon, A., and R. Gabbard. 1998. Hypermedia as an Educational Technology: A Review of the Quantitative Research Literature on Learner Comprehension, Control, and Style. *Review of Educational Research*, 68(3), 322–349.

Edwards, P. 1995. *The Closed World: Computers and the Politics of Discourse*. Cambridge, MA: MIT Press.

Efthimiadis, E. N. 1992. Interactive Query Expansion and Relevance Feedback for Document Retrieval Systems. Ph.D. dissertation, City University, London.

Efthimiadis, E. N. 1993. A User-Centered Evaluation of Ranking Algorithms for Interactive Query Expansion. In R. Korfhage, E. Rasmussen, and P. Willett, eds., *Proceedings of the 16th International Conference of the Association of Computing Machinery, Special Interest Group on Information Retrieval (Pittsburgh, PA, June 1993)* (pp. 146–159). New York: ACM Press.

Egan, D. E. 1988. Individual Differences in Human-Computer Interaction. In M. Helander, ed., *Handbook of Human-Computer Interaction* (pp. 543–568). Amsterdam: Elsevier.

Egan, D. E., J. R. Remde, T. K. Landauer, C. C. Lochbaum, and L. M. Gomez. 1989. Behavioral Evaluation and Analysis of a Hypertext Browser. In K. Bice and C. Lewis, eds., *CHI '89: Human Factors in Computing Systems Conference Proceedings* (pp. 205–210). New York: ACM Press.

Ehn, P. 1988. *Work-Oriented Design of Computer Artifacts*. Stockholm: Arbetslivscentrum.

Europe and the Global Information Society. 1994. *Recommendations to the European Council* ("the Bangemann Report"). Brussels: European Council.

Foley, J. D., A. Van Dam, S. K. Feiner, and J. F. Hughes. 1990. *Computer Graphics: Principles and Practice* (2nd ed.). Reading, MA: Addison-Wesley.

Fox, E. A., ed. 1993. *Sourcebook on Digital Libraries: Report for the National Science Foundation*. TR-93-35. Blacksburg, VA: Computer Science Department, Virginia Polytechnic Institute and State University. Available by anonymous FTP from ⟨directory pub/DigitalLibraryonfox.cs.vt.edu⟩ or at ⟨http://fox.cs.vt.edu/DLSB.html⟩.

Fox, E. A., R. M. Akscyn, R. K. Furuta, and J. J. Leggett. 1995. Digital Libraries. *Communications of the ACM*, 38(4) (special issue), 22–28.

Gilbreth, L. M. 1921. *The Psychology of Management: The Function of the Mind in Determining, Teaching and Installing Methods of Least Waste*. New York: Macmillan.

Gilliland-Swetland, A. 1998. Defining Metadata. In M. Baca, ed., *Introduction to Metadata: Pathways to Digital Information*. Los Angeles: Getty Information Institute.

Glass, A. L., K. J. Holyoak, and J. L. Santa. 1979. *Cognition*. Reading, MA: Addison-Wesley.

Greenberg, D. 1998. Camel Drivers and Gatecrashers: Quality Control in the Digital Research Library. In B. L. Hawkins and P. Battin, eds., *The Mirage of Continuity: Reconfiguring Academic Information Resources for the 21st Century* (pp. 105–116). Washington, DC: Council on Library and Information Resources and the Association of American Universities.

Harter, S. P., and C. A. Hert. 1997. Evaluation of Information Retrieval Systems: Approaches, Issues, and Methods. In M. E. Williams, ed., *Annual Review of Information Science and Technology* (vol. 32, pp. 3–94). Medford, NJ: Information Today.

Hert, C. A. 1996. User Goals on an Online Public Access Catalog. *Journal of the American Society for Information Science*, 47(7), 504–518.

Hildreth, C. R. 1993. An Evaluation of Structured Navigation for Subject Searching in Online Catalogues. Ph.D. dissertation, Department of Information Science, City University, London.

Hill, L. L., L. Carver, M. Larsgaard, R. Dolin, T. R. Smith, J. Frew, and M. A. Rae. 2000. Alexandria Digital Library: User Evaluation Studies and System Design. *Journal of the American Society for Information Science*, 51(3), 246–259.

Hirsh, S. G. 1998. Relevance Determinations in Children's Use of Electronic Resources: A Case Study. In R. Larson, K. Petersen, and C. M. Preston, eds., *ASIS '98: Proceedings of the 61st American Society for Information Science Annual Meeting (Pittsburgh, PA, October 24–29, 1998)* (vol. 35, pp. 63–72). Medford, NJ: Information Today.

Ingwersen, P. 1984. Psychological Aspects of Information Retrieval. *Social Science Information Studies*, 4(2/3), 83–95.

Ingwersen, P. 1996. Cognitive Perspectives of Information Retrieval Interaction: Elements of a Cognitive IR Theory. *Journal of Documentation*, 52(1), 3–50.

Koberg, D., and J. Bagnall. 1972. *The Universal Traveler: A Soft-Systems Guidebook to Creativity, Problem-Solving and the Process of Design*. Los Altos, CA: Kaufman.

Lancaster, F. W. 1968. *Information Systems: Characteristics, Testing, and Evaluation*. New York: Wiley.

Lancaster, F. W., T. H. Connell, N. Bishop, and S. McCowan. 1991. Identifying Barriers to Effective Subject Access in Library Catalogs. *Library Resources and Technical Services*, 35(2), 377–391.

Lancaster, F. W., and E. G. Fayen. 1973. *Information Retrieval On-Line*. Los Angeles: Melville.

Landauer, T. K. 1995. *The Trouble with Computers: Usefulness, Usability and Productivity.* Cambridge, MA: MIT Press.

Larson, R. R. 1991a. Between Scylla and Charybdis: Subject Searching in the Online Catalog. In *Advances in Librarianship* (vol. 15, pp. 175–236). San Diego: Academic Press.

Larson, R. R. 1991b. Classification Clustering, Probabilistic Information Retrieval and the Online Catalog. *Library Quarterly*, 61(2), 133–173.

Larson, R. R. 1991c. The Decline of Subject Searching: Long-Term Trends and Patterns of Index Use in an Online Catalog. *Journal of the American Society for Information Science*, 42(3), 197–215.

Leazer, G. H. 1994. A Conceptual Schema for the Control of Bibliographic Works. In D. L. Andersen, T. J. Galvin, and M. D. Giguere, eds., *Navigating the Networks: Proceedings of the ASIS Mid-Year Meeting (Portland, Oregon, May 21–25, 1994)* (pp. 115–135). Medford, NJ: Learned Information.

Leazer, G. L., A. J. Gilliland-Swetland, and C. L. Borgman. 2000. Classroom Evaluation of the Alexandria Digital Earth Prototype (ADEPT). In *Proceedings of the American Society for Information Science Annual Meeting (Chicago, November 2000)* (vol. 37, pp. 334–340). Medford, NJ: Information Today.

Leazer, G. L., A. J. Gilliland-Swetland, C. L. Borgman, and R. Mayer. 2000. Evaluating the Use of a Geographic Digital Library in Undergraduate Classrooms: The Alexandria Digital Earth Prototype (ADEPT). In *Proceedings of the Fifth ACM Conference on Digital Libraries (San Antonio, TX, June 2–7, 2000)* (pp. 248–249). New York: ACM Press.

Lesk, M. E. 1997. *Practical Digital Libraries: Books, Bytes, and Bucks.* San Francisco: Morgan Kaufman.

Leventhal, L., A.-M. Lancaster, A. Marcus, B. Nardi, J. Nielsen, M. Kurosu, and R. Heller. 1994. Designing for Diverse Users: Will Just a Better Interface Do? In C. Plaisant, ed., *CHI '94, Human Factors in Computing Systems, Conference Companion (Boston, MA, April 24–28, 1994)* (pp. 191–192). New York: ACM Press.

Levy, D. M., and C. C. Marshall. 1995. Going Digital: A Look at the Assumptions Underlying Digital Libraries. *Communications of the ACM*, 38(4), 77–84.

Lohse, G. L., and P. Spiller. 1998. Electronic Shopping. *Communications of the ACM*, 41(7), 81–87.

Lucier, R. E. 1995. Building a Digital Library for the Health Sciences: Information Space Complementing Information Place. *Bulletin of the Medical Library Association*, 83(3), 346–350.

Lyman, P. 1996. What Is a Digital Library? Technology, Intellectual Property, and the Public Interest. In *Daedalus, Journal of the American Academy of Arts and Sciences*, 125(4), 1–33. Republished in S. R. Graubard and P. LeClerc, eds. 1998. *Books, Bricks, and Bytes: Libraries in the Twenty-first Century* (chs. 2, 4, 7). New Brunswick, NJ: Transaction.

Lynch, C. A. 1993. *Accessibility and Integrity of Networked Information Collections.* Background Paper No. BP-TCT-109. Washington, DC: Office of Technology Assessment.

Lynch, C., and H. Garcia-Molina. 1995. Interoperability, Scaling, and the Digital Libraries Research Agenda. ⟨http://www.hpcc.gov/reports/reports-nco/iita-dlw/main.html⟩.

Lynch, C. A., and C. M. Preston. 1990. Internet Access to Information Resources. In M. E. Williams, ed., *Annual Review of Information Science and Technology* (vol. 25, pp. 263–312). Amsterdam: Elsevier.

Lynch, C. A., and C. Preston. 1991. Evolution of Networked Information Resources. In M. E. Williams, ed., *National Online Meeting 1991: Proceedings of the 12th National Online Meeting (New York, NY, May 7–9, 1991)* (pp. 221–230). Medford, NJ: Learned Information.

Lynch, C., A. Michelson, C. Summerhill, and C. Preston. 1995. *Information Discovery and Retrieval*. Washington, DC: Coalition for Networked Information. ⟨ftp.cni.org⟩.

Markey, K. 1984. *Subject Searching in Library Catalogs: Before and after the Introduction of Online Catalogs*. Dublin, OH: OCLC Online Computer Library Center.

Markey, K. 1986. Users and the Online Catalog: Subject Access Problems. In J. R. Matthews, ed., *The Impact of Online Catalogs* (pp. 35–69). New York: Neal-Schuman.

Markey, K., and A. N. Demeyer. 1986. *Dewey Decimal Classification Online Project: Evaluation of a Library Schedule and Index Integrated into the Subject Searching Capabilities of an Online Catalog: Final Report to the Council on Library Resources*. Dublin, OH: OCLC Online Computer Library Center.

Marshall, C. C. 1998. Making Metadata: A Study of Metadata Creation for a Mixed Physical-Digital Collection. In *Digital Libraries '98: Proceedings of the 3rd ACM Conference on Digital Libraries (Pittsburgh, PA, June 23–26, 1998)* (pp. 162–171). New York: ACM Press.

Matthews, J. R., G. S. Lawrence, and D. K. Ferguson. 1983. *Using Online Catalogs: A Nationwide Survey*. New York: Neal-Schuman.

McGarry, D., and E. Svenonius. 1991. More on Improved Browsable Displays for Online Subject Access. *Information Technology and Libraries*, 10(3), 185–191.

Meadow, C. T., B. A. Cerny, C. L. Borgman, and D. O. Case. 1989. Online Access to Knowledge: System Design. *Journal of the American Society for Information Science*, 40(2), 99–109.

Meadows, A. J. 1974. *Communication in Science*. London: Butterworths.

Meadows, A. J. 1997. *Communicating Research*. San Diego: Academic Press.

Micco, M. 1991. The Next Generation of Online Public Access Catalogs: A New Look at Subject Access Using Hypermedia. In D. A. Tyckoson, ed., *Enhancing Access to Information: Designing Catalogs for the 21st Century* (pp. 103–132). New York: Haworth Press.

Myers, B. A., and M. B. Rosson. 1992. Survey on User Interface Programming. In *Proceedings of the ACM Computer-Human Interaction Conference (Monterey, CA, May 3–7, 1992)* (pp. 195–202). New York: ACM Press.

Namioka, A., and D. Schuler, eds. 1990. *PDC '90: Participatory Design Conference Proceedings (Seattle, WA, March 31–April 1, 1990)*. Palo Alto, CA: Computer Professionals for Social Responsibility.

National Research Council; Commission on Physical Sciences, Mathematics, and Applications; Computer Science and Telecommunications Board; Committee on Information Technology Literacy. 1999. *Being Fluent with Information Technology*. Washington, DC: National Academy Press. ⟨http://www.nap.edu⟩.

National Science Foundation. 1999. *Digital Libraries Initiative Phase 2 Home.* ⟨http://www.dli2.nsf.gov⟩.

Nielsen, J. 1993. *Usability Engineering.* Boston: Academic Press.

Norman, D. A. 1988. *The Psychology of Everyday Things.* New York: Basic Books.

O'Brien, A. 1994. Online Catalogs: Enhancements and Developments. In M. E. Williams, ed., *Annual Review of Information Science and Technology* (vol. 29, pp. 219–242). Medford, NJ: Learned Information.

Perry, J. W., A. Kent, and M. M. Berry. 1956. *Machine Literature Searching.* New York: Interscience.

Phelps, T. A., and R. Wilensky. 2000. Multivalent Documents. *Communications of the ACM,* 43(6), 83–90.

Polya, G. 1957. *How to Solve It* (2nd ed.). Garden City, NY: Doubleday/Anchor.

President's Information Technology Advisory Committee. 1999. *Information Technology Research: Investing in Our Future.* Report to the President, February 24. National Coordination Office for Computing, Information, and Communications. ⟨http://www.ccic.gov/ac/report⟩.

Priss, U., and J. Old. 1998. Information Access through Conceptual Structures and GIS. In R. Larson, K. Petersen, and C. M. Preston, eds., *ASIS '98: Proceedings of the 61st American Society for Information Science Annual Meeting (Pittsburgh, PA, October 24–29, 1998)* (vol. 35, pp. 91–99). Medford, NJ: Information Today.

Reich, R. B. 1992. *The Work of Nations.* New York: Vintage.

Reitman, W. 1964. Heuristic Decision Procedures, Open Constraints, and the Structure of Ill-Defined Problems. In M. W. Shelley and G. L. Bryan, eds., *Human Judgements and Optimality.* New York: Wiley.

Robertson, S. E., and M. M. Hancock-Beaulieu. 1992. On the Evaluation of IR Systems. *Information Processing and Management,* 28(4), 457–466.

Rosenberg, J. B., and C. L. Borgman. 1992. Extending the Dewey Decimal Classification via Keyword Clustering: The Science Library Catalog Project. In *Proceedings of the 54th American Society for Information Science Annual Meeting (Pittsburgh, PA, October 26–29, 1992)* (vol. 29, pp. 171–184). Medford, NJ: Learned Information.

Salton, G. 1992. The State of Retrieval System Evaluation. *Information Processing and Management,* 29(7), 646–656.

Saracevic, T. 1975. Relevance: A Review of and a Framework for the Thinking on the Notion in Information Science. *Journal of the American Society for Information Science,* 26(6), 321–343.

Sawyer, P., A. Flanders, and D. Wixon. 1996. Making a Difference: The Impact of Inspections. In *Proceedings of the Conference on Human Factors in Computing Systems, Association for Computing Machinery* (pp. 375–382). New York: ACM Press.

Schamber, L. 1994. Relevance and Information Behavior. In M. E. Williams, ed., *Annual Review of Information Science and Technology* (vol. 29, pp. 3–48). Medford, NJ: Learned Information.

Schauble, P., and A. F. Smeaton, eds. 1998. *An International Research Agenda for Digital Libraries*. Summary Report of the Joint NSF-EU Working Groups on Future Developments for Digital Libraries Research, DELOS Workshop on Emerging Technologies in the Digital Libraries Domain, European Consortium for Informatics and Mathematics, Brussels, Belgium, 12 October 1998. Le Chesnay, France: ERCIM. ⟨http://www-ercim.inria.fr⟩.

Shneiderman, B. 1980. *Software Psychology: Human Factors in Computer and Information Systems*. Boston: Little, Brown.

Shneiderman, B. 1987. *Designing the User Interface: Strategies for Effective Human-Computer Interaction*. Reading, MA: Addison-Wesley.

Shneiderman, B. 1992. *Designing the User Interface: Strategies for Effective Human-Computer Interaction* (2nd ed.). Reading, MA: Addison-Wesley.

Shneiderman, B. 1998. *Designing the User Interface: Strategies for Effective Human-Computer Interaction* (3rd ed.). Reading, MA: Addison-Wesley.

Shneiderman, B., D. Byrd, and W. B. Croft. 1997. Clarifying Search: A User-Interface Framework for Text Searches. *D-Lib Magazine*, 3(1). ⟨http://www.dlib.org/dlib/January97/retrieval⟩.

Simon, H. 1973. The Structure of Ill Structured Problems. *Artificial Intelligence*, 4, 181–201.

Smith, T. R., D. Andresen, L. Carver, R. Donlil, C. Fischer, J. Frew, M. Goodchild, O. Ibarra, R. B. Kemp, R. M. Larsgaard, B. S. Manjunath, D. Nebert, J. Simpson, A. Wells, T. Yang, and Q. Zheng. 1996. A Digital Library for Geographically Referenced Materials. *IEEE Computer*, 29(5) (special issue on Digital Libraries), 54–60.

Tague-Sutcliffe, J. M., ed. 1996. *Journal of the American Society for Information Science*, 47(1) (special Topic issue on Evaluation of Information Retrieval Systems), 1–105.

Taylor, A. G. 1984. Authority Files in Online Catalogs: An Investigation of Their Value. *Cataloging and Classification Quarterly*, 4(3), 1–17.

Tillett, B. B. 1991. A Taxonomy of Bibliographic Relationships. *Library Resources and Technical Services*, 35, 150–159.

Walker, S. 1988. Improving Subject Access Painlessly: Recent Work on the Okapi Online Catalogue Projects. *Program*, 22(1), 21–31.

Walker, S., and M. M. Hancock-Beaulieu. 1991. *OKAPI at City: An Evaluation Facility for Interactive IR*. British Library Research Report 6056. London: British Library.

Waters, D. J. 1998. What Are Digital Libraries? *CLIR (Council on Library and Information Resources) Issues*, no. 4. ⟨http://www.clir.org/pubs/issues/issues04.html⟩.

Wilensky, R. 2000. Digital Library Resources as a Basis for Collaborative Work. *Journal of the American Society for Information Science*, 51(3), 228–245.

Zhao, D. G., and A. Ramsden. 1995. Report on the ELINOR Electronic Library Pilot. *Information Services and Use*, 15(3), 199–212.

6

The People in Digital Libraries: Multifaceted Approaches to Assessing Needs and Impact

Gary Marchionini, Catherine Plaisant, and Anita Komlodi

Digital libraries (DL) serve communities of people and are created and maintained by and for people. People and their information needs are central to all libraries, digital or otherwise. All efforts to design, implement, and evaluate digital libraries must be rooted in the information needs, characteristics, and contexts of the people who will or may use those libraries. Like most principles, the devil is in the details— in implementing and applying the principle to practical problems.

Human-centered digital library design is particularly challenging because human information behavior is complex and highly context dependent and because the digital library concept and technologies are rapidly evolving. Two important aspects of human-centered design are assessing human information needs and the tasks that arise from those needs and evaluating how the digital library affects subsequent human information behaviors.

Given the evolving nature of digital library development, solutions to these challenges must be process-oriented and iterative rather than product-oriented and summative. Given the complexity of human information needs and the uncertainty about the effects of new systems, multiple data views are essential to guide design and to help us understand the impact of digital libraries. This chapter focuses on two elements of design—information needs assessment and ongoing evaluation of impact.

Multifaceted approaches to needs assessment and evaluation of digital libraries are illustrated using three case studies with particular emphasis on a user needs assessment conducted as part of a project to develop prototype interface designs for the Library of Congress National Digital Library Program. The human-centered design principle links three clusters of constructs or facets—(1) people and their needs, characteristics, and contexts; (2) design, implementation, and evaluation; and (3) digital libraries.

Human Information Needs, Characteristics, and Contexts

The term *people* is used here to include the individuals, groups, and communities that have a stake in a digital library. Individuals' information needs have long been studied by researchers in marketing, education, and information science. A substantial history exists of studies of the information needs that people bring to libraries (e.g., Krikelas 1983; Marchant 1991; Paisley 1980; Wilson 1981). Brenda Dervin and Michael Nilan's (1986) review of the information needs literature dichotomizes system-oriented and user-oriented approaches to determining information needs. They criticize the system-oriented approach as too narrow to actually identify user needs and propose an approach that attempts to assess people's information needs directly.

Needs assessment research in information science recognizes that there are different levels of needs that users may not be able to articulate. For example, Robert Taylor (1962) specified visceral, conscious, formalized, and compromised levels of needs. Nicholas Belkin (1980) noted that users often bring anomalous states of knowledge to a search task and that needs change as information-seeking progresses. Highly personalized needs must eventually be translated into executable tasks.

Practical design aims to support a common set of these tasks. A fundamental goal of needs assessment is to identify large numbers of unique needs and map these into common classes of needs that may be met with standardized task procedures. A related goal is to develop systems that assist and guide people in mapping their personal needs onto system-supported tasks. One reason for the popularity of hypertext selection and the browsing mechanisms available on the World Wide Web (WWW) is that people are able to personalize these mappings experimentally, albeit laboriously.

That individuals vary on a host of physical, mental, and emotional characteristics is a defining condition of humanity. Much of psychology is devoted to identifying the essential dimensions of human behavior (for example, the theories of multiple intelligence presented by Gardner 1983 or Sternberg 1985). A much-promoted but seldom realized aim of design in a democratic information society is universal access. It is axiomatic that designing for universal access is much more difficult than designing for specific populations because the entire range of human characteristics must be supported. Thus, assessing needs and designing a national digital library service requires examination of many communities and will likely lead to multiple system solutions.

Table 6.1
Human-Centered Design and Evaluation Questions

Designers	Evaluators
Who are the users?	Who is impacted?
Who are the potential users?	Who and what may influence impact?
What are the common needs?	What are the indicators of impact?
How can those needs be mapped onto tasks?	How can indicators be measured?
How will the new system change needs (and tasks)?	How do impacts influence future generations and systems?

In addition to the needs of individuals and groups who make use of information in DLs, the needs of the providers and managers also influence design and evaluation. Many groups and individuals (such as digital librarians, taxpayers, political leaders, and philanthropists) have needs that must also be taken into consideration in digital library design. Individuals are embedded in many different communities, and communities are embedded in larger social and cultural contexts. When it comes to human behavior, these contexts are inescapable and confound efforts to artificially isolate specific variables for assessment.

The variety of stakeholders and contexts exacerbates the inherent complexity of assessing human information needs and the impact of systems designed to meet them. Designers and evaluators who wish to take a human-centered approach are thus challenged to specify which people will be served, what levels and types of information needs will be supported, and what contextual influences will be at play. These challenges are summarized in table 6.1, which expresses questions from the perspectives of designers and evaluators. Clearly, this is an overstated dichotomy: designers are concerned with evaluation questions, and evaluators must consider the design questions. In practice, design team members collaborate to address these questions, and in some cases individuals serve in both roles.

Design, Implementation, and Evaluation

Design, implementation, and evaluation processes marshal intellectual and physical capital to yield tangible, usable, and testable products. The design process is of primary concern to architects, engineers, and inventors. There is a significant body of literature devoted to theory (e.g., Braha and Maimon 1997; Simon 1996), history (e.g., Petroski 1996), and practice (e.g., Brooks 1975; Norman 1988). Computer system designers have begun to consider the physical and psychological human factors associated with system usage. Recognition of the importance of user-centered design for systems used by general populations is also growing. This approach is increasingly adopted by practicing software designers. Deborah Hix and H. Rex Hartson (1993) offer many practical suggestions for assessing the tasks that users bring to computer systems. Ben Shneiderman (1998) provides a rationale for mapping user tasks and needs onto the syntax and semantics of interface designs. Gary Marchionini (1995b) provides a framework for mapping users, tasks, and information need settings onto interface designs.

Implementation issues follow design and much of the work libraries do to digitize collections, provide access, and ensure interoperation advances practice by demonstrating "how to" procedures. Ideally, the design specifications are perfect, the work goes smoothly, the project comes in on time and on cost, and it includes all and only the functionality defined in the specifications. In practice, systems contain a variety of workarounds, add new features, and do not include all functions in the specifications. In spite of efforts to build a science of design, iterative design informed by evaluation feedback is more typical.

Evaluation may be a research genre aimed at assessing classes of techniques or methods (e.g., Suchman 1967 for social programs; Flagg 1990 for educational technology) or a systematic assessment of a specific product for the purposes of improvement (e.g., Nielsen 1993). Summative product testing is another form of evaluation that is not applicable to complex and evolving concepts like digital libraries. A human-centered approach to design, implementation, and evaluation is fundamentally complicated by the variability in human characteristics and behavior. Stephen Harter and Carol Hert (1997) present a recent review of evaluation research in information retrieval.

Evaluation of a digital library may serve many purposes ranging from understanding basic phenomena (such as human information-seeking behavior) to assessing the effectiveness of a specific design to ensuring sufficient return on investment. Human-centered evaluation serves stakeholders ranging from specific users and librarians to various groups to society in general. Additionally, evaluation may target different goals ranging from increased learning and improved research to improved dissemination to bottom-line profits. Each of the evaluation goals may also have its own set of measures and data collection methods. Finally, the evaluation must have a temporal component that can range from a very short term through multiple generations. One approach to dealing with evaluation complexity is presented in the Perseus DL case below.

Digital Libraries: Integrating People, Information, and Systems

Digital libraries are the logical extensions and augmentations of physical libraries in the electronic information society. Extensions amplify existing resources and services, and augmentations enable new kinds of human problem solving and expression. As such, digital libraries offer new levels of access to broader audiences of users and new opportunities for the information science field to advance both theory and practice (Marchionini 1998). A substantial body of literature relates to digital libraries, including many conference proceedings, such as the Association of Computing Machinery's annual DL series. Special issues of journals are numerous, including *Journal of the American Society for Information Science* (1993), *IEEE Computer* (1995), *Communications of the ACM* (1995, 1998), and *Information Processing and Management* (1999). Gary Marchionini and Edward Fox (1999) introduce one special journal issue by framing digital library design space with community, technology, service, and content dimensions; they argue that most research and development projects to date have been devoted to technology and content. Thus, the bulk of the work in DLs has focused on extending access beyond the physical walls of libraries and on extending citizen access to government-produced information. As extensions, we should be able to access more relevant information faster and with less expense. The augmentation of community and of information services remains an important challenge for the years ahead.

As DLs are actually developed, used, and improved, design guidelines will slowly evolve through experience and reflection. Needs assessment and evaluation offer several special challenges. An inherent limitation in directly assessing the human needs for an innovation is the fact that potential users must imagine what the innovation can and will do for them. This is very difficult to do, and innovators often justify adopting a "Build it and they will come" (BITWC) policy based on their own imaginations of needs and applications. If the engineering is good and the marketing successful, people will recognize the system's value and adopt it. Information technology history is filled with cases of top-down BITWC design success and failure.

Alternatively, designers can study users continually and involve them at all stages of the design and evaluation process, thus ensuring a ready-made market. This systematic bottom-up approach can produce lowest-common-denominator solutions and, in the worst case, may exhaust time and resources before any solution can be built. Clearly, some middle ground is needed for DL design. Holding on to high-level visions that are guided by astute observations of human behavior and are coupled with systematic and iterative assessments seems to be the right approach.

Because digital libraries are extensions and augmentations of physical libraries, needs assessment and evaluation may be modeled initially and generally on physical libraries. But in starting with the general goals, stakeholders, methods, and outcomes of physical libraries and related information technology services, designers and evaluators must be alert to new applications and goals, new user communities that may emerge, changing needs and abilities of existing user communities, new technological developments, changing information processes and capabilities, and new possibilities for data collection and manipulation.

The design and evaluation of DLs is driven by high-level visions but is mainly a bottom-up process that synthesizes specific instances and cases based on systematic probes of authentic environments with results from controlled investigations in simulated environments (such as laboratories). These approaches are equally expensive and mutually complementary. Authentic environments are context-rich but are therefore complex and not under the designers' control. Additionally, the environments must exist: there must be a working design to study. Thus, to design new environments, comparable built worlds must be investigated and results applied to new designs. Prototypes and laboratory studies offer good control over specific technical variables but give only glimpses of authentic environments. It seems clear that multifaceted approaches to determining user information needs and evaluating DLs must be used and the results integrated to inform design as an ongoing process.

The integration is not algorithmic. It is, however, systematic, interpretive, and driven by high-level goals. This integration is analogous to medical imaging techniques (such as the CAT scan) that aggregate a plethora of data slices so that diagnosticians may interpret holistic organ status. Of course, the data cases in DL design and evaluation are less precise and come from several different sources, making the interpretations and conclusions more time consuming and more dependent on inference. For complex phenomena such as DLs where human characteristics, the world's knowledge, and sophisticated information technology and social systems intersect, it is clear that principles and guidelines are synthesized over time rather than hypothesized and demonstrated.

Design and evaluation must be customized because every DL is situated in a context defined by community policies, human needs and characteristics, and technical constraints. As Aristotle noted in *Ethics* (1985): "In practical science, so much depends on particular circumstances that only general rules can be given." To identify some of the general rules for DL design and evaluation and to demonstrate how multifaceted data streams can be synthesized, we turn now to three cases. The first case (Perseus) illustrates an iterative and multifaceted approach to evaluation of DLs. The second case (Baltimore Learning Community) highlights specific high-level goals that drive design decisions. The final case (Library of Congress National Digital Library) illustrates a multifaceted approach to user needs assessment.

The Perseus Project Evaluation

The Perseus Project began in 1987 to develop a corpus of multimedia materials and tools related to the ancient Greek world. The mission of this project was driven by the perceived needs of students and faculty to have improved access to primary source materials and to have juxtaposed linguistic and visual resources to better learn and understand culture. This evolving digital library (⟨http://www.perseus.tufts.edu⟩) began as a HyperCard-based CD-ROM library of Greek texts and English translations; images of vases, sculpture, and sites; maps and drawings of Greek sites; and a variety of retrieval and philological tools. The DL transitioned to the WWW in 1995. From the first days of the project, an evaluation team worked to address a set of research questions related to learning, teaching, scholarly research in the humanities, and electronic publishing. The evaluation effort has continued for a decade, and there are many published reports on the project, as well as the evaluation (several evaluation reports are available at ⟨http://www.perseus.tufts.edu/FIPSE⟩; see also Marchionini and Crane 1994). The discussion here provides an overview of how the evaluation was initially framed and how it evolved over the years as the digital library was developed, used, and expanded.

In the original four-year evaluation plan, four goal sets were identified—learning, teaching, system (performance, interface, and electronic publishing), and content (scope, accuracy). Three characteristics of the computational medium that we believed would add particular value were identified for special emphasis—access, learner control, and collaboration. Based on these goal sets and media characteristics, a hierarchical set of ninety-four questions was developed to guide the overall evaluation.[1] Four sets of stakeholders were identified for study, including students, instructors, project staff, and classics researchers. A set of data collection methods was adopted, based on the identified goals and stakeholders. These methods fell into three general classes—observations, interviews, and document analysis.

Observations included baseline notes made by evaluators during classes or in laboratories, structured checklists and forms completed in these same settings, audio recordings of people thinking aloud while using the system, and transaction logs of

Perseus usage. Interviews included one-on-one and group interviews and discussions, all of which were audiotaped and transcribed. Various written questionnaires were completed by students, instructors, and scholars as part of the interview process. Document analysis included examination of the system software and documentation, syllabi and assignments created by instructors, and student responses to assignments (essays, journals, and hypermedia paths). Observations and interviews were done at a variety of sites that included major research universities (both public and private), small liberal arts schools, high schools, and museums.

Over the first few years of the evaluation, the scope of work and available resources caused the evaluation team to decide to invest most effort in the learning and teaching goals and on students and instructors as stakeholders. This is a practical example of prioritizing evaluation goals. Additionally, more data collection techniques were added, such as gathering written comments of visitors to a museum exhibit (at the National Gallery of Art in Washington) where Perseus was available as an adjunct to a sculpture exhibit. As the Perseus corpus grew in size and especially as it migrated to the WWW, the evaluation team was able to examine more longitudinal effects. By the late 1990s, evidence was gathered to support the claim that Perseus was having systemic effects on the field of classical studies.

The Project's Main Findings

The main findings over the first eight years of the Perseus Project evaluation can be summarized in four categories—amplification and augmentation, physical infrastructure, conceptual infrastructure, and systemic change.

Amplification and Augmentation   The Perseus DL amplifies and augments teaching and learning. Amplification takes several forms. First, more texts were available for students, including some that did not exist in print form. Additionally, more images and maps were available than department slide and map libraries typically offer. Second, an integrated corpus allows text-oriented courses to add image-based content and vice versa. Another type of amplification often noted is that content may be accessed more quickly and easily than physical versions in libraries (this is a kind of mechanical advantage).

Augmentation is evidenced by instructors who introduce new activities that are otherwise impossible. For example, the philological tools allowed instructors in class to illustrate points with word analyses or to visually and easily correlate geographic characteristics and textual passages. In addition, entirely new courses were created that integrated the varied and multiple resources in the DL. More important, Perseus empowered new kinds of student learning, such as sophisticated philological investigations by students who knew no Greek, visual investigations of themes, and new discoveries by students alone or as part of a class.

Physical Infrastructure   Using the Perseus DL requires substantial physical infrastructure investment. At every site evaluated, hardware and network problems caused frustration for faculty and students, substantial economic and human resources were necessary to make Perseus available in classrooms and laboratories, and laboratory staff had to be trained to support faculty and student access.

These challenges are faced by all educational institutions introducing technology into instruction and reflect the larger learning curve investments taking place in all disciplines at the close of the twentieth century. One effect apparent when Perseus was used through the WWW was the relative ease of use when compared to the HyperCard-based version. Students did not have to learn a new system but used the familiar interfaces of Web browsers. One result that recurred over the years is that self-reports on the system interface and learning effects were highly correlated, whereas demographics, computer experience, and frequency of Perseus use were not statistically correlated with learning effects.

Conceptual Infrastructure   The Perseus DL demands new conceptual infrastructures for teaching and learning. Instructors must learn to teach with the DL, students must learn how to learn with it, and both these requirements involve substantial amounts of time. Many instructors noted the large investments in time required to create new assignments and Perseus-augmented lectures. Likewise, some students complained about the length of time needed to learn to use and access the system and to find information. Several instructors noted that students took longer to complete assignments than anticipated and that classroom use often took longer than planned, since interesting alternatives or additional examples could easily be pursued with the system.

Instructors generally should take into consideration that the novelty and amount of work an innovation demands may lower student course evaluation results during the early years of adoption. Several instructors noted that Perseus raised their levels of expectation about the scope of material accessible to students. Likewise, students at schools where Perseus was used in multiple courses came to expect that such resources would always be available for use in their courses.

A number of opportunities and challenges related to teaching emerged. The traditional dilemma of how best to mix open-ended and guided instructional activities was exacerbated by the many possible uses that Perseus offers. Other considerations were how to use class time best and what content was displaced when Perseus-based content is introduced into a course or curriculum. Perseus allowed instructors to model how they do their own research; the risks and time required to model research should be considered. Instructors also had to learn to evaluate electronic assignments. One instructor noted that more extensive feedback was made possible through having assignments and comments in electronic form, since he could leverage all the advantages of word processing while grading. Certainly, instructors and administrators must understand that iterative planning and implementation cycles are required over years rather than weeks or semesters, and appropriate allowances, resources, and rewards must be available.

Opportunities and challenges related to learning were also varied. Students were certainly motivated by Perseus, especially by the images. They were observed to work harder and better when their assignments were put on the Web. The persistence of the assignment beyond the end of the course and the "publication" of the work are likely explanatory factors in this regard. Some students reported being overloaded by the amount of available content. Likewise, some students were overwhelmed in lectures that included many Perseus examples and multiple verbal themes. The learning curve necessary to use Perseus tended to be more problematic in large general studies courses than in advanced courses for classics majors, who tended to recognize the time it took to learn to use the tool as an investment to be amortized over multiple courses.

Systemic Change   Perseus is bringing systemic changes to the multiple fields within classics. There were well over fifty courses included on the Perseus Web site, representing more than a dozen colleges and twice that many instructors. These courses use Perseus in a variety of ways and illustrate the penetration of Perseus into the classics curriculum internationally. In some universities, multiple instructors use Perseus for many of their courses. Several instructors noted that Perseus use led students, faculty, and administrators to see classics as technologically "plugged-in," leading to increased recognition and resources on campus.

While it is too soon to generalize, the new courses created based on Perseus tend to integrate textual and visual materials and illustrate ways to break down barriers between distinct areas such as philology and art history. New classics faculty position announcements list computer experience as requisites. Popular textbooks now include Perseus companion paths, and a spin-off company, Classical Technology Systems, provides training and support.

The Perseus Project continues to attract funding for expanded work. Workshops and papers related to Perseus are standard fare at professional conferences in the classics fields, as well as education. By 1998, the Perseus DL was responding to approximately 25,000 hits per day (unique requests excluding GIFs) and had become an electronic gateway to a suite of digital resources well beyond the original ancient Greek culture corpus and tools. Many Perseus tools and techniques have been integrated into new projects at Tufts University (including Rome and ancient science) and elsewhere. The Perseus DL impacts a large and diverse community beyond university classics courses. It serves as a stable and authoritative resource for other publications (for example, several commercial online encyclopedias refer to the Perseus DL), as well as for distance education and other nontraditional education venues.

Perseus and the "Big Picture" in DL Evaluation

Perseus evaluation results address many of the high-level goals related to teaching and learning that were set out in the original evaluation plan more than ten years ago. The evaluation adapted to new technologies and content and to the new ways that instructors and learners found to apply this DL to their needs. It was opportunistic in that it took advantage of new venues (museums and high schools) and new data collection techniques (such as the transaction logging scripts built for the HyperCard version and the less user-specific but more broad-based transaction logs of the WWW). It took a multifaceted, bottom-up approach to evaluation that integrated many specific data collection efforts and was guided by high-level general questions.

The longitudinal dimension extends the evaluation space to include a temporal dimension. Figure 6.1 illustrates this expansion by depicting a wide range of stakeholders across time (Marchionini 1995a). Such a framework gives a "big picture" flavor to digital library evaluation research and can easily be adapted for other stakeholder sets or evaluation goals. In the figure, time intervals are immediate (days or weeks), annual (year), short-term (two to five years), midterm (five to fifteen years), generational (twenty to thirty years), and very long-term (over fifty years). The curves in each cell represent tradeoff balances between tangible and intangible costs and added values of the DL. In the figure, all of the curves show equivalent

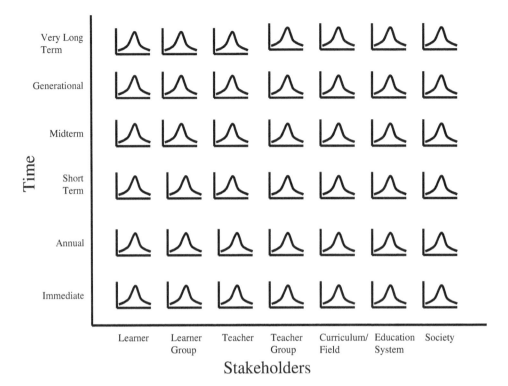

Figure 6.1
Temporal framework for evaluating digital library effects on educational stakeholders

balances. Specific DL evaluation efforts can aim at determining what cost-benefit tradeoffs are evident for any stakeholder or time subset. In the Perseus example, we might depict the immediate benefits to learners and instructors as being more costly than traditional resources but showing better payoffs in the short- or midterm rows. Such situations would be visualized by the peaks of curves occurring in different locations in different rows and columns.

Baltimore Learning Community Project

The Baltimore Learning Community (BLC) project leverages DL resources and technology to support a community of teachers who create and share outcome-oriented instructional modules. Beginning with a seed collection of video material from Discovery Communications, Inc., the BLC project team worked with teachers to create a small repository of multimedia materials, tools for accessing those materials, and tools for building and using instructional modules. Teachers are encouraged to share the modules they create, together with reflections on how the modules were used in classes. In addition, teachers may contribute specific resources to the BLC repository. Together with face-to-face meetings that require multiple participants in each school and Internet communication facilities, these ideas, raw materials, and modules were central components for the vision of community building and sustainability. The BLC DL provides support for teachers to

• Find multimedia resources in the database by a variety of entry points such as topic, instructional standards, media, and source;

• Use those resources and others found on the WWW to create instructional modules keyed to State of Maryland instructional outcomes;

• Present the modules in classes;

• Contribute new resources to the BLC repository, including project-defined metadata;

• Contribute instructional modules to the repository; and

• Communicate with other teachers in the BLC community.

Although the original vision may be characterized as BITWC, informed by state and local curriculum guides, the design process involved teachers after funding was secured, and this involvement has increased as the project evolved. The basis for design is rooted in assumptions about teaching practice and needs. Teacher needs are actualized in curriculum standards and guidelines. It is unmistakable that teachers in public schools are strongly influenced by local and state curriculum guidelines and assessment procedures.

Our intention from the start was to assist teachers in addressing these needs. The mechanism was providing teachers with access to high-quality multimedia materials tied to curriculum rubrics and encouraging them to leverage these materials to build a community where teaching practice and lessons are shared to everyone's benefit. This approach to meeting common needs through shared resources and experiences is the basis for the *sharium* concept—that is, a distributed electronic problem-solving space with rich information resources and tools, where people can work independently or collaboratively to solve problems (Marchionini 1999). The BLC evaluation plan documents how teaching practice evolves in such an environment and, eventually, how student learning is affected.[2]

The system design used included a selection and form-based approach to lesson construction; alternative interface options for teachers to facilitate following personal styles as well as working at home or at school; a dynamic query interface style to facilitate resource exploration; and different video surrogates to save teacher preview time. The system design evolved as teachers expressed their needs more sharply as they gained experience. Based on teacher feedback, several changes to the design have been made:

• A simple slide-show presentation component was added to the original presentation component that included details on pedagogy (for example, objectives, groupings, and instructional strategy).

• The full Maryland School Performance Assessment Program Outcomes Model and Content Indicators for social studies and science were made available from the module construction tool.

• A feature for printing locally required lesson plan formats from the modules was added.

• The resource explorer tools were embedded in the module construction tool rather than remaining separate functions.

In the fourth year of acquisition, indexing, interaction, and use, the system continues to evolve. Overall, progress has been slow and arduous. There are several indications, however, that a community is forming. First, a few teachers have contributed original materials. For example, photographs of the Baltimore harbor were taken, indexed, and contributed. Second, more than 100 modules have been created and added to the repository although teacher reflections, and commentaries on usage have seldom been included. Third, most teachers have made significant strides in incorporating the Internet into their classrooms. In most cases, this happened without using the BLC repository directly, although they often create modules that incorporate WWW materials they find. A final indicator is that a fourth middle school and a group of more than a dozen teachers have joined the project.

On the research and development side, the advanced interfaces have been incorporated into a Web-based application and a series of empirical studies of video surrogates have been conducted (e.g., Komlodi and Marchionini 1998). It is clear that the process of community building in schools and across schools is a very slow process. Building the DL has proven to be much easier than building a community of instructional practice. Although some of the difficulties teachers have in creating and sharing materials are due to technical issues (including slow access, system crashes, and multiple platform incompatibilities), the challenges of teaching in inner-city middle schools, administrative pressures to raise test scores, and lack of incentives for out-of-class time investments also deter forward progress. Few teachers have actually used the modules created by other teachers. There is a flurry of activity immediately after the annual summer institute and less activity through the school year. With the exception of a few teachers who are actively creating and using modules, most teachers ignore e-mail and choose not to attend after-school development sessions (small stipends are provided). It seems clear that the demands of students, of administrators, and of the school system climate get much more attention than the project.

It is likely that this is as it should be. DLs and technology are not panaceas for solving the problems of public education. It will take time for communities to form around the resources and tools the DL research and development community provides. This case illustrates that content and technologies are not sufficient to create sustainable communities in highly structured environments such as public schools. Funding bodies and designers must show patience in assessing impact in such environments.

Library of Congress National Digital Library Program

This case study reports on the user needs assessment that was conducted for the Library of Congress National Digital Library Program to develop prototype user interfaces for the National Digital Library (NDL). The design process and the various prototypes are described in Catherine Plaisant, Gary Marchionini, Tom Bruns, Anita Komlodi, and Laura Campbell (1997) and Marchionini, Plaisant, and Komlodi (1998). The following discussion will focus on the extensive information needs assessment that was undertaken before system design was begun. Based on the perceived needs of the populace to have better access to the Library of Congress's (LC) treasures, a multifaceted and flexible approach to estimating the needs of the citizenry was taken.

The user needs assessment of NDL was conducted over a relatively short period of time (four months). Flexibility in choosing techniques that were practical, given the allotted time, proved especially important in identifying potential users and tasks. For example, online surveys of users of the existing NDL Web interface were not used for two reasons. First, such surveys reach only self-selected users who have already found the NDL and chosen to complete the questionnaire; and second, they require a long process of authorization, review, and implementation that was prohibitive in this case.

As we collected information about users' characteristics and needs, we identified and examined corresponding user interface challenges. A taxonomy of users and tasks was proposed. It is highly likely that this taxonomy will change as the NDL evolves, since it is rooted in data that come from a mix of current users of the physical LC, early adopters of the nascent implementation, and nonusers of the NDL who were asked to speculate about using it. This taxonomy and a general list of interface challenges became important reference materials for the team designing the NDL prototypes.

The LC NDL Context and High-Level Goals

The Library of Congress has taken a leadership role in the development of a National Digital Library. The LC's role in this effort is to digitize five million items from 200 Americana collections over a five-year period (see ⟨http://memory.loc.gov/ammem/amabout.html⟩). In this initiative, the NDL Program planned activities for preparing, digitizing, archiving, and providing access to these historical and contemporary materials; managing the overall project; and collaborating with other institutional partners. The challenges of providing access are addressed by efforts to develop an easy-to-use, yet powerful, human-computer interface. It is our view that a poor user interface is actually worse than denying equitable access since it frustrates people and wastes their time. To guide the development of the interface, a user needs assessment was undertaken to determine what tasks and search strategies users bring to the library today; what new types of users, tasks, and strategies the NDL will attract; and how these user needs may best be served by the NDL interface. Because the NDL Program was new, there was no extant user community. To assess the needs of potential users, a variety of data collection methods was employed, and a flexible approach to integrating and interpreting data was adopted.

Needs Assessment Stakeholders

Three general user communities were considered at the beginning of the project (others emerged as the needs assessment progressed). The user community comprising LC staff is expert in both library systems and content, represents heavy users of the existing systems, is knowledgeable about articulated (formalized) user needs, and is able to devote short (usually less than two hours) amounts of time to individual needs. The second user group included scholars and other visitors to LC. This community has high levels of domain expertise and knowledge of library systems. They are also able to devote large but specific amounts of time (days or weeks) to their individual information needs. The third group, visitors and prospective users, has varied levels of domain expertise and low levels of library systems knowledge. They are able to devote only short time periods to their information needs or explorations.

Data Collection Procedures

Three types of data collection were undertaken. Selected LC reading rooms were visited, and staff hosts were interviewed. Written questionnaires were created and distributed to important prospective user communities. Third, relevant LC documents were examined.

Reading Room Visits and Interviews

This was considered to be the most informative component of the assessment, since the librarians in the reading rooms are themselves important users of the existing system and also have extensive knowledge about current user needs at the LC. In addition, many staff members are either directly or indirectly involved in the development and maintenance of the NDL and have stakeholder needs that must be captured. Two or more project team members participated in visits to nine different reading rooms. For each visit, information was gathered with regard to three primary facets of the user needs assessment.

The first facet emphasized content—that is, what materials are housed in the reading room, what indexes or finding aids existed, and which materials were scheduled for inclusion or possible inclusion in the NDL. Users were the focus of the second facet. Information was gathered about the types of users who typically used the reading room, the types of information needs they brought to the Library, what levels of searching skill they typically had, how these users might benefit from the NDL, and what new users might be attracted. The third facet, strategies, focused on issues relating to how users conducted searches, what types of search tools were available in the reading room, and how the reading room staff assisted users. A fourth category was added for miscellaneous data, any notes on interface implications, and additional notes specific to that reading room.

The notes were transcribed by one of the interviewers and sent via e-mail to the other team members who participated in the visit. These team members augmented and edited the electronic notes based on their own written notes, and the final electronic version was sent for comment (member checking) via e-mail to the reading room staff member who had hosted the visit. After staff had made comments, corrections, or additions, a final set of notes was prepared.

Questionnaires for Potential Users

Because K–12 teachers and students are featured prominently in plans for the NDL, a limited survey of educators was conducted. Rather than an expensive and time-consuming random sample, we took advantage of existing contacts to survey a convenient sample of educators. Two groups were selected. The first group was a set of twenty-four teachers from different areas of the United States who participated in NDL orientation sessions. The second group of educators was a set of twenty-seven school library media specialist supervisors in the state of Maryland. Both of these groups were administered a questionnaire that included items on types of materials used, search strategies applied, system characteristics preferred, and possible future applications of an NDL collection. The difficult task of assessing the needs of the general citizen was addressed by a survey of parents and workers in a large daycare center. This facility serves a diverse community of blue-collar and white-collar workers in a midwestern U.S. city.

Document Analysis

Documents represent a library's expression of procedures, policies, and responses to user needs. Three types of LC documents were examined as part of the needs assessment to better understand user and system needs:

• *Reading room handouts and brochures*   During reading room visits, many handouts were obtained; these handouts reinforced the notes taken during visits in that, in some cases, they illustrate user needs that are so pervasive that special publications or finding aids are created to address them.

• *User study reports prepared by LC staff*   These two studies (Library of Congress 1993, 1995) reported on the experiences of different user groups in different environments with the pilot American Memory project, a precursor to the NDL using similar materials. These reports identified many interface issues.

• *User e-mail commentaries or inquiries about the NDL*   E-mail messages sent by early users of the infant NDL, as well as staff responses, were analyzed and categorized.

Results of Reading Room Visits and Interviews: Content

The NDL is not an academic exercise that begins with well-articulated user needs and then adds content that is tailored and organized to meet those needs. Instead, it is a large real-world effort that is rooted in a complex cultural context. Content for the NDL is selected based on a variety of economic, legal, social, and political exigencies. As the digitization plan continues to evolve, we can anticipate that materials added to the NDL will be as varied as the holdings of the Library of Congress itself. This implies that interfaces to the NDL must be driven by content decisions as well as users' needs.

Challenges are posed by the variety of content digitized from the collections of the different reading rooms. For each visit, we used notes to compile a list of content-related interface challenges to be addressed in the prototypes. Table 6.2 gives samples extracted from notes and lists of challenges.

Next, all challenges were summarized into a set of general content-related interface challenges. First, a variety of materials will become part of the NDL. The materials vary by topic, size, and format and also by degree of cataloging. Some materials are cataloged at the item level, some are cataloged only at the collection level, and some are uncataloged. This state of affairs presents huge challenges to LC staff as they work to serve user needs, as well as to users who are trying to find information in the NDL. It is highly unlikely that new cataloging efforts can be undertaken to include those items not currently cataloged (in the Geography and Map Reading Room alone, only approximately 250,000 items are cataloged out of 4 million).

It is also unlikely that catalog records for specific items in collections cataloged at the aggregate level can be created or that existing catalog records can be edited to reflect any special requirements of the NDL (such as georeferencing for maps). Thus, the first challenge is to develop a conceptual interface design—an organizational framework with appropriate rules for applying the framework. The framework must characterize for users the granularity, size, and nature of objects in the NDL. This challenge cuts across all reading rooms and has several components. The interface must communicate to the user the following:

• The contents of the entire NDL,

• The level of representation for a displayed object (bibliographic record to collection, series, or item),

Table 6.2
Samples from Reading Room Notes Related to the "Content" Facet of the Assessment and Corresponding Interface Challenges

Examples of Visits	Extracts from Notes	Interface Challenge Examples
American Folklife Center	The collection is largely uncataloged, much of it is unpublished. Originally consisting of folk songs, it now includes cultural documentation in various media. Most collections are multiformat. Most include recorded sound.	Browsing of materials more important than catalog search; accessing audio (searching capability as well as download times); showing multiformat items (i.e., coordinating a sound track and textual field notes)
Geography and Map Reading Room	Items include panoramic maps (cataloged), county atlases (cataloged), railroad maps, East European maps (not cataloged, heavily used by genealogists), Civil War maps, Sanborn fire insurance maps (about 700,000 items), and selected maps of American history.	Showing very large maps at varying resolutions; specifying areas and regions on maps; dealing with the problem of place-name ambiguity
Prints and Photographs Room	Items for NDL include Washington, Lincoln, Jefferson, papers; WPA life histories; Whitman notebooks; Margaret Mead collection; some finding aids for other collections. Finding aids are important but vary in detail.	Making a full-text search; specifying limits of search within a manuscript, across a collection, etc.; distinguishing finding aids levels and primary materials

• The alternative levels of representation available for a displayed object (bibliographic record only, thumbnail or other extract, primary object), and

• The nature of a displayed object (secondary or primary, format(s), concomitant or linked objects).

This challenge must be met within the current levels of indexing. The Library takes advantage of user browsing behavior in uncataloged collections or in collections that are described only at a group level to create records subsequently for individual items selected by patrons and therefore presumed to be "high demand." These include, for example, photographs found and copied by patrons in Prints and Photographs. Similarly, the NDL presents an opportunity to add additional cataloging information to items that users find while browsing the digital collection. Another element of this challenge is to integrate searching of catalog records with searching of SGML finding aids available for some objects, such as manuscripts.

A second challenge is to communicate to the user what items are *not* in the NDL. At one level, this places the NDL within the context of the entire LC itself (such as copyrighted materials, three-dimensional objects, and so on) and in the context of the world of information available in other institutions and on the entire Internet. On another level, this assists users in planning for visits or becoming aware of related information at LC or other institutions.

A third challenge is to support users in the NDL without human intervention. Although some level of reference service will be necessary in the NDL, not every patron can expect the level of human support that current visitors to LC need, receive, and expect.

A fourth challenge is to create an interface that is accessible to a user with state-of-the-market technology. The interface (and underlying retrieval system) cannot be based on assumptions about state-of-the-art hardware. For example, in 1995 terms, state-of-the-art hardware included 20-inch high-resolution displays, very high-speed connections, huge amounts of memory, and specialized input and output devices. And state-of-the-art software would mean the latest operating system or Web browser.[3] Just as the overall system must leverage compression and advanced retrieval algorithms, the interface must inform the user about temporal demands for data transfer, provide posting information about result set sizes, and provide some level of explanation for ranked results. In addition, the interface must be "growable" to keep pace with the evolution of hardware and software.

A fifth challenge is to invent new techniques to search for multimedia objects and to integrate those techniques into the interface (such as visual and audio query languages, and surrogate viewers).

Together with the specific interface challenges listed in the reading room summaries above, the NDL content begins to define both a development and a research agenda for more general digital library interface design. Although these challenges were addressed at preliminary levels in the resulting prototypes, they all will remain problems for the DL research and development field for many years to come.[4]

Results of Reading Room Visits and Interviews: Users and Strategies

The Library of Congress is mandated to serve Congress and its staff first. It has traditionally been a library of last resort for other citizens in that its collections are so large and specialized that only well-prepared researchers can take full advantage of them. Although many casual visitors come to the LC to see and experience it as an institution, it is a research library and operates to serve those prepared to work in a complex and scholarly environment. For example, citizens under the age of sixteen are not permitted to use the Library of Congress. The NDL thus inaugurates a fundamental change in the service mission of the LC in that a much broader user community is addressed. Table 6.3 provides samples of types of users and their search strategies, as well as some of the interface challenges that arise from use (taken from the reading room visit notes).

Many interface challenges are posed by the assessment of users and the strategies used in the reading rooms. The most fundamental challenge is how to serve the wide range of users who will visit the NDL. This includes the entire range of U.S. citizenry:

• Users of different ages,
• Users representing the entire spectrum of education levels,
• Users with a range of cultural and ethnic perspectives,
• Users with special physical and cognitive needs,
• Users who vary in their experiences with computer technology,
• Users with a large variety of experiences specific to their visit to the NDL,
• Users who vary in experiences in the domain of the information problem they bring to the NDL, and
• Users who vary in their experiences with libraries and research collections.

In addition to the challenge of individual user characteristics, the interface must support the wide variety of information needs that users bring to the NDL. Because we are focused on formalized and compromised needs, we use the term *task*. From a user's perspective, these tasks vary on five nonorthogonal dimensions:

• Complexity (the number of concepts involved and how abstract they are),
• Specificity (how confident the user is about the accuracy and completeness of results, ranging from a particular fact to interpretations),
• Quantity (the amount of information expected or required to meet the need),
• Criticality (how important it is to the user to meet the need), and
• Timeliness (how long users expect or are willing to spend in meeting the need).

The NDL interface must help users easily communicate some of these characteristics of their task to the system. Other general challenges echo the content challenges

Table 6.3
Samples from Reading Room Notes Related to the "Users and Strategies" Facets of the Assessment and Sample Interface Challenges

Examples of Visits	Extracts from Notes	Interface Challenge Examples
Geography and Map Reading Room	One-third of users (estimate) are looking for genealogical information. An estimated one-fourth of users are contractors looking for environmental information. Specialized maps are often sought (e.g., railroads, land use, hot topics). Users rely heavily on reference librarians to get started and are often not cartographically or geographically literate.	Specifying areas; transferring and displaying potentially huge files
Newspaper and Current Periodical Reading Room	Many users are first-time visitors to the LC; about 20 percent are regular users who come for sustained scholarly visits. Reference interviews are important. Since much of collection is copyrighted, finding aids and a few specialized items will be in NDL. Holdings information is as important as pointer information.	Helping NDL users quickly understand that few of the primary materials are online
Law Library	Primary users are congressional staff. People in the wrong place are directed to their local libraries; there are good law materials on the Internet so an analogous strategy can be used in NDL. Novices are pointed to law encyclopedias. Main entry points are time, country, type of material (law, court decision).	Determining when to point users elsewhere
Manuscript Reading Room	About 80 to 90 percent are academics visiting for a few days to more than a month. Part of the entry registration process for visitors is a reference interview with a librarian. Most users have to work through multiple levels of guides to get to primary materials and are heavily dependent on librarian assistance. Entry points are mainly name of person, and some subject access that points to finding aids.	Helping NDL users quickly understand that few of the primary materials are online and that many levels of search must be worked through

presented earlier. The interface should help users distinguish primary and secondary materials (including multiple layers of each). It should help users make links among items across different collections and reading rooms. The interface must also capture the essential elements of the reference interview so that users can find what they need without human intervention.

In sum, the reading room interviews reinforced well-known general challenges for interface design but also raised concrete cases specific to multimedia DLs that strive to serve general populations. The interviews also helped to introduce the needs assessment project team to the rich culture and collections of the LC.

Questionnaire Results

Teachers and School Library Media Specialists

Responses from the teachers and school library media specialists were very similar to each other and provided insights into the great diversity of technology penetration in U.S. schools in 1995. Thirteen of the twenty-four questionnaires (54 percent) sent to teachers were returned. Eleven of the twenty-seven questionnaires (41 percent) sent to school library media specialist supervisors in the state of Maryland were returned. The two-page teacher questionnaire had four general demographic questions, three Likert-scaled questions on types of materials, five scaled questions on search strategies, five scaled questions on student search strategies, five scaled questions on computer system characteristics, three scaled questions on access to the NDL, and one open-ended question about desired useful materials for teaching. The school media specialist questionnaire was the same except that "media center" was substituted for "classroom" and "teachers in your school" for "students."

The data for the two groups showed very similar patterns. For preferred information-seeking patterns, both groups reported that they often used all the strategies listed in the questionnaire—browse a list of potential terms, browse items with embedded links, navigate a hierarchical set of menus, search using controlled vocabulary, and search using natural language. Both groups were similarly generous in estimating students' use of the entire variety of information-seeking strategies. Likewise, teachers and school library media specialists were unified in demanding a system that could be used in school and at home and one that was easy to learn and use. There was less unanimity within or across the groups about the need for access to large amounts of primary materials. Even less agreement was reported on the need for access to multimedia materials, perhaps reflecting concerns about the computing infrastructure in schools at that time. Both groups were far more concerned about professional use of the NDL than they were about personal uses.

Although both groups rated primary documents as important, teachers were more uniformly consensual in rating them as extremely important. Teachers rated finding aids as somewhat less important than school library media specialists (most of whom are trained as librarians). It is interesting to note that teachers rated teacher guides as somewhat less important than did the school library media specialists.

At the time of the survey (fall 1995), there was a wide range of equipment available in these schools ranging from DOS and Apple II computers to Windows and Macintosh machines. This wide range of equipment was problematic, as significant numbers of workstations in these schools were not capable of using the NDL. It was somewhat surprising that all but two respondents said that there was some kind of Internet access in their schools, most often in the school library media center. Perhaps even more surprising, eight of the thirteen teachers and six of the eleven school library media specialist supervisors reported having Internet access at home. Although Internet access was not defined as an IP-capable connection, some type of remote access capability was available either in school or at home to most of this very select group of educators.

The final open-ended question asked what types of materials respondents would find most useful for themselves or their teachers. Not surprisingly, the teachers were more verbose in describing specific materials they could use in class, in many cases citing specific NDL collections that presumably they had encountered in their LC training. Examples of suggested materials from the teachers' questionnaire responses included

• Images of the past and of current interest linked to curriculum, interesting primary sources and documents, sounds of historical events, and music;

• Major American figures from all areas of life (pictures, writing, inventions, personal items such as diaries);

• Life in other times (Edison film of New York harbor, sharecroppers, immigrants, farm life);

• Documents that helped shape our nation;

• Links to related sites (archives, Smithsonian); and

• Photos related to the Depression, American Indians, black studies, popular culture, and world wars.

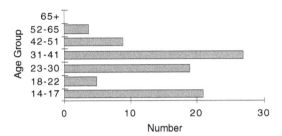

Figure 6.2
Age distribution (N = 85)

Teachers also emphasized that ease of access to all content was vital. Suggestions from the school library media specialists were somewhat different and included

· Downloadable public domain information (indexes, finding guides),

· Lesson plans for all subject areas,

· Support materials for lessons (text, sound, images),

· Text to accompany images (that can be printed and used away from terminal),

· Social studies, American history documents (text, visual, census statistics), and

· Source information on specific subject area topics, teaching strategies, and techniques for given objectives/outcomes.

Day Care Center

The results of the day care center questionnaire stand in somewhat sharp contrast to the school-based data. The two-page questionnaire was administered in fall 1995 and had five demographic questions, five questions on computer usage, three questions on library usage, and twelve questions on expectations about the NDL. Eighty-five questionnaires were returned and provide a 1995 snapshot of the views of citizens remotely located from the Library of Congress. About two-thirds of the respondents were female, and the largest number were in the twenty-three to forty-one age group (another large group was composed of high school students who worked at the center). Figures 6.2 and 6.3 illustrate these data. This distribution is representative of working parents with day-care age children.

There was a wide range of educational completion levels in the group, and most respondents expressed an interest in learning more about computers. The responses to computer usage questions illustrate the gap between willingness to learn and

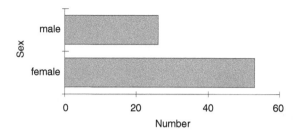

Figure 6.3
Sex distribution (N = 79)

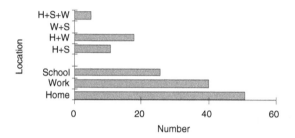

Figure 6.4
Location of computer use

actual experience. Figure 6.4 shows where respondents accessed computers. Ten participants did not respond to this question. As they were asked to check all that apply, the total number of responses was 117. It is interesting to note that 51 respondents reported using a computer at home (60 percent), a figure that exceeded national survey data on home computer penetration nationally at that time: 37 percent of U.S. homes had computers in early 1995 (Fox 1995, p. 9). Forty respondents (47 percent) reported using computers at work. Twelve of the eighteen (67 percent) high school student respondents reported using computers at school. Eighteen respondents (21 percent) reported using computers both at home and work; eleven (13 percent) reported using computers both at home and school; no respondent reported using computers at work and school; and five respondents (6 percent) reported using computers in all three locations.

Figure 6.5 depicts how often respondents use computers. Twenty-nine (35 percent) of the respondents reported using computers on a daily basis; twenty (24 percent) reported using computers several times a week; fifteen (18 percent) reported

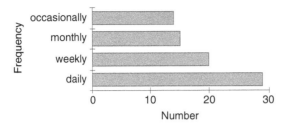

Figure 6.5
Frequency of computer use

using computers several times a month; fourteen (16 percent) reported using computers once in a while; and seven (8 percent) did not respond to this question. Seventeen respondents selected reasons for not using computers, with access (six respondents) and knowledge (five respondents) the most frequent selections. Twenty-five (29 percent) of the respondents reported having access to some online service. These data suggest that computer usage was pervasive, with well over half of the respondents using computers in a variety of settings multiple times per week, and almost one-third reporting some type of online access.

A very different portrait emerged from the responses related to the World Wide Web. Although thirty-two (38 percent) of the respondents had heard of the WWW, only nine (11 percent) had actually used it. Of those who did use it, only three were high school students; so the adults were beginning to use the WWW. Although this very low penetration of the WWW suggests that NDL access today would reach a small portion of these citizens, the fact that almost one-third now have some online access and over half are regular computer users suggest that WWW access will follow quickly as the NDL continues to evolve.

The questionnaire was also designed to determine the basic facility of the respondents in library use. The results (see figure 6.6) illustrate that most (59 percent) of the respondents use the library on occasion (less than several times a month). In response to the question about why they use libraries, responses varied across the categories, with school (47 percent) and reference (45 percent) garnering the most frequent usages (see figure 6.7). It is interesting that thirty-three respondents (39 percent) selected leisure reading as a usage; of all these usages, leisure reading is perhaps least likely to be affected by NDL availability.

The main objective of the questionnaire was to determine what needs exist or may evolve for the NDL. This was also the most difficult objective to achieve since sub-

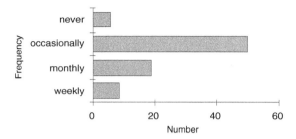

Figure 6.6
Frequency of library use (N = 84)

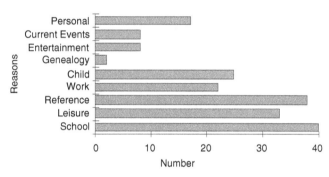

Figure 6.7
Reasons for library use

jects were asked to articulate needs related to a new entity about which they knew very little, if anything. Thirty-six (42 percent) of the respondents said they expected to use the NDL for the same reasons they currently use libraries; however, the fact that forty-one (48 percent) selected the answer "don't know" illustrates the lack of knowledge about the NDL. To address the difficulty of eliciting information about an unknown entity, the questionnaire was designed to be open-ended in this section. Although we recognized that this would make responses more difficult and that many participants would perhaps thus skip these questions, we wanted to provide the broadest possible scope of responses.

For the question about why subjects would like to use the NDL, most respondents (64 percent) wrote nothing or "NA/none." Several of the respondents made generic comments about easy, fast access to information or about gaining more information or knowledge. A few made general comments about using it for school, and a few noted the importance for children. Five respondents gave specific topics or types of information (World War II, genealogy, history from a different perspective, educational materials for teachers, and maps and photographs). Two respondents noted that they would use the NDL to find information not available in their local libraries. Thus, only a small number of respondents were able to give any reasons beyond generic library use for the NDL.

A question about search strategy (title, author, subject, and browse) was left unanswered by fifty-three subjects (62 percent). Most who did respond noted subject (eleven), author (five), or some combination of subject/author/title (eleven). Thus, the respondents mainly expected to use the NDL as they do finding aids in physical libraries. To the question about other preferable ways to find information, ten wrote something about computers or other electronic tools (CD-ROM), and three listed a librarian. When asked if they thought they could find information easier if it were available electronically, most respondents (84 percent) answered yes, showing confidence about both their ability to use computers to find information and high expectations of the use of computers in searching for information.

Faster and easier access to information was given as an example in the questionnaire items on expected advantages and disadvantages of DLs.[5] Forty-seven (55 percent) of the respondents indicated one or both of these reasons. Two of those respondents also noted that there could be more information, another noted that it would be faster since they could send information to their local printer, and another noted that there could also be many more cross-references. Twenty-two (26 percent) gave no response. Five respondents said there would be more information. Six said they could do work from home or work. Two noted that they could find information beyond that available in their local libraries. Two thought that they could get more precise ("exactly what I want") information; and one noted that it would have no advantages since the computers in libraries were too confusing to use already.

The disadvantage question gave "slower, more difficult access to information" as an example. Six respondents noted that access might be slower, seven said that it might be difficult to use the technology, three noted generic difficulty, and three noted difficulties related to finding information. Forty-five (53 percent) of the respondents did not answer or said no disadvantages. Two respondents were concerned with access to terminals. Five noted potential problems with technology failure; three noted concerns about system overload (busy lines). One noted potential information overload, one was concerned that not everything needed would be available, three were concerned about costs, one noted that no librarian would be available to help, and one thought the entire idea was a waste of tax dollars.

The final open-ended question asked participants to list ways they might use NDL materials. Thirty-eight (45 percent) of the respondents did not write anything or wrote "NA." Fifteen (18 percent) specifically noted schoolwork or projects; seven (8 percent) noted to help children (one said younger brother); eight (9 percent) said reports, presentations, or projects; and four said for general reference. The remaining responses related to some sort of personal knowledge: five noted general personal information; five noted travel or vacation information; three noted general knowledge acquisition; two said knowledge of history; two said genealogy; and one each noted music/literature, multiple sclerosis, and art inspiration. One other respondent indicated potential use of the NDL for entertainment, and one said "just like a library."

Although it was difficult for the respondents to speculate about the pros and cons of an NDL they knew little about, the comments given do cover the commonly expressed advantages of digital libraries. They also present a wide range of concerns about technical and intellectual challenges to using the NDL.

Summary of Questionnaire Analyses

At the time of this assessment, the general populace showed interest in computers and the information resources networks offer, but they had limited firsthand experience with the WWW. Home computer use was high, with some sort of online access becoming more common. Schools used a wide range of computers, many have some type of online capacity in the building, but few had it in classrooms. The day-care results show that few respondents gave specific reasons for using the NDL, and none cited specific collections they wanted to access. The media specialists gave some specific reasons but no specific collections. The teacher respondents who had participated in LC training cited specific materials. These results reinforce the theme of diversity in users, needs, and computational settings and suggest that the NDL must be purposefully introduced to potential users.

LC Document Analysis

Reading Room Handouts

The first documents examined were those available in the reading rooms as handouts for visitors. A rich set of knowledge is captured in these handouts, and the NDL interface must provide some of their functionality. Some of the handouts related to physical space and working hours—functions less critical in the NDL but still needed in terms of services and the interplay between digital and physical collections. Others provided descriptions of the collection and policies for using the collection, and these were clearly applicable to the NDL. A challenge was finding ways to integrate the many different reading room descriptions into a common introduction for the integrated NDL. Documents that help users actually conduct research or otherwise use the collection will also be as valuable for NDL users as they currently are for LC reading room users. Finally, the reading rooms have created specialized documents related to particular user needs or hot topics (such as current events and anniversaries). The existence of these documents illustrates the need for some similar service in the NDL interface.

All these documents reinforce the complex nature of the individual reading rooms and the complexity of the aggregate LC. Preliminary versions of the interface can simply point to electronic versions for these documents. However, specialized electronic versions that emerge over time can be integrated into the system help and guided tour components.

LC User Studies

The two LC user studies we utilized offer a rich set of specific interface recommendations. One represented a user evaluation of the digital American Memory collection (Library of Congress 1993). Some of the impressions and commentary at the early American Memory test sites strictly reflect individual preferences. However, some trends were common across many sites:

• Students' preference for buttons over pull-down menus,

• Poor understanding of search strategy (such as misunderstanding Boolean AND and a lack of systematic reformulations or use of results),

• Complaints about the time needed to use the system (both physical access due to a single workstation being available and response time),

• Teachers' desire for linkages between the primary material and their curricula (a common use for the content and system was as an enrichment adjunct rather than as an integral part of the course), and

• The feeling that it is essential that materials can be printed and/or saved to disk.

The Prints and Photographs WWW Public User Interviews (Library of Congress 1995) identified a variety of navigational problems. For example, the lack of basic graphical user interface literacy made it difficult to use a mouse, to scroll, or to use pull-down menus. Moving among levels of representation (between a full image, thumbnail, and brief display) was also difficult, and losing buttons after scrolling was frustrating.

Search strategy problems that were identified included confusion regarding relevance ranking of results, a need for easy-to-use alphabetical subject headings, a need to limit hit lists to 100 to avoid having too much information to look through, and a need for posting data (frequency of occurrence) after all moves. Both sets of results highlight challenges for the NDL interface. In addition to reinforcing the importance of computer experience, these findings illustrate the need for crisp and intuitive dynamics between screens and windows so that users can focus on different levels of representation easily or possibly juxtapose them. The findings also indicate the need for judicious use of any scrolling functions and the need for clear yet powerful search articulation and results display.

E-mail Messages

The sample of representative electronic mail messages answered by LC staff provides an interesting mosaic depicting people who are actually using the infant NDL and the types of information needs and problems they bring to the LC. Based on content analysis of user messages, staff responses, and a brief classificatory commentary provided by staff for the exchanges, the messages fall into three broad categories.

The "system" category includes several types of messages. People sent messages of praise with compliments about the NDL. They complained or inquiried about server crashes or telecommunications problems. They sent technical questions (such as when having difficulty launching video players on a client) or suggestions (for example, for alternative data formats). Suggestions about content or adding specific collections were also sent in. One interesting request was made for a zipped index of the NDL that could be downloaded and used locally to save connect charges. Patrons also e-mailed in corrections, such as to correct a mislabeled photograph.

In another category, e-mail classified as identifying "user needs" messages included reference questions (for example, census data for New York State from 1600 to 1900 and how to find it). Other user needs messages included various types of requests, such as those for special collections, for specialized services (online photograph copy requests), and for help. In the latter instances, users wanted help in narrowing ongoing searches; other users were confused about what to ask or where their mail was going.

The third category includes a variety of "miscellaneous" messages, such as how to contribute specific items to the LC (donating a photograph), requests that the LC link to users' Web sites, and requests by users seeking jobs to work on the NDL.

General communication also filled in this category. Some users wanted to maintain a dialog, responding with thanks and commentary to a reply to their earlier messages.

Taken together, these documents capture a wide range of knowledge about the LC, the people who use it, and the types of information needs they bring to it. The existing documents must be captured and integrated into the NDL. Analysis of future e-mail will surely contribute important data as the system evolves, and new systematically scheduled user studies are highly recommended.

User Type and Task Taxonomy

A central theme emerging from this investigation is diversity. There were wide ranges in users, the tasks they bring to the NDL as manifestations of their needs, the technological settings they work in, and the wide variety of content that makes up the NDL. To integrate the results from the different components of the investigation, we revisited the nature of users and needs in light of the types of user communities identified before data collection began. It was clear that the three communities (LC staff, current scholars, and prospective NDL users) were too broad to capture the diversity or to guide fully the interface design.

A more fine-grained classification can be made by combining the analytical framework of users and tasks (Marchionini 1995b) with the empirical data from this investigation. The analytical framework crosses users, tasks, and the personal situations that motivate the search. *Users* have a host of individual characteristics, preferences, and experiences that are not orthogonal. They differ in physical, cognitive, and social (personal) attributes. They differ in their experiences in the domain of knowledge related to their information need and in their experiences in using library systems, information technology, and research techniques. As noted above, there are also a number of nonorthogonal dimensions associated with the information-seeking *task*—complexity, specificity, quantity, and timeliness.

Analytically, we can define scales for each dimension (low, average, high) and populate the resulting matrix with examples or cases. Assuming three points per dimension, we would have 27 cells for user characteristics and 243 cells for tasks. Crossing the two matrices would yield 6,561 cells to fill. Adding the many types of motivations and situations that contextualize specific instances of information seeking similarly expands the theoretical possibilities. Clearly, this approach is impractical, due to the large number of variations for which interface features are considered. Additionally, it is both minimal and simplistic in terms of user characteristics.

Table 6.4
User Taxonomy

---

1. *LC staff*   High motivation, medium domain knowledge, high library system knowledge, high focus, and limited time allocations

2. *Hobbyists* (e.g., genealogy, Civil War, railroads, other examples)   High motivation, typically high domain knowledge, a range of library system knowledge, high focus, and high time allocations

3. *Scholars* (e.g., historians, sociologists, anthropologists, authors)   High motivation, high domain knowledge, high library system knowledge, high focus, and high time allocations

4. *Professional researchers* (e.g., picture researchers)   High motivation, medium domain knowledge, average to high library system knowledge, very high focus, and medium time allocations

5. *Rummagers (browsers)* (e.g., Ph.D. students looking for topics, scholars looking for new topics)   High motivation, medium domain knowledge, range of library system knowledge, low focus, and medium to high time allocations

6. *Object seekers* (e.g., some authors, CD-ROM/multimedia developers, TV/video producers, and instructional materials developers)   High motivation, range of domain knowledge, low library system knowledge, high focus, and low to medium time allocations

7. *Surfers* (e.g., those who are curious, those who bump into the NDL, etc.)   Low motivation, low domain knowledge, low library system knowledge (but may be high computing system knowledge), low focus, and very low time allocations

8. *Teachers K–16*   Medium motivation, medium to high domain knowledge, low to medium library system knowledge, medium focus, and low time allocations

9. *Students K–16*   Low to medium motivation, low domain knowledge, low library system knowledge, low to medium focus, and low to medium time allocations

---

A more realistic approach collapses some dimensions in light of the empirical evidence collected in reading rooms, surveys, and document examinations. Based on the evidence, factors that characterize users include

- Motivation (the personal situation that brings one to the Library, including quantity, criticality, and timeliness),
- Domain knowledge (related to the particular need),
- Library system knowledge (including information technology),
- Focus (a combination of complexity, quantity, and specificity), and
- Time allocated (combining timeliness and criticality).

These factors were applied to the different types of users and user needs described in the reading room visits. Such an approach yielded the nine user classes listed in table 6.4. These classes are not exhaustive, nor are they mutually exclusive.

Table 6.5
Interface Challenges Identified by the User's Need Assessment and Addressed by the Prototypes

Interface challenges: Content

General interface challenges across reading rooms:
Developing a framework characterizing for users the granularity, size, and nature of objects in the NDL across all the reading rooms
Communicating to the user what items are *not* in the NDL
Helping the user to see that the NDL has multiple, but not uniform, access points
Creating an interface that is accessible to users with state-of-the-market technology
Inventing new techniques to search for multimedia objects and to integrate those techniques into the interface (e.g., visual and audio query languages)

Specific interface challenges from different reading rooms:
Audio access (searching as well as download times)
Access to multiformat items (i.e., a sound track and textual field notes)
User specification of areas/regions on maps
Place name ambiguity
Identifying and representing links
Distinguishing documents that help one do genealogical research from the primary materials
Integrating full-text and controlled vocabulary searching both across and within collections
Distinguishing finding aids levels and primary material
Need for linkages from browsable covers to bibliographic records and microfilm text available at the LC
Image searching (including displaying series of related images and images in challenging formats such as panoramas and oversized posters)—possible use of P&P thesaurus
Enabling researchers to absorb enough of the context for historical images and captions (why images were made) to deal sensitively with content that might otherwise be deemed offensive.

Moreover, any individual belongs to a class for each information need (in different visits or sessions, users may fall into different classes). It is highly likely that this taxonomy will change as the NDL evolves since it is rooted in data that come from a mix of current users of the physical LC, early adopters of the nascent implementation, and nonusers asked to speculate about using it.

This taxonomy was very useful to the prototyping phase of the NDL project by providing detail to the general design vision and acting as a "reality check" for design discussions. Three additional uses are apparent for continued development of DL interfaces. First, the taxonomy can guide the development of features that substitute for the reference interview. Since a human resource will not be readily avail-

Table 6.5
(continued)

Interface challenges: Users and strategies

General interface challenges across reading rooms:
Serving a wide range of users
Serving a wide variety of information needs
Helping users distinguish primary and secondary materials (including multiple layers of each)
Helping users make links among items across different collections and reading rooms
Capturing the essential elements of the reference interview so that users can find what they need without human intervention
Specific interface challenges from different reading rooms:
Specifying geographical areas
Potentially huge files to transfer and display
Determining when to point users elsewhere
Helping LC NDL users to quickly understand that few of the primary materials are online
Helping LC NDL users to quickly understand that there are many levels of search to work through
Identifying copyrighted materials (so patrons do not find pointers to them and expect they can come and copy them at LC)
Supporting hot topics, specialized exhibits
Overcoming patrons lack of knowledge about media (e.g., how pictures were produced at different times)

---

able, the system must provide ways for users to articulate specific needs (queries) as well as contextual information (granularity and scope of need). A set of user templates varying according to the parameters above was sketched early in the prototyping phase of work, although that work was not pursued as other features with higher priority were addressed. Second, the taxonomy can provide the basis for variations in discussions of interface features such as help, tours, and tutorials. Third, it can provide the basis for testing prototypes with scenarios and representative tasks.

Design Implications

This needs assessment identified many design challenges specific to the LC's NDL program, a number of which have been described above. Table 6.5 summarizes these challenges, which can also be read as system requirements. As the design of the

NDL proceeded, these challenges were prioritized, and specific prototype features were created. We chose to address the general challenges of "content" and "users and strategies" in the prototypes we implemented.

These are strong requirements, and the degree of success varied according to how one prioritized the challenges. The general user needs identified in this assessment oriented the design team in understanding both the needs of potential users and the characteristics of the rich and varied collections of materials scheduled for the NDL. In addition to contributing to the eventual prototypes, the multifaceted approach to assessing needs was applied and extended in other projects in different public access settings Hert and Marchionini (1998) built on these methods to develop a user-task taxonomy for federal government statistical Web sites.

Conclusion

The case studies presented in this chapter illustrate three different situations in which multifaceted and flexible approaches to human-centered DL design and eval-uation were taken. Each case included prioritizations and design decisions that were woven into the fabric of interface prototypes, systems, and our understanding of human information behavior. They are juxtaposed to highlight some of the general approaches to DL design and evaluation, while demonstrating the unique challenges of each setting. Practical design is a compromise between the top-down BITWC and bottom-up, organically grown system. Our central assumption is that practical design is guided by general design visions that are informed by multifaceted, ongo-ing assessments of user needs and system impact.

The Perseus DL evaluation illustrates how multifaceted approaches, over time, can yield evidence of significant effects across a broad community of users. In this case, the user needs were perceived by the DL designers, the primary user popula-tions are university students and professors, and the main design goals were to provide large quantities of primary text and image materials with little or no inter-pretation. The evaluation effort was begun at the inception of the project. It used a multiplicity of methods to collect data on a large set of evaluation questions that were tied to a matrix of stakeholder, DL mission, and technical dimensions. Early results demonstrated typical effects of mechanical advantage and the difficulties of learning to teach with technology. They also yielded some examples of new kinds of learning, but it was only through longitudinal study that systemic change in larger sets of stakeholders (for example, the field of classics) began to emerge. Without a steady and varied stream of data, these larger effects would likely not have been demonstrated except through some future historical efforts.

The Baltimore Learning Community case illustrates the difficulty of influencing complex human behavior (like teaching) through external interventions, regardless of the information or technical innovation. In this case, user needs were determined by state and local administrative curricula and assessment programs. The primary user population is middle school science and social studies teachers and their students. The predominant design goals were to provide relatively small quantities of multimedia materials that teachers could map to well-defined curriculum guidelines in creating lesson plans. The experience of this case demonstrates that advanced technical solutions and high-quality content are not sufficient to initiate or sustain community in settings where day-to-day practice is strongly determined by personal, social and political constraints.

The Library of Congress NDL case details how one multifaceted user needs assessment was conducted and used to guide the creation of user interface prototypes. User needs were largely unknown, and the first phase of the design project was devoted to investigating potential needs in a variety of settings. This DL aims to serve the entire U.S. population. The overarching design goal was to provide universal access to historical materials in a variety of formats and at the entire range of granularities from bibliographic records and finding aids to full manuscripts and images. The user needs assessment yielded a set of user types and interface design challenges that guided the development of eventual prototypes and may find use in other DL settings.

Although these three DLs aim to serve different user populations and had distinct design goals, there are several general design and evaluation principles that resonate across the cases. First, DL designers and evaluators must know the users. This general rule is followed through needs assessments, and the different cases illustrate that needs may be externally or internally motivated, well defined or ill defined, and narrow or wide-ranging. The three cases demonstrate the benefits of using direct methods such as interviews and observations along with indirect methods such as document analyses or interviews with intermediaries. It is only through understanding user characteristics and needs that DL designers can build tools to help users map information needs onto DL tasks and evaluators can develop a good set of questions to guide research.

Second, it should be clear that DL design and evaluation are processes that aim to create and understand complex products. As such, they proceed in stages over multiple iterations. DLs are constantly evolving to meet the needs of users who grow and learn and technologies that advance exponentially. Design and evaluation efforts must be embedded into the overall DL management culture and applied on an ongoing basis.

Third, in addition to the complexities of change noted above, DLs must serve people with a wide range of characteristics and needs. To do so requires that designers and evaluators create and use multiple and flexible systems and tools. Designers must create overviews that allow DL patrons to quickly understand what is and is not available in the collection. We must provide alternative entry points and search tools so that diverse user needs and experiences can be accommodated. Evaluators must seek data from all fronts and create ways to integrate those data streams to make judgments about progress and next steps. Just as the blind men could understand only parts of the elephant they experienced firsthand, our understanding of DL design and evaluation processes will require us to examine multiple views through multiple lenses.

Notes

1. Learning had four general questions, twenty-four subquestions, and two more specific questions. Teaching had three general questions and nine subquestions. System had three general questions, thirteen subquestions, and twenty-six more specific questions. Content had three general questions and twelve subquestions. As an example, the first general learning question was "What tactics and strategies do students employ at particular junctures in the Perseus environment?" A subquestion under that question was "What proportion of time do students spend in the primary text?"

2. See ⟨http://www.learn.umd.edu⟩ for background on the BLC, including a demonstration of selected system components and a set of reports on the project. See Ernestine Enomoto, Victor Nolet, and Gary Marchionini (1998) for preliminary evaluation results.

3. A more parsimonious position is to assume a state-of-the-installed-base, which seems entirely too constraining for an evolving, forward-looking project like the NDL.

4. See ⟨http://www.cs.umd.edu/projects/hcil/ndl/index.html⟩ for reports and various prototypes.

5. In asking open-ended questions about phenomena that respondents may have little experience with, a questionnaire design decision is whether to give examples or not. Examples explicate and focus the question but also may lead and bias responses. We decided to use speed and ease attributes as examples of both possible advantages and disadvantages of NDLs. Participants were asked to speculate on advantages and disadvantages that might be expected with NDL use.

References

Aristotle. 1985. *Nicomachean Ethics*, trans. Terence Irwin. Indianapolis: Hackett.

Belkin, N. 1980. Anomalous States of Knowledge as a Basis for Information Retrieval. *Canadian Journal of Information Science*, 5, 133–143.

Braha, D., and O. Maimon. 1997. The Design Process: Properties, Paradigms, and Structure. *IEEE Transactions on Systems, Man, and Cybernetics—Part A: Systems and Humans*, 27(2), 146–166.

Brooks, F. 1975. *The Mythical Man-Month: Essays on Software Engineering*. Reading, MA: Addison-Wesley.

Dervin, B., and M. Nilan. 1986. Information Needs and Uses. In M. E. Williams, ed., *Annual Review of Information Science and Technology* (vol. 21, pp. 3–33). White Plains, NY: Knowledge Industries.

Enomoto, E., V. Nolet, and G. Marchionini. 1998. Video and Dynamic Query Capability in Schools: Implications for Learning in a Networked Community. *Bulletin of Science, Technology, and Society*, 18(6), 432–440.

Flagg, B., ed. 1990. *Formative Evaluation for Educational Technologies*. Hillsdale: Erlbaum.

Fox, R. 1995. Newstrack. *Communications of the ACM*, 38(11), p. 9.

Gardner, H. 1983. *Frames of Mind: The Theory of Multiple Intelligence*. New York: Basic.

Harter, S., and C. Hert. 1997. Evaluation of Information Retrieval Systems: Approaches, Issues, and Methods. In M. E. Williams, ed., *Annual Review of Information Science and Technology* (vol. 32, pp. 3–94). Medford, NY: Information Today.

Hert, C., and G. Marchionini. 1998. Information Seeking Behavior on Statistical Websites: Theoretical and Design Implications. In C. Preston, ed., *Proceedings of the 61st Annual Meeting of the American Society for Information Science (Pittsburgh, Oct. 25–29)* (pp. 303–314).

Hix, D., and H. Hartson. 1993. *Developing User Interfaces: Ensuring Usability through Produce and Process*. New York: Wiley.

Komlodi, A., and G. Marchionini. 1998. Key Frame Preview Techniques for Video Browsing. In C.-W. Karat, A. Lund, J. Coutaz, and J. Karat, eds., *Digital Libraries '98: Proceedings of the Third ACM Conference on Digital Libraries (Pittsburgh, June 23–26)* (pp. 118–125). New York: ACM.

Krikelas, J. 1983. Information-Seeking Behavior: Patterns and Concepts. *Drexel Library Quarterly*, 1911, 5–20.

Library of Congress. 1993. American Memory User Evaluation 1991–1993: Site Summaries. An Addendum to the Main Report. Report prepared by the American Memory User Evaluation Team, Washington, DC.

Library of Congress. 1995. *P&P World Wide Web Public User Interviews*. Washington, DC: Prints and Photographs Staff.

Marchant, M. 1991. What Motivates Adult Users of Public Libraries? *Library and Information Science Research*, 133, 201–235.

Marchionini, G. 1995a. The Costs of Educational Technology: A Framework for Assessing Change. In Hermann Mauer, ed., *Proceedings of ED-MEDIA 95 World Conference on Educational Multimedia and Hypermedia (Graz, Austria, June 17–21)* (pp. 33–38). Charlottesville, VA: AACE.

Marchionini, G. 1995b. *Information Seeking in Electronic Environments*. New York: Cambridge University Press.

Marchionini, G. 1998. Research and Development in Digital Libraries. In A. Kent, ed., *Encyclopedia of Library and Information Science* (vol. 63, pp. 259–279). New York: Marcel Dekker.

Marchionini, G. 1999. Augmenting Library Services: Toward the Sharium. In S. Sukimoto, ed., *Proceedings of International Symposium on Digital Libraries 1999 (Tsukuba, Japan, September 28–29)* (pp. 40–47).

Marchionini, G., and G. Crane. 1994. Evaluating Hypermedia and Learning: Methods and Results from the Perseus Project. *ACM Transactions on Information Systems*, 12(1), 5–34.

Marchionini, G., and E. Fox. 1999. Progress toward Digital Libraries: Augmentation through Integration. *Information Processing and Management*, 35(3) (special issue on Progress toward Digital Libraries), 219–225.

Marchionini, G., C. Plaisant, and A. Komlodi. 1998. Interfaces and Tools for the Library of Congress National Digital Library Program. *Information Processing and Management*, 345, 535–555.

Nielsen, J. 1993. *Usability Engineering*. Boston: Academic Press.

Norman, D. 1988. *The Design of Everyday Things*. New York: Doubleday.

Paisley, W. 1980. Information Work. *Progress in Communication Sciences*, 2, 113–165.

Petroski. 1996. *Invention by Design: How Engineers Get from Thought to Thing*. Cambridge, MA: Harvard University Press.

Plaisant, C., G. Marchionini, T. Bruns, A. Komlodi, and L. Campbell. 1997. Bringing Treasures to the Surface: Iterative Design for the Library of Congress National Digital Library Program. In S. Pemberton, ed., *CHI 97 Proceedings (Atlanta, March 22–27)* (pp. 518–525). New York: ACM.

Shneiderman, B. 1998. *Designing the User Interface: Strategies for Effective Human-Computer Interaction* (3rd ed.). Reading, MA: Addison-Wesley.

Simon, H. 1996. *The Sciences of the Artificial* (3rd ed.). Cambridge, MA: MIT Press.

Sternberg, R. 1985. *Beyond IQ: A Triarchic Theory of Human Intelligence*. New York: Cambridge University Press.

Suchman, E. 1967. *Evaluative Research: Principles and Practice in Public Service and Social Action Programs*. New York: Russell Sage Foundation.

Taylor, R. 1962. The Process of Asking Questions. *American Documentation*, 13(4), 391–397.

Wilson, T. 1981. On User Studies and Information Needs. *Journal of Documentation*, 371(1), 3–15.

7

Participatory Action Research and Digital Libraries: Reframing Evaluation

Ann Peterson Bishop, Bharat Mehra, Imani Bazzell, and Cynthia Smith

Introduction

What particular issues arise in digital library research when intended users are on the wrong side of socioeconomic, information, and digital divides? How can digital library evaluation account more fully for the practices and consequences of use for marginalized members of society, such as the poor and people of color? The evaluation of information technology is too often characterized by features that we feel are especially detrimental to those outside the social mainstream: it relies on external standards of expertise, treats users as subjects or objects of the evaluation, and pays only indirect attention to social impacts associated with use, often striving to take a neutral and objective view.

We hold in abeyance the question of whether such an evaluation stance is ever appropriate, and look instead at what is happening in one case where we are trying to take a different course. In the Afya (Swahili for "health") project, we are establishing a collaboration between African American women and other community partners in building a collection of digital tools and resources that women will find both usable and useful in their efforts to nurture a healthy lifestyle. Our goal is to build capacity for creating and sharing health information across the social, cultural, economic, and technology divides that separate black women from health and information service providers. Ultimately, we are striving for social transformation on a communitywide basis.

To support these aims, evaluation in the Afya project strives to

• Incorporate local knowledge held by marginalized groups,

• Gain the participation of marginalized groups early in the design and development process, and

• Build capacity and achieve constructive social outcomes.

Our work in the Afya project reconsiders traditional approaches to information technology evaluation. To engage with the social practice of disenfranchised users, we are incorporating ideas and techniques from the domain of participatory action research, a domain that both claims social practice as its fundamental object of study and explicitly pursues an agenda focused on improving conditions for disenfranchised members of society.

Community Health, African American Women, and Digital Libraries

The health of community residents depends on social and cultural factors at the local level (such as politics, economics, beliefs, traditions, and prejudices) and on the activities of multiple community institutions (McKenzie, Pinger, and Kotecki 1999). In the United States, "the gap in health status and mortality between those commanding, and those who lack, economic power and social resources continues to widen" (Moss and Krieger 1995, p. 302). A landmark 1985 study demonstrated that African American women were at significantly greater risk of death from cancer, diabetes, and hypertension (U.S. Department of Health and Human Services 1985) than were white women, and improvements since then have been modest (U.S. Department of Health and Human Services 1990). Better health care for African American women remains high on the federal agenda (U.S. Department of Health and Human Services 2001) and the inequitable delivery of health-care services to minorities continues to receive attention in the press (White 1999).

African American women often are considered a hard-to-reach population by health-care institutions, and they face particular problems in finding and using health information. Health behavior change models typically used by health-care professionals are often ineffective in promoting a healthy lifestyle among African American women because they fail to fit their sociocultural context (Ashing-Giwa 1999). For example, efforts to promote prenatal care for black women may fail unless support services like transportation, child care, and literacy training are provided (Floyd 1992). Health information is often difficult for black women to understand and use because it is viewed as both irrelevant to their needs and laden with medical jargon (Gollop 1997). Those with higher incomes tend to acquire information at a faster rate, so knowledge gaps increase over time in communities where black women occupy the lower end of the socioeconomic spectrum. And these gaps in knowledge lead to negative health outcomes (Tichenor, Donohue, Donohue, and Olin 1970, pp. 159–160, cited in Viswanath, Finnegan, Hannan, and Luepker 1991).

Can the creation of digital libraries related to African American women's health help resolve these problems? Digital collections and services in health domains include community-based health information networks (Payton and Brennan 1999), interactive health monitors (Berlin and Schatz 1999), and online health resources found at sites sponsored by not-for-profit community networks (Cohill and Kavanaugh 2000; Gurstein 2000; Schuler 1996) or commercial health organizations. Collections of Web resources are organized by public libraries or health-related organizations, including pharmaceutical companies. Online systems, however, may duplicate the mazes and turf battles associated with offline community information (Dervin 1976; Dewdney and Harris 1992; Pettigrew and Wilkinson 1996). Moreover, health information on the Internet is often poorly organized, inaccurate, and difficult to read and understand (Berland et al. 2001).

National studies of the digital divide highlight disparities in information technology access and use along socioeconomic lines. Census data reveal that computer ownership and Internet use is less prevalent among marginalized groups such as the poor, ethnic minorities, and women (NTIA 2000). As pointed out in a Benton Foundation (1998, p. 5) report: "The information poor will become more impoverished because government bodies, community organizations, and corporations are displacing resources from their ordinary channels of communication onto the Internet." While the Internet has proved a useful tool for collective action among African American women in some instances (Mele 1999) and the number of women online has surged recently, women's organizations and groups have typically been behind the curve in technology adoption and use of computers (Balka and Doucette 1994).

Lack of access to technology is only one factor that is delaying the spread of computer and Internet use among marginalized groups. Equally important is the lack of "useful content on the Internet—material and applications that serve the needs and interests of millions of low-income and underserved Internet users" (The Children's Partnership 2000, p. 5). Web sites with health information geared specifically to African American women have appeared only recently. Perhaps most important, information access and use models and digital library design and evaluation assumptions may fail to fit the sociotechnical context of black women. The Benton Foundation (1998, p. 12) found that "creative ways will have to be found to make computer networking more a part of the social lives" of traditionally underserved audiences. We know little about the situational or contextual factors that prompt community members to seek information online and about how it helps in their daily lives (Pettigrew, Durrance, and Vakkari 1999).

While previous research has identified community information needs and exchange practices of marginalized groups (see, for example, Agada 1999; Chatman 1996; Metoyer-Duran 1993; Uehara 1990), it has not directly pointed the way to participatory models for involving local residents or for crossing institutional boundaries and tiers in the development of digital community information resources. Traditional conceptual frameworks and methods for the design and evaluation of information systems seem ill suited to this task. The active and direct participation of marginalized community members is needed to create usable and useful online information services suited to the social context of their everyday lives and to the achievement of their goals.

This chapter refocuses digital library evaluation around action, empowerment, and social change. In the Afya project, evaluation is being reframed in an attempt to mobilize the local knowledge of African American women, to expand the utility of evaluation as a concept and process employed throughout the system lifecycle, and to build the capacity of community members. Our work depends on tapping into the knowledge held by local African American women. They are the experts when it comes to identifying their own problems and needs in the delivery of health services and information. Their help is needed to make decisions regarding the sociocultural relevance of digital library content, desired functionality, and interface preferences.

Reconsiderating Digital Library Evaluation

Social Responsibility in the Information Professions

We argue that greater attention needs to be paid to the concerns of marginalized members of society in the evaluation of digital libraries. One of the defining characteristics of a profession is the responsibility of its practitioners to examine the effects of their activities on society and to strive for positive social outcomes. Virtually every profession related to the growth and exchange of knowledge has strong advocates for social responsibility. For example, Jay Rosen (1995, p. 18) makes a case for "public journalism," arguing that "journalists should do what they can to support public life. The press should help citizens participate and take them seriously when they do. It should nourish or create the sort of public talk that might get us somewhere, what some of us would call a deliberative dialogue. The press should change its focus on the public world so that citizens aren't reduced to spectators in a drama dominated by professionals and technicians."

One example of public journalism practice cited by Rosen is a local newspaper's attempt to ground its campaign coverage in a list of issues identified as important by residents. Campaign speeches were mapped against this citizens' agenda so that it was easy for residents to tell what was said about their concerns. Here, as in the Afya project, we see the link between attending to the needs of local residents and gaining their active participation in shaping information generation and flows.

The profession of librarianship is fundamentally grounded in a sense of social responsibility, expressed prominently in advocacy for intellectual freedom and broad, equitable access to information resources. There is a strong tradition of outreach to underserved groups in public librarianship (e.g., Bundy 1980; De la Pena McCook 2000). In Great Britain, attention to social inclusion in all public library services, including access and use of computers, is a matter of national policy (Muddiman et al. 2000). The Computer Professionals for Social Responsibility's (CPSR 2002) mission statement includes evaluation as a fundamental need that is grounded in concerns about the impact of information systems on society: "As technical experts, CPSR members provide the public and policymakers with realistic assessments of the power, promise, and limitations of computer technology. As concerned citizens, we direct public attention to critical choices concerning the applications of computing and how those choices affect society."

In the Afya project, we seek approaches to the evaluation of digital libraries that are rooted in social responsibility. Our research tries to integrate information system evaluation methods with a philosophy and epistemology oriented to social justice.

Need for a New Epistemology: Legitimizing "Local" Knowledge

In addition to foregrounding social responsibility, the reconsideration of digital library evaluation presented in this chapter focuses on new ways of thinking about how knowledge is created and used by those with a stake in the process and outcomes of evaluation—social science researchers, digital library designers, librarians, and marginalized members of society. On the institutional level, we see a growing call for universities, which form the seat of traditional expertise for research and professional practice related to digital libraries, to become more fully engaged in society (Kellogg Commission on the Future of State and Land-Grant Universities 1999). Donald A. Schön (1995, p. 34) argues that universities should admit new forms of scholarship, like action research, and a new "epistemology of reflective practice" that must "account for and legitimize not only the use of knowledge produced in the academy, but the practitioner's generation of actionable knowledge."

We see a need for digital library evaluation that derives from the diverse knowledge held by multiple practitioners (where "practitioners" is understood broadly as those engaged in the practice under investigation) in the arena of community health—librarians, health-care providers, and community residents. Actionable knowledge comes from examining the current practice of diverse participants in community health and reflecting on how these groups can work together to improve the delivery of health care and health information.

Especially germane to the Afya project is the negotiation of whose knowledge is legitimate when formal, expert systems clash with informal, local knowledge (Star and Bowker 1999). The "funds of knowledge" approach in education seeks innovation in classroom teaching by drawing on the knowledge, skills, and social networks found in households that are challenged by difficult social, political, and economic circumstances (Moll 1994; Moll, Amanti, Neff, and Gonzales 1992). In this approach, an attempt is made to integrate—in a more reciprocal and holistic manner—the substance and flows of folk and formal knowledge.

Following John Dewey (1916), Michael Glassman (2001, p. 5) explains the importance of the individual's unique set of locally grounded intellectual tools, defined as "the socially developed tools such as morals, ideals, values, and customs that serve as reference points for the individual as she attempts to navigate life situations." He argues that a "disturbed equilibrium" occurs when diverse local realities conflict and that this necessary grounding leads to true learning and change in a democratic society. Here, then, we see the positive side of identifying and representing different stores of local knowledge in the creation a digital library, even when they clash. Authentic learning on a communitywide basis demands that differences in the intellectual tools of community members be exposed. Arriving at a more socially just equilibrium requires adjustments in practice that recognize the validity of knowledge arising from the values, goals, and experiences of marginalized groups.

Socially Grounded Evaluation of Digital Libraries

How do we create digital libraries that will be both usable and useful for traditionally marginalized groups in society, including women, people of color, and the poor? How do we uncover and bridge varying sets of intellectual tools in evaluation? Close studies of social practice are becoming increasingly common in research being done on information systems use, and a number of researchers provide frameworks for coupling consideration of social practice with social values.

Bonnie Nardi and Vicki O'Day (1999) introduce the concept of information ecologies to describe the way in which people, practices, technologies, and values are intertwined in information use. Considering digital libraries from this perspective (O'Day and Nardi, chapter 4), they also make a connection between understanding social values and practice and enhancing system usability. They argue for the need to look closely and holistically at particular use communities and environments, so that technological innovations that seem fine when assessed in isolation do not turn out to be flawed in real use settings.

Bertram C. Bruce and Maureen P. Hogan (1998) present a particularly cogent statement regarding the nexus of social practice, values, and outcomes in the design and evaluation of information technology. They argue that researchers need to undertake situated studies that examine how technologies are realized in given settings and how ideology comes into play in situations where technology and humans interact. In the arena of the digital divide, the crucial question is whether marginalized users are included in the mix of people studied and whether system success is gauged against their needs, interests, and capabilities. As Bruce and Hogan (1998, pp. 275, 271) note, information system access and use are always "partial, restricted, and stratified," but we need to pay closer attention to whom the technology marks as "full, marginal, or nonparticipating."

Here we see a strong emphasis on change—on exploring how technology changes what people do and how people change what technology does in particular settings of use. Situated evaluation provides a new framework for understanding sociotechnical innovation, one that focuses analysis on actual practices of use and assumes that "the object of study is neither the innovation alone nor its effects, but rather, the realization of the innovation—the innovation-in-use" (Bruce and Rubin 1993, p. 215). Michael B. Twidale, David Randall, and Richard Bentley (1994, p. 441) also call for a reframing of evaluation to account for how an information system is realized in a specific situation of use. They argue that evaluation "can be understood as a process which should saturate and be constitutive of the design process precisely because the 'context of use' is central" to the analysis of information systems. They consider the problems of evaluating systems in use and conclude that ethnographic insights are central to integrating evaluation into design.

Reframing evaluation in this fashion brings a representation of the digital library in use to those responsible for making decisions about the digital library's contents, functionality, interface, and access. At the heart of the matter is finding ways to bridge the distance between system designers and users so that the digital libraries can be evaluated with authentic representations of the context of use for marginalized members of society. In the Afya project, we are developing a socially grounded approach to evaluation that considers social values, practice, and consequences;

includes participation of intended users throughout design and implementation in a manner that grants greater parity to the local knowledge held by users; and strives for constructive change in the lives of marginalized members of society. The tradition of participatory action research provided us with relevant perspectives from social and educational program evaluation. It also gave us a model for information technology evaluation for situations in which use and nonuse are assumed to have critical consequences for disenfranchised community members.

Participatory Action Research in Digital Library Evaluation: Empowerment and Social Change

Action Research

Action research is familiar territory in educational and social program evaluation, urban planning, and community development. It claims social practice as its object of study and the transformation of social practice, along with the social institutions and relationships that support it, as its primary aim (Kemmis and McTaggart 1988). With social practice as the target, it does not isolate any action—such as digital library use—from its social context and ideological roots, no matter how technically or cognitively based that action may seem. In action research, social practice is not observed and interpreted according to externally imposed goals and theories by researchers whose sphere of action lies outside that of the practitioners whose social practice is under investigation. Rather, those directly involved in an activity set the course for what needs to be examined and changed to improve their actions and circumstances.

Action permeates both the methods and results of action research and is inseparable from them. Methods are those actions performed by practitioners as they conduct, observe, and reflect on their current mode of social practice. Specific techniques typically include diaries and recorded observations of practice, interviews, and content analysis of texts associated with practice in some way. Another primary activity is tinkering with the current mode of action and then assessing and reflecting on the outcomes. In other words, an innovation is entered into use and is observed and reflected on, discussed, and assessed. Since the methodology of action research is grounded in the actions of those engaged in the social practice being studied, its goal is "authentic insights, grounded in participants' own circumstances and experiences," and it "tends to be informal and convivial, rather than formalistic and overtly 'theoretical'" (Carr and Kemmis 1983, p. 179).

Results are similarly intertwined with action. Changes in social practice (in the knowledge, values, and actions of participants in the practice) are the primary form that research outcomes take. Exploring and tinkering with practice leads to new insights and relationships that take root in practice. Confronting one's own ideology and value system is a necessary starting point for understanding one's behavior. Understanding the value systems of those with whom one's actions intersect is also necessary in forging the limits of change. Thus, those engaged in action research also strive to engage the full range of relevant stakeholders, widening the circle of reflection and change.

In the social practice of digital library design and evaluation, the range of people who can be considered practitioners (or stakeholders in digital library practice) include programmers and interface designers, content developers, social scientists, potential users, and people responsible for implementing and managing digital library services. Action research in this context involves examination and reflective analysis of each person's role in the evaluation process; results encompass improved social practice associated with digital library use.

Action research that explicitly seeks to build capacity and attain immediate improvements in local conditions or marginalized groups at the community level is often called participatory action research (Reardon 1998, p. 59):

Participatory action research focuses on the information and analytical needs of society's most economically, politically, and socially marginalized groups and communities, and pursues research on issues determined by the leaders of these groups. It actively involves local residents as co-investigators on a equal basis with university-trained scholars in each step of the research process, and is expected to follow a nonlinear course throughout the investigation as the problem being studied is "reframed" to accommodate new knowledge that emerges.

In participatory action research, capacity building takes the form of promoting learning among all participants in an investigation. Scholars help marginalized groups improve their ability to frame and investigate their own concerns. Local residents help scholars improve the relevance and accuracy of their research.

Participatory evaluation (along with its near relative empowerment evaluation) extends participatory action research principles and methods into the domain of evaluation. It reflects concerns for both social justice and the improved utilization of evaluation results. Learning and empowerment are explicit goals of both the conduct and product of participatory evaluation, which is seen as a developmental process (Whitmore 1998). There is thus a broader view of what constitutes legitimate knowledge and how it is generated for and through action. Key participants are those with little power who will be affected by the evaluation. Their knowledge and how it can be engaged in improving social practice are central to evaluation (Pursley 1996, pp. 32–33):

Participants are active partners who gain insight into their own situation, and acquire the information, skills, and other resources necessary to change it.... People know more about their own needs than anyone else, and therefore, their knowledge and experience are valuable resources in coping with and changing their environment.

What emerges from these participatory approaches is a theoretical framework for investigating social practice that embraces social values, consequences, and change.

In the last decade, the social sciences have increasingly critiqued how race is treated in evaluation (Madison 1992), providing additional insights into the role of marginalized community members in research affecting them. Regardless of the nature of the research approach adopted in a particular evaluation, concern is growing that when evaluators are too far removed from the authentic experiences of women, the poor, and people of color, they are unable to conduct valid assessments of their experiences. The results they produce can be inaccurate or incomplete and do little to shift existing social practice and power relationships. Too often, professional control of the research agenda, budget, and data results in the return of very few benefits to marginalized members of society. Because evaluators lack real interest in and meaningful contact with "subjects" and refuse to articulate power-based explanations for current social practice, their evaluations of marginalized communities portray their inhabitants as pathological and ignore their indigenous culture (Stanfield 1999). A parallel emphasis on strengths drives community development work that counters the dominant "deficit" model with the argument that all communities have assets that can be tapped to build local capacity (Kretzmann and McKnight 1993).

Both the process and product of evaluation must respect multiple perspectives. Robert E. Stake (1990) has coined the term *responsive evaluation* for a research process that emphasizes the issues, context, and standards of a diverse set of stakeholders. In concert with the claim of action research that all participants examine and assess their own social practice, Michael Quinn Patton (1999, p. 437) calls for evaluators to reflect more critically on their own knowledge bases and values: How does the lens of race, gender, or socioeconomic status shape the understanding and actions of researchers? What methods and measures capture fairly the experiences of people of color and the poor? When are evaluative judgments conditioned by personal politics?

Application of Participatory Action Research to Information System Development
Library and information science has a rich tradition of research on the information needs and behaviors of marginalized members of society, some of which focuses on the design and evaluation of information technology and services (see, e.g., Mueller and Schement 1996; Spink, Jaeckel, and Sidberry 1997). Brenda Dervin (1980, 1989, 1992) is especially vocal and innovative in striving for theories and methods that permit the authentic representation of marginalized society members and can lead to system improvements for them. In general, however, this work does not place primary emphasis on action to achieve immediate social change or on building the capacity of marginalized members of society, nor do the people who serve as subjects play a leading role in all phases of the research.

Participatory design—a well-accepted approach to information systems development that typically includes evaluation in authentic use settings—shares much common ground with action research in that both emphasize the social responsibility of professionals and the empowerment of those with little control over their situation. Arising from concerns about the social impact of technology in the workplace, participatory design produces ethnographically informed results geared toward ensuring that the introduction of computers will enrich work environments and benefit workers (Greenbaum and Kyng 1991).

Participatory design recognizes that social practice and technology are mutually constituted. Working in this approach, one looks for fruitful ways to bring users into the design process. Perhaps most important, it raises the issue of whether information technology professionals can "recognize and affirm the validity of perspectives other than their own, and value expertise that comes from experience, not just the knowledge that is attested by academic credentials" (Muller and Kuhn 1993, p. 26). Participatory design practices are far from monolithic, yet a few broad distinctions can be drawn between participatory design and action research as applied to information system development. Participatory design, as its name suggests, aims first at technology innovation rather than social change. In practice, it has been applied more often to workplace technologies rather than those encountered in other spheres of life.

Some recent studies associated with the development of different kinds of information systems seem to blend participatory design and action research. A particularly strong current of participatory action research exists in the domain of geographic information systems (GIS). The gap is especially evident between the technical expertise of those who create and use GIS and the technical skills of most community members affected by GIS. Further, implications of system use and non-use for marginalized groups are especially critical: decisions based on GIS data and analyses have a direct effect on the local conditions and livelihoods of community members.

In its first white paper on GIS and society, the University Consortium on Geographic Information Systems (UCGIS) proposed a set of research priorities, developed at a 1995 specialist meeting sponsored by the National Center for Geographic Information and Analysis. These included the following research questions, which convey a strong social stance and a sociotechnical approach to research on information systems associated with marginalized community members (Harris and Weiner, 1996, n.p.):

In what ways have particular logics and visualization techniques, value systems, forms of reasoning, and ways of understanding the world been incorporated into existing GIS techniques, and in what ways do alternative forms of representation remain to be explored and incorporated?

How has the proliferation and dissemination of databases associated with GIS, as well as differential access to these databases, influenced the ability of different social groups to utilize this information for their own empowerment?

How can the knowledge, needs, desires, and hopes of non-involved social groups adequately be represented as input in a decision-making process, and what are the possibilities and limitations of GIS technology as a way of encoding and using such representations?

What possibilities and limitations are associated with using GISs as participatory tools for democratic resolution of social and environmental conflicts?

It is easy to see how a similar set of questions could be developed for digital libraries: Whose knowledge is valued, and how is it represented in digital libraries? Who is empowered by digital libraries? How can we gain the participation of marginalized groups (those likely to be noninvolved as users and designers) in digital library development? To what extent can digital libraries be used to make social conditions more equitable and just and resolve conflicts in a more democratic and participatory manner?

Keeping these issues in mind, we present several examples of participatory action research applied to the development of information systems to illustrate how this approach has been applied across system genres, from GIS to community networks to the creation of a computer-based "street library."

In workshops with local leaders and community members in Kiepersol, South Africa, researchers hoping to support grassroot initiatives for the redistribution of natural resources collected oral histories related to forced removals, perceptions of land quality and borders, and other aspects of land use (Weiner, Warner, Harris, and Levin 1995). This local knowledge was integrated in a GIS containing traditional information, such as data on hydrology, topography, climate, and soil quality. The participatory process linked feedback from the village workshops to map construction, articulation of policy issues and positions, and the dissemination of research results. The goal of the project was to construct a GIS that was "capable of representing not just one (official) version of reality, but multiple realities capturing the everyday life experiences of diverse social groups" (Weiner et al. 1995, p. 33). For example, the official limit of a 12 percent slope for plowable land encoded in the traditional system was based on the constraints of mechanized agriculture. Villagers, on the other hand, reported that hand hoeing and animal traction permitted planting on steeper slopes.

In his book on community networking, Douglas Schuler (1996) includes an appendix that describes the participatory action research associated with the development of community environmental indicators. This research was conducted by local residents associated with the Seattle Community Network. Participatory action research was also employed to study the computing needs of neighborhood-based organizations across Ohio. Randy Stoecker (1996, n.p.) reports that this new breed of collaborative inquiry and action produces better results that are more sensitive to both the needs and the strengths of poor communities: "Grass roots participants and researchers each bring expertise to the table and the inclusion of everyone's expertise in the research process ultimately produces not only more useful research but more accurate research."

The issues of gaining participation of marginalized society members, social equity and empowerment, and incorporating local knowledge are highlighted in an action research project whose aim was to introduce computers in New York City neighborhoods as part of a broader initiative to foster creativity, learning, and peace among very poor children and their families (Tardieu 1999). Researchers collaborated with children to develop a "neighborhood encyclopedia," with the text produced online and then printed and mounted in a giant wooden model of a book. Children contributed entries on whatever was important to them, such as why firefighters are heroes and how pretty their moms are. As work developed on the encyclopedia, children and parents invented a number of new uses for it, such as reestablishing contact with a friend forced to relocate to a welfare hotel. Other uses included sharing ideas about the nature of home and family with people around the world and providing hope and improved social relationships for a woman whose children had been removed to foster care.

The divides that marginalize some people in our society are not simply digital; they are also social and political. In considering the equitable development of digital libraries, we can look to research frameworks that recognize social values and their role in shaping practice and that explicitly aim at constructive social change through the participation and empowerment of marginalized groups. As these examples demonstrate, participatory action research provides a framework in which the social consequences of digital library use are palpable considerations from the outset and in which evaluation becomes a developmental process that contributes to new learning among all research participants. Digital library developers and users have much to gain from each other if the distance between them is closed and the legitimate knowledge of both is recognized and integrated.

Reframing Evaluation in the Afya Project

Afya Project Overview
In the Afya project, we reframe the concept of digital library evaluation by adopting the basic tenets of participatory action research. The Afya project is designed to engage African American women in assessing and improving their access to health information and services. At the same time, it nurtures their interest, proficiency, and participation related to information technology and the Internet. Through the Afya project, we are striving to develop a digital library that promotes social justice through communitywide alliances that model democratic and participative relationships. Thus, Afya is concerned, fundamentally, with developing new social technologies (ways that people communicate and collaborate) as well as new digital tools and resources (Bishop, Bazzell, Mehra, and Smith 2001a).

At the heart of the Afya project is SisterNet, a grassroots organization of African American women whose members are committed to achieving better health through improvements in lifestyle, behavior, and support systems. SisterNet was founded in Champaign, Illinois, by Imani Bazzell and has developed numerous projects and opportunities designed to nurture both healthy lifestyles and community activism. The Afya project is viewed as an essential part of a political strategy to resist oppression and shape a livable community. Faculty and students affiliated with the

University of Illinois Graduate School of Library and Information Science are involved in all aspects of the research, and students from several other units on campus participate also.

Afya unites SisterNet members with others who are concerned about African American women's health and have an interest in developing Web-based resources and services to form a community digital library. Prairienet, the local community network, provides computer technology and training. Like most community networks, Prairienet (⟨http://www.prairienet.org⟩) is a not-for-profit organization that contributes to community development by offering local digital content, Internet access, public access computers, and user training and support. Finally, local health-care institutions and libraries as well as the local community college have contributed resources. Thus, individual community members participating in the project include faculty and students from several institutions and disciplines, SisterNet's director and members, Prairienet Web site designers and trainers, and staff from local health-care facilities and libraries.

Afya's members are working to change social practice associated with the design and delivery of health information services for African American women in the community. The goals are to

· Increase computer access and literacy among African American women,

· Improve the quantity and quality of health information from local providers,

· Establish and institutionalize ongoing information provision (in digital and print formats) from black women,

· Improve relationships between providers and black women, and

· Facilitate the development of a strong social network for the exchange of support and information among SisterNet members.

We began with a series of discussion groups—three with African American women (grouped loosely by age) and one with health and information providers—to learn about the barriers to good health for black women. We also discussed community members' interests and capabilities related to taking action on the problems they identified. From discussion group transcripts, we culled a set of scenarios that represented individual community members' experiences related to the provision of health and information services for African American women.

We set project goals based on results from the discussion groups and began planning for two basic kinds of community action aimed at addressing the problems and opportunities represented by our set of scenarios. The first kind of action was to develop a collection of digital information resources and services to be provided through a SisterNet Web site. The second was to develop community-based activities associated with building and sharing expertise among SisterNet members and health and information service providers.

SisterNet members then participated in workshops—that is, situated evaluation sessions where they assessed existing Web-based communication services, including locally developed resources that are accessible through Prairienet. The workshop participants evaluated online services according to how well they addressed the situation described in either a previously generated scenario or a new one that they themselves provided. In a final workshop, women evaluated the prototype SisterNet Web site that had been developed by students in collaboration with Bazzell. Here again, scenarios helped in design and evaluation. The Web site designers used both our growing scenario set and related evaluation findings of SisterNet members to make decisions about the content, functionality, and interface design of our prototype Web site.

In developing Afya's community action plan (specific community-based activities designed to achieve project goals), we first reviewed suggestions contributed in our discussion groups and situated evaluation sessions. The resulting list was distributed at a SisterNet event billed as a "Friday Night Salad Extravaganza," where we shared favorite salad dishes and nutrition tips. Attendees checked off Afya project activities they thought were most important and indicated activities in which they personally wanted to participate. Sign-up sheets for the final set of proposed activities were presented at SisterNet's annual community conference in fall 2000. Only activities that SisterNet members signed up for are being pursued in the final community action plan.

Incorporating Local Knowledge in Digital Library Evaluation

African American women in the community contribute local knowledge and expertise to the development of digital library services through their participation in all phases of the Afya project. Imani Bazzell coauthored the Afya project proposal, and Cynthia Smith, a SisterNet member and master of arts student in social work, was hired as a research assistant. Other SisterNet members are involved throughout as community action researchers. In this role, they pinpoint key problems in community health-care services, evaluate existing digital services, and frame project goals and activities to address the problems they identified.

Another way that local knowledge has been kept front and center as we design SisterNet's digital collection of health information and plan communitywide activities is through our continual, though fluid, reliance on scenarios (for a more complete discussion of our use of scenarios, see Bishop, Mehra, Bazzell, and Smith 2001b). Scenarios are use-oriented design representations that model users' tasks to guide system development and evaluation. In contrast to top-down, technology-driven, abstract, and formal approaches, scenarios provide concrete descriptions of use, focus on particular instances, elicit envisioned outcomes, and are open-ended, rough, and colloquial (Carroll 1995, p. 5). Users become empowered as initiators in the analysis of information about their expectations and requirements and are not treated as mere informants in the design process. This is especially important in developing a more complete picture of the context of information-seeking and technology use for marginalized groups who are often on the fringes of system design and evaluation.

Our scenarios are simple, unstructured anecdotes, narratives that bring together values and practices related to the use of health services and information. Our growing set of scenarios is helpful as an evaluation tool because it provides authentic insights into local women's experiences related to health, information, and technology. These insights can then be used to assess digital library services. Scenarios provided by health and information workers and by SisterNet members noted the lack of useful and usable health information—information that is conveniently accessible, jargon free, relevant, and culturally appropriate. The problem is complex partly because it is embedded in the social practices and ideologies of knowledge generation and transfer in the arena of health care. No matter how limited or how sweeping, improvements can come only from achieving change in existing ideology and practice.

One important purpose of the scenarios is to reveal social problems that hinder the exchange of health information. For example, the following scenarios generated in our discussion groups depict the need to reduce mistrust in relationships between African American women and providers. They also highlight the importance of personal social networks in the exchange of health information:

It's hard to know if it is really racism or if the health-care providers are being pushed. The fact that if a black woman presents with abdominal pains the first thing they want to do is run a series of venereal disease tests on her, whether that is the issue or not.

They [doctors] bank on our level of ignorance. I'm not a doctor. I don't know anything about what's going on. But they walk in there, and they start spurting off these words in their lingo, and they're saying this and that. Talk to me in layman's terms because I don't know what's going on. They really do rely on us not to ask too many questions. Luckily for me, like I said, I have an aunt who has been through everything possible, and she asks more questions. That's what's really good about having people like that in your lives.

A digital library devoted to African American women's health can attempt to address these problems, though technology alone cannot remedy them. Nonetheless, participatory action research provides an evaluation framework that moves the concerns of marginalized groups onto the agenda and at least registers them as criteria against which digital library success will be evaluated. Through their scenarios, SisterNet members have redirected Afya's notions about the content of the SisterNet Web site that will serve as Afya's digital library. The scenarios validated local women's interest in health problems that commonly affect African American women, such as diabetes and heart disease. But local women advocated for content that had not previously occurred to Afya project staff, such as tips on building self-confidence in personal interactions related to health care. Some of the content recommended by SisterNet members represents local knowledge. Women advocated for the inclusion of "home remedies" and "motherwit" in SisterNet's digital library of health resources.

SisterNet members' scenarios have also provided guidelines for assessing digital library functionality and articulating how the digital library would fit with existing social practice and interactions. The two scenarios cited above, for example, suggest that an online discussion forum might allow more experienced women to provide support and information to others. In a workshop session devoted to evaluating existing Internet communication services related to African American women's health, SisterNet members provided additional scenarios that described the circumstances in which they would find such services useful. One woman described how hard it was for her to find sports partners. She said she would use a bulletin board on the SisterNet Web site to look for tennis partners and to organize a softball team. Another woman described how busy she was, managing a retail outlet for her clothing designs and developing various community programs, such as after-school activities for young teens. She commented that she would like to consult a calendar of events on the SisterNet Web site when she found herself with a few free hours on a Sunday afternoon to see if any SisterNet activities were taking place that day.

The local knowledge of SisterNet members is also being tapped to help evaluate specific interface design choices. For example, in a working session devoted to critiquing the SisterNet Web site prototype, one woman talked about using fabric that often adorned tables in SisterNet gatherings as part of the Web site background. She noted that the fabric had become a symbol of the group's identity and an easily recognized marker for SisterNet activities. When she saw the digitized fabric swatch on the Web site, she said it was a great choice. As in their face-to-face gatherings, "Women will know immediately they've come to the right place" when they see that fabric.

Evaluation as Part of Design

The second aspect of reframing evaluation that we are learning about in the Afya project is how evaluation can be used as a concept and technique throughout the lifecycle of system design and implementation. In Afya, evaluation is cast as the fundamental task and responsibility of the SisterNet members who serve as our community action researchers, an activity that began in our needs assessment discussion groups, where women presented their assessment of community health care and health information provision.

As noted above, to help in the design of SisterNet's Web site, we held three "situated user evaluation" workshops at which a few SisterNet members acted as community action researchers to evaluate existing local and national Web resources related to African American women's health. Each workshop was devoted to a single type of Web resource: national Web sites, local offerings on Prairienet, and Internet communication services (listserves, newsgroups, chat rooms). In each session, we first demonstrated Web resources and services related to African American women's health and then asked SisterNet members to critique them in terms of relevance, appeal, and ease of use by referencing scenarios contributed by their peers.

In the workshop on national Web sites related to African American women's health, SisterNet members first selected one of the scenarios developed from focus groups, such as an elderly woman's desire to learn more about side effects of her diabetes medication or a young woman's concern that she couldn't deal with the stress of combining parenthood, employment, and pursuit of a college degree. Next, they critiqued a site from the perspective of how well it addressed the problem described in that scenario. This evaluation technique limited the authenticity of the use situation since the SisterNet evaluators were not assessing Web resources according to a real problem they had but rather were making judgments based on their affinity with problematic health situations typical of their peer group. We chose this course because we felt women might feel too awkward about describing personal health problems to strangers.

For the second site-evaluation task, however, they picked a Web site that they wanted to explore, at which point we elicited their own scenarios by asking questions such as "What prompted you to pick this one? How is it relevant to you? Do you think you'll come back here? Under what circumstances?"

We acknowledge several benefits of casting SisterNet members as evaluators throughout the Afya project. Doing so gives us a filter for interaction with SisterNet members that is biased toward design. It tends to channel their comments toward key design decisions related to content, functionality, interface choices, and other aspects of deployment and training. Our emphasis on evaluation by SisterNet members also recognizes explicitly the importance of what they know and what role they play in the design process. In legitimizing their local knowledge, we are conferring value to their views and building trust. Shifting the power dynamic in the relationship between potential users and system designers makes authentic insights easier to come by.

We find that the labels we attach to people's roles are important. In typical need-assessment focus groups and usability sessions, people are reduced to being merely users who are forced into a situation in which their identities are diminished to fit the scope of their interaction with information systems. As subjects, they participate in exercises from which they may gain little and in which their actions and responses are controlled by researchers. With our situated user evaluation sessions, we attempted something more like an exchange of expertise, an experience in which mutual learning occurs. In our situated evaluation sessions, SisterNet community action researchers were called the *evaluators* and project participants who elicited and recorded their assessments were called *evaluation mentors*. In exchange for their expert knowledge on how valuable and usable existing Web resources are to them, women come away with newly acquired expertise related to computers and the Internet and some additional information on what Web sites are available and how they can find them. They took home brief printed guides and received Prairienet memberships as concrete benefits from their participation in the Afya project.

Evaluation as a Tool for Capacity Building and Social Change

The most obvious way in which evaluation in Afya is tied to transformation in social relationships and immediate improvement in local conditions is in the development of a community action plan as the primary outcome of the project's first stage (see table 7.1). The specific activities that we are pursuing all have at root the goal of building the capacity of people across the community who share a concern

Table 7.1
Afya Community Action Plan

Establish an action circle to develop a Web site featuring jargon-free, culturally appropriate health information for our physical, emotional, spiritual, and intellectual well-being; chat and bulletin board space; news; tips; and public policy information.

Establish an action circle to develop a SisterNet Technology/Internet Guide.

Identify and assist other local African American women's organizations interested in developing their own Web sites.

Develop and offer a series of free computer training workshops targeting beginner, intermediate, and advanced needs.

Identify and train resource and referral agents (online and offline) to promote healthy lifestyles and use of technology.

Establish an action circle to develop five SisterNet Resource Centers featuring relevant books, magazines, pamphlets, and Internet access in comfortable and convenient locations for African American women.

Establish an action circle to develop and design visually appealing, easy-to-read health promotional materials on various topics and distribute them to numerous drop-off sites.

Establish an action circle to design a Bill of Rights (for health and technology) poster for public distribution and display.

Establish an action circle to identify leadership opportunities for women as SisterNet representatives on relevant committees and task forces related to health information and access and to technology information and access (e.g., libraries, public health district, social service agencies, community organizations, private providers).

Establish an action circle to organize an African American women's health fair designed to increase knowledge about health concerns and resources and provide opportunities for interaction between health information and service providers and community women.

for African American women's health and improving relationships among them. Our community action plan also frequently emphasizes constructive social outcomes over technical results.

The plan we developed for community action in Afya outlines ways that capacity building can occur in many quarters. Both women and providers of health and information services acknowledged that providers need help in making their services more relevant to African American women. In the Afya project, local black women are contributing their expertise to help build the capacity of providers. Further, we've chosen to use SisterNet's action circle model for implementing the Afya community action plan, with its built-in bias toward capacity building and its practical approach to facilitating direct action. SisterNet action circles are *ad hoc* in nature. They form when three or more women identify a problem and are willing to research and develop a plan to resolve it. Action circles build knowledge and experience as skills and information are shared and developed. They are also designed to achieve constructive social transformation in that they further the interests of African American women by foregrounding their concerns, insights, and relationships.

Capacity building and social change are part of the grand scheme in Afya and of each individual activity. Some of the scenarios point to the desire of SisterNet members to contribute directly to constructive social change. An elderly woman participating in our community discussion groups said she would "try to be a part of anything that I know will help even after I'm gone.... I've got more time on my hands than I have money, and I would like to make my time valuable and do something worthwhile, more than just sit at home and sew and look at television." The health Web site review exercise that Imani Bazzell developed as part of the SisterNet computer training workshop directs participants to explore a given health site on the Web, find something of interest that will help them make a positive change in their lives, and then report back to all workshop participants on what they learned. Thus, along with Web browsing skills, SisterNet members find, share, and discuss action-oriented health information. They also gain confidence in public speaking through their informal oral reports to their workshop cohorts.

The SisterNet Health Fair conducted with support from the Afya project contributed to improving the capacity of participants by including a focus on how to find and use health information. The Health Fair was held in a local public library and included booths and activities set up by a wide range of community health and information providers. It also helped redefine community relationships in a manner that was recommended by both health information and service providers and women in our Afya discussion groups.

Providers were especially eager to develop ways to bring women and providers together in an informal and enjoyable environment where they could get to know each other outside of the impersonal situations that typify exchanges in hospitals and libraries. Another detrimental aspect of official exchanges is that the balance of power dramatically favors the experts—health-care and information professionals. The Health Fair provided an avenue for the kind of informal exchanges that all stakeholders in community health sought. For example, the University of Illinois Health Sciences librarian set up a booth that provided information on how community members could access and use her library's resources, including books on women's health and the PubMed database of medical literature. She got to know women from the community better, increase the visibility of her library, and demonstrate that it was a congenial environment that had something relevant to offer African American women.

The issue of who benefits from evaluation has been an important consideration throughout Afya as we strive to engage women in practices that will build their capacity. The project is continually reframed, and community women's role negotiated, so that they gain directly in ways meaningful to them. Early on, for example, we realized that we needed to revise the original project proposal that Afya community action researchers should participate in all project activities over the course of a year while receiving only a very small honorarium (compared to what other project staff were earning). Instead, we realized that recruiting women to participate in each activity separately was a more practical approach that acknowledged the value of their participation, given their busy lives, and allowed them to decide which activities were really of interest to them.

Conclusions

The Afya project exemplifies participatory action research in association with evaluation of digital libraries. In Afya, we hope to develop digital library services related to community health that African American women in our community will find usable, useful, and congenial. We hope to contribute to a reframing of digital library evaluation as a systematic and collaborative approach to inquiry that gives people the means to take action toward improving contemporary social conditions (Stringer 1999, p. 17; Whyte 1991). We explore techniques such as scenario building and situated user evaluations in an attempt to mobilize the local knowledge of SisterNet members, foster evaluation as a concept and process useful throughout the system life cycle, and build the capacity of a diverse set of community members as we move toward broader social change.

Both our method and our results are instantiated through action so that knowledge is returned to practice. Women take action as evaluators of digital library services as a technique for identifying appropriate criteria and guidelines for the digital library we are building. Their participation as evaluators helps them, in turn, to build their capacity as critical users of digital health information so that their future practice is improved. By placing social practice at center stage, we also generate and strive to act on findings that have more to do with social than technical problems. The boundaries of the digital library we are creating encompass social values and relationships, and we look for ways to improve these aspects of the practices associated with digital library use.

What distinguishes action research from other research approaches is that the driving force comes from people's desire to understand and change their *own* practice. By stepping back enough to make their own actions the subject of critical self-reflection, discussion, and experimentation, those engaged in a particular practice work toward constructive change. Here we indeed see Schön's new epistemology, where reflection bridges knowledge and action, and change occurs by gaining the participation of all practitioners (librarians, SisterNet members, doctors, Web site designers, and workshop trainers) in a cycle of action and reflection. Our community action plan creates opportunities for SisterNet members and providers to practice together activities and relationships that are more socially just.

In the Afya project, we are trying out a blend of the theories, methods, and philosophies associated with arenas that, although critical to the evaluation of digital libraries for marginalized members of society, seem too seldom allied—participatory action research, information system design and evaluation, and the traditional stance of social responsibility in the information professions. Because participatory action research claims social practice as its primary object of study, it provides an approach that seems particularly well suited to socially grounded digital library evaluation. Information system evaluation too often lacks the attention to social values, practice, and outcomes inherent in participatory action research. The library profession prides itself on its tradition of social inclusion and outreach to under-served segments of society, but its ethical stance of neutrality and service mentality can limit its capacity to promote social change (Durrance 2001).

Our work in the Afya project is helping us learn about digital library evaluation as a socially grounded activity that is based in professional social responsibility, attends to details of social practice, and can lead to positive social consequences. An evaluation approach that is situated in the social practice of use and closes the distance among digital library practitioners is always helpful, but we feel it is crucial for developing collections and services intended for use by marginalized groups in a world scarred by social, economic, and digital divides.

Acknowledgments

We are grateful to the U.S. Institute of Museum and Library Services, especially Joyce Ray and Beverly Sheppard, for their support of the Afya project, which has gone beyond financial backing. Nancy Van House and Chip Bruce offered helpful comments on early drafts of this chapter. Thanks to SisterNet members for what they have taught us about digital library evaluation.

References

Agada, J. 1999. Inner-City Gatekeepers: An Exploratory Survey of Their Information Use Environment. *Journal of American Society for Information Science*, 50, 74–85.

Ashing-Giwa, L. 1999. Health Behavior Change Models and Their Socio-Cultural Relevance for Breast Cancer Screening in African American Women. *Women and Health*, 28(4), 53–71.

Balka, E., and L. Doucette. 1994. The Accessibility of Computers to Organizations Serving Women in the Province of Newfoundland: Preliminary Study Results. *Arachnet Electronic Journal on Virtual Culture*, 2(3). ⟨http://www.monash.edu.au/journals/ejvc/balka.v2n3⟩.

Benton Foundation. 1998. *Losing Ground Bit by Bit: Low-Income Communities in the Information Age*. Washington, DC: Benton Foundation.

Berland, G. K., M. N. Elliott, L. S. Morales, J. I. Algazy, R. L. Kravitz, M. S. Broder, D. E. Kanouse, J. A. Mu–oz, J. A. Puyol, M. Lara, K. E. Watkins, H. Yang, and E. A. McGlynn. 2001. Health Information on the Internet: Accessibility, Quality, and Readability in English and Spanish. *Journal of the American Medical Association*, 285, 2612–2621.

Berlin, R. B., and B. R. Schatz. 1999. Internet Health Monitors for Outcomes of Chronic Illness. *Medscape General Medicine (MedGenMed)*, 6 sections. ⟨http://www.medscape.com⟩.

Bishop, A. P., I. Bazzell, B. Mehra, and C. Smith. 2001a. Afya: Social and Digital Technologies That Reach across the Digital Divide. *First Monday*, 6(4). ⟨http://www.firstmonday.org/issues/issue6_4/bishop/index.html⟩.

Bishop, A. P., B. Mehra, I. Bazzell, and C. Smith. 2001b. Scenarios in the Design and Evaluation of Networked Information Services: An Example from Community Health. In C. R. McClure and J. C. Bertot, eds., *Evaluating Networked Information Services: Techniques, Policy, and Issues* (pp. 45–66). Medford, NJ: Information Today.

Bowker, G. C., W. Turner, S. L. Star, and L. Gasser. 1997. Introduction. In G. C. Bowker, S. L. Star, W. Turner, and L. Gasser, eds., *Social Science, Technical Systems, and Cooperative Work: Beyond the Great Divide* (pp. xi–xxiii). Mahwah, NJ: Erlbaum.

Bruce, B. C., and M. P. Hogan. 1998. The Disappearance of Technology: Toward an Epistemological Model of Literacy. In D. Reinking, M. L. Labbo, and R. Kieffer, eds., *Handbook of Literacy and Technology: Transformations in a Post-Typographic World* (pp. 269–281). Hillsdale, NJ: Erlbaum. ⟨http://www.lis.uiuc.edu/chip/pubs/disappearance.shtml⟩.

Bruce, B. C., and A. D. Rubin. 1993. *Electronic Quills: A Situated Evaluation of Using Computers for Writing in Classrooms*. Hillsdale, NJ: Erlbaum.

Bundy, M. L. 1980. *Helping People Take Control: The Public Library's Mission in a Democracy*. College Park, MD: Urban Information Interpreters.

Carr, W., and S. Kemmis. 1983. *Becoming Critical: Knowing through Action Research* (Rev. ed.). Victoria, Australia: Deakin University.

Carroll, J. M. 1995. Introduction: The Scenario Perspective on System Development. In J. M. Carroll, ed., *Scenario-Based Design: Envisioning Work and Technology in System Development* (pp. 1–17). New York: Wiley.

Chatman, E. A. 1996. The Impoverished Life-World of Outsiders. *Journal of the American Society of Information Science*, 47(3), 193–206.

Cohill, A. M., and A. L. Kavanaugh, eds. 2000. *Community Networks: Lessons from Blacksburg, Virginia* (2nd ed.). Boston: Artech House.

Computer Professionals for Social Responsibility. 2002. *Mission Statement*. Palo Alto, CA: CPSR. ⟨http://www.cpsr.org/cpsr/about-cpsr.htm/#mission⟩.

De la Pena McCook, K. 2000. *A Place at the Table: Participating in Community Building*. Chicago: ALA Editions.

Dervin, B. 1976. The Everyday Needs of the Average Citizen: A Taxonomy for Analysis. In M. Kochen and J. C. Donohue, eds., *Information for the Community* (pp. 19–38). Chicago: American Library Association.

Dervin, B. 1980. Communication Gaps and Inequities: Moving toward a Reconceptualization. In B. Dervin and M. Voigt, eds., *Progress in Communication Sciences* (vol. 2, pp. 73–112). Norwood, NJ: Ablex.

Dervin, B. 1989. Users as Research Inventions: How Research Categories Perpetuate Inequities. *Journal of Communication*, 39, 216–232.

Dervin, B. 1992. From the Mind's eye of the User: The Sense-Making Qualitative-Quantitative Methodology. In J. D. Glazier and R. R. Powell, eds., *Qualitative Research in Information Management* (pp. 61–84). Englewood, CO: Libraries Unlimited.

Dewdney, P., and R. M. Harris. 1992. Community Information Needs: The Case of Wife Assault (in Six Ontario Communities). *Library and Information Science Research*, 14(1), 5–29.

Dewey, J. 1916. *Democracy and Education*. New York: Free Press.

Durrance, J. 2001. Personal conversation.

Floyd, V. D. 1992. Too Soon, Too Small, Too Sick: Black Infant Mortality. In R. L. Braithewaite and S. E. Taylor, eds., *Health Issues in the Black Community* (pp. 165–177). San Francisco: Jossey-Bass.

Glassman, M. 2001. Dewey and Vygotsky: Society, Experience, and Inquiry in Educational Practice. *Educational Researcher*, 30(4), 3–14.

Gollop, C. J. 1997. Health Information-Seeking Behavior and Older African American Women. *Bulletin of the Medical Library Association*, 85(2), 141–146.

Greenbaum, J., and M. Kyng, eds. 1991. *Design at Work: Cooperative Design of Computer Systems*. Hillsdale, NJ: Erlbaum.

Gurstein, M., ed. 2000. *Community Informatics: Enabling Communities with Information and Communications Technologies*. Hershey, PA: Idea Group.

Harris, T. M., and D. Weiner. 1996. *GIS and Society: The Social Implications of How People, Space, and Environment Are Represented in GIS*. Technical Report 96(7). Santa Barbara, CA: National Center for Geographic Information and Analysis.

Kellogg Commission on the Future of State and Land-Grant Universities. 1999. *Returning to Our Roots: The Engaged Institution*. Washington, DC: National Association of State Universities and Land-Grant Colleges.

Kemmis, S., and R. McTaggart, eds. 1988. *The Action Research Planner* (3rd ed.). Victoria, Australia: Deakin University.

Kretzmann, J. P., and J. L. McKnight. 1993. *Building Communities from the Inside Out: A Path toward Finding and Mobilizing a Community's Assets*. Chicago: ACTA.

Madison, A. M., ed. 1992. *Minority Issues in Program Evaluation*. San Francisco: Jossey-Bass.

McKenzie, J. F., R. R. Pinger, and J. E. Kotecki. 1999. *An Introduction to Community Health*. Boston: Jones and Bartlett.

Mele, C. 1999. Cyberspace and Disadvantaged Communities: The Internet as a Tool for Collective Action. In M. A. Smith and P. Kollock, eds., *Communities in Cyberspace* (pp. 291–310). London: Routledge.

Metoyer-Duran, C. 1993. *Gatekeepers in Ethnolinguistic Communities*. Norwood, NJ: Ablex.

Moll, L. C. 1994. Mediating Knowledge between Homes and Classrooms. In D. Keller-Cohen, ed., *Literacy: Interdisciplinary Conversations* (pp. 385–410). Cresskill, NJ: Hampton.

Moll, L. C., C. Amanti, D. Neff, and N. Gonzales. 1992. Funds of Knowledge for Teaching: Using a Qualitative Approach to Connect Homes and Classrooms. *Theory into Practice*, 31, 132–141.

Moss, N., and N. Krieger. 1995. Measuring Social Inequalities in Health: Report on the Conference of the National Institutes of Health. *Public Health Reports*, 110(3), 302–305.

Muddiman, D., S. Durrani, M. Dutch, R. Linley, J. Pateman, and J. Vincent. 2000. Open to All? The Public Library and Social Exclusion. Vol. 1, Overview and Conclusions. *Library and Information Commission Research Report 84*. London: Resource: The Council for Museums, Archives, and Libraries.

Mueller, M. L., and J. R. Schement. 1996. Universal Service from the Bottom Up: A Study of Telephone Penetration in Camden, New Jersey. *Information Society*, 12, 273–292.

Muller, M. L., and S. Kuhn. 1993. Participatory Design. *Communications of the ACM*, 36(6), 24–28.

Nardi, B., and V. O'Day. 1999. *Information Ecologies*. Cambridge, MA: MIT Press.

National Telecommunications and Information Administration (NTIA). 2000. Falling through the Net: Toward Digital Inclusion. Washington, DC: NTIA. ⟨http://www.ntia.doc.gov/ntiahome/fttn99/contents.html⟩.

Patton, M. Q. 1999. Some Framing Questions about Racism and Evaluation: Thoughts Stimulated by Professor Stanfield's "Slipping through the Front Door." *American Journal of Evaluation*, 20(30), 437–444.

Payton, F. C., and P. F. Brennan. 1999. How a Community Health Information Network Is Really Used. *Communications of the ACM*, 42(12), 85–89.

Pettigrew, K. E., J. C. Durrance, and P. Vakkari. 1999. Approaches to Studying Public Library Networked Community Information Initiatives: A Review of the Literature and Overview of a Current Study. *Library and Information Science Research*, 21(3), 327–360.

Pettigrew, K. E., and M. A. Wilkinson. 1996. Control of Community Information: An Analysis of Roles. *Library Quarterly* 66(4), 373–407.

Pursley, L. A. 1996. Empowerment and Utilization through Participatory Evaluation. Ph.D. dissertation, Cornell University, Department of Human Service Studies. UMI 9608262.

Reardon, K. M. 1998. Participatory Action Research as Service Learning. In R. A. Rhoads and J. P. F. Howard, eds., *Academic Service Learning: A Pedagogy of Action and Reflection* (pp. 57–64). New Directions of Teaching and Learning 73. San Francisco: Jossey-Bass.

Rosen, J. 1995. A Scholar's Perspective. In D. Merritt and J. Rosen, eds., *Imagining Public Journalism: An Editor and Scholar Reflect on the Birth of an Idea* (pp. 14–26). Roy W. Howard Public Lecture in Journalism and Mass Communication Research 5, April 13, 1995. Bloomington: School of Journalism, Indiana University.

Schön, D. A. 1995. The New Scholarship Requires a New Epistemology. *Change*, 27(6), 27–34.

Schuler, D. 1996. *New Community Networks: Wired for Change*. New York: ACM Press.

Spink, A., M. Jaeckel, and G. Sidberry. 1997. Information Seeking and Information Needs of Low-Income African American Households: Wynnewood Healthy Neighborhood Project. In C. Schwartz and M. Rorvig, eds., *ASIS '97: Proceedings of the Sixtieth ASIS Annual Meeting* (vol. 34, pp. 271–279). Medford, NJ: Information Today.

Stake, R. E. 1990. Responsive Evaluation. In H. J. Walberg and G. D. Haertel, eds., *International Encyclopedia of Educational Evaluation* (pp. 75–77). Oxford: Pergamon Press.

Stanfield, J. H. 1999. Slipping through the Front Door: Relevant Social Scientific Evaluation in the People of Color Century. *American Journal of Evaluation*, 20(3), 415–431.

Stoecker, R. 1996. Putting Neighborhoods On-Line; Putting Academicians in Touch: The Urban University and Neighborhood Network. Paper presented at the International Meeting of the Community Development Society, Melbourne, Australia. ⟨http://uac.rdp.utoledo.edu/docs/uunn/cdsppr.htm⟩.

Stringer, E. T. 1999. *Action Research* (2nd ed.). London: Sage.

Tardieu, B. 1999. Computer as Community Memory: How People in Very Poor Neighborhoods Make a Computer Their Own. In D. A. Schon, B. Sanyal, and W. J. Mitchell, eds., *High Technology and Low-Income Communities: Prospects for the Positive Use of Advanced Information Technology* (pp. 289–312). Cambridge, MA: MIT Press.

The Children's Partnership. 2000. *Online Content for Low-Income and Underserved Americans: The Digital Divide's Frontier: A Strategic Audit of Activities and Opportunities*. Santa Monica, CA: The Children's Partnership.

Tichenor, P. J., G. A. Donohue, J. M. Donahue, and C. N. Olien. 1970. Mass Media Flow and Differential Growth in Knowledge. *Public Opinion Quarterly*, 34, 159–170.

Twidale, M. B., D. Randall, and R. Bentley. 1994. Situated Evaluation for Cooperative Systems. In R. K. Furuta and C. Neuwirth, eds., *CSCW '94: Proceedings of ACM 1994 Conference on Computer-Supported Cooperative Work* (pp. 441–452). New York: ACM Press.

Uehara, E. 1990. Dual Exchange Theory, Social Networks, and Informal Social Support. *American Journal of Sociology*, 9(3), 521–527.

United States Department of Health and Human Services. 1985. *Report of the Secretary's Task Force on Black and Minority Health*. Washington, DC: U.S. Department of Health and Human Services.

United States Department of Health and Human Services. 1990. *Healthy People 2000*. DHHS Publication No. PHS 91-50213. Washington, DC: U.S. Government Printing Office.

U.S. Department of Health and Human Services. 2001. *Healthy People in Healthy Communities: A Community Planning Guide Using Healthy People 2010*. Washington, D.C.: U.S. Government Printing Office.

Vishwanath, R., J. Finnegan, P. Hannan, and R. Luepker. 1991. Health and Knowledge Gaps: Some Lessons from the Minnesota Heart Health Program. *American Behavioral Scientist*, 34(6), 712–726.

University Consortium for Geographic Information Science. 1996. Research Priorities for Geographic Information Science. *Cartography and Geographic: Information Systems*, 23(3), 115–127.

Weiner, D., T. A. Warner, T. M. Harris, and R. M. Levin. 1995. Apartheid Representations in a Digital Landscape: GIS, Remote Sensing and Local Knowledge in Kiepersol, South Africa. *Cartography and Geographic Information Systems*, 22(1), 30–44.

White, J. E. 1999. Prejudice? Perish the Thought. *Time*, March 8, 1999, 36.

Whitmore, E. 1988. Participatory Approaches to Evaluation: Side Effects and Empowerment. Ph.D. dissertation, Department of Human Service Studies, Cornell University.

Whitmore, E., ed. 1998. *Understanding and Practicing Participatory Evaluation*. San Francisco: Jossey-Bass.

Whyte, W. F., ed. 1991. *Participatory Action Research*. Newbury Park, CA: Sage.

8

Colliding with the Real World: Heresies and Unexplored Questions about Audience, Economics, and Control of Digital Libraries

Clifford Lynch

Introduction

Without becoming entangled in the still unresolved (and at least for now perhaps, unresolvable) debate about "what is a digital library" (DL) (see Borgman 1999; Levy 2000), I believe that three general kinds of services or systems are emerging that might be considered digital libraries. The first are commercial information services targeted at the work of particular professions or disciplines; perhaps the clearest examples are Lexis, Nexis, and Westlaw, though I later argue that many others are less directly commercially framed and marketed but are nonetheless emerging within the commercial marketplace. The second group contains research prototypes such as those that developed during the NSF/ARPA/NASA Phase I Digital Libraries Initiative grant program. The third are extensions of the research or academic library (or occasionally the public library) that incorporate extensive network-based collections and services. These include the efforts of virtually all major research libraries to deploy what they are most commonly calling a digital library component. These examples demonstrate that we actually had more than thirty years of experience with various types of digital libraries before the term came into popular use.

A few additional categories might plausibly be recognized as digital libraries. Technically and in terms of content, they have the most in common with some of the extensions of the traditional library, but they also have important differences. These include long-standing federally funded projects such as the work of the National Library of Medicine,[1] which serves a national and international scientific community (and more recently a national consumer community as well), or the U.S. Department of Education's Educational Resources Information Center (ERIC) program.[2] Another category includes the more recent developments in state-based digital library projects such as SAILOR in Maryland, the California Digital Library (to the extent that it moves beyond its roots at the University of California), Ohio-link, or the programs in the Commonwealth of Virginia. These introduce new complexities because their funding and missions are shaped largely by political processes and because, in part as a result of these processes, they incorporate elaborate consortia of stakeholder and participant institutions. Because of these complexities and their still evolving character, they are discussed only in passing in the rest of this chapter. But it is important to recognize that they exist and are likely to expand and reach ever broader audiences over the next few years. They will be important venues for future research, particularly because of the scale of investment, the complexity of the efforts, and their potential impact in terms of the number of people they are intended to serve.

Again, beyond questions of the nature and definition of digital libraries, one key axis of variation that is important throughout this chapter is the extent to which the digital library provides an environment for actually doing active work rather than just locating and reviewing information that can support work processes.

In digital libraries, we see a continuum from personal monolithic information access to analysis (for example, in geospatial information systems) to distributed collaboration in an information-rich environment (such as a colaboratory). Historically, libraries have tended toward the first end of this continuum—the provision of simple information access—and not surprisingly, the digital library systems they are deploying also usually gravitate toward this end of the spectrum. Some (but certainly not all) operators of research and commercial systems have been more receptive to including collaborative work tools. This continuum is important because the further digital libraries move toward creating collaborative work environments (rather than just representing reference tools that might be embedded in or sit alongside such environments), the greater their potential for significantly changing (it is hoped improving) the ability of their user communities to accomplish work. Their potential for ultimately transforming both the practice of the user community and the nature of the work also grows, but the more they move in this direction, the further they move away from the traditions of the libraries that are funding and developing many of them.

Three of the most critical factors associated with digital libraries are (1) sustainability of sources of economic support and, where relevant, business models; (2) governance and control of the digital library's design, operation, and ongoing development; and (3) the composition of the audience, both intended and actual.

These factors are significant in the design and evolution of any digital library. A central theme of this chapter is how these factors apply to various types of digital libraries and how they are interrelated.

Much of the attention elsewhere in this book has been focused on digital library research prototypes that provide simple (or perhaps simplistic) answers to questions about economics and sustainability: the problems are just ignored. The project is funded by grants for a fixed duration, after which more grants are sought or the project ends. The answers to questions about control and governance are also often simple: it's a research prototype that is run by a research group and doesn't need governance.

Indeed, it's not an organization created for the long term; at most, an advisory committee is established. The answers to questions about audience and adapting to audience needs are more complex, and two models are popular. The first is the "field of dreams" (build it and they will come). The second emphasizes intense user studies (including studies of individual users, communities, and other social groups and incorporating sociological and anthropological perspectives) and iterative user-centered design and evaluation—what I term, for want of a better phrase, *socially grounded engineering*[3] techniques. Of course, in a research project one is free to target any user community that makes sense in the context of the research; this can be done without too much regard for politics or economics.

As other chapters in this book have shown, there's a tremendous amount to be learned about socially grounded engineering, design, and evaluation in these settings. Indeed, we still have very limited understanding of what users want, what they do, how to provide for their needs, and what we learn by studying the research prototypes. All of this will inform future development of digital libraries. And in cases where socially grounded engineering is a real commitment included as part of project planning and implementation (as opposed to an add-on reluctantly accommodated by the engineering people really running the project), it can make a significant difference in the characteristics of the system that is produced.

However, I believe that an exclusive focus on research prototypes is highly misleading, particularly in relation to the character and utility of socially grounded engineering approaches and to the value, support, and underwriting of user studies. The other two classes of digital libraries (what might reasonably be thought of as real-world digital libraries) come with a very different set of political, economic, governance, and cultural dynamics that can call into question, and even subvert, many of the socially grounded engineering approaches. Socially grounded design isn't going to work effectively when institutions are driving the development of digital libraries and when the social design doesn't take institutional needs, objectives, biases, and constraints into account but focuses narrowly on system users. Such socially grounded design may not be allowed to have much influence on system evolution because the users of the system are not the ones developing it or paying for its development. Issues about how the system reduces cost, shifts pressures, or helps to justify and enhance other investments by the institution may be more important.

Socially grounded engineering studies users, their needs, and their practices. Real-world systems sometimes aren't being designed for users but for institutional customers acting as intermediaries (such as spokespersons or representatives) for sometimes captive user populations (as in the university case) or even possibly imaginary users (as in the failed commercial service case).

Ann Bishop suggested a helpful distinction between authentic use situations (real users of experimental systems) and authentic (that is, production) systems in her comments on an early draft of this chapter. My concern is with authentic systems rather than authentic use situations. My intent is in no way to belittle the importance of user studies or to suggest that they are unsound but rather to suggest that, for authentic systems, those in charge may not assign much importance to studies of authentic use situations. Indeed, for such authentic systems, Ann raised the important question (which I can't answer and, as far as I know, is unexplored) about which kinds of social studies are likely to be funded and which are likely to have a real and helpful impact on the ongoing development of systems.

Relatively little attention, as far as I know, has been focused on these real-world situations, and my purpose here is to try to illuminate some of the unique characteristics of real-world digital libraries and the way that they differ from research prototypes, placing emphasis on some of the fundamental tensions and disconnects. I want to examine some of the problems and limitations of the socially grounded engineering approach in these contexts and ask questions about what kind of socially grounded design is relevant for these types of systems.

The emphasis here will be primarily on the case of digital libraries evolving from traditional libraries in academic settings. This is the situation that I know best, and it is a richly complex one where the institutional context fundamentally reshapes many of the considerations of socially grounded design. Commercial digital libraries will be used as a counterpoint to library-situated digital libraries. This will illustrate some implications of removing many of the constraints that characterize the library-based model, while still leaving intermediaries, rather than end-users, in control of the development agenda.

I will perhaps exaggerate occasionally to make certain points; the reality of the long-term development and evolution of any real system is always more complex and ambiguous and represents a complicated and unique political balance among the interested parties. In particular, even in highly intermediated situations, the users often have a habit of ultimately reasserting control: commercial systems succeed or fail in the marketplace, and libraries are ultimately accountable to their users, though sometimes it takes a managerial or political crisis to reinforce this accountability.[4] I hope this chapter will help the collective understanding by providing another—perhaps less scholarly and idealized—perspective on the factors that shape how digital libraries come into being and subsequently evolve.

A Few Disturbing Observations from Early Digital Library Experience

We are beginning to get some insights into how digital libraries might actually work based on actual experience. The answers are very diverse, depending on what purposes and audiences the digital libraries in question serve. And these insights are disturbing to those who approach the construction of digital libraries with preconceived notions and vested interests. They are particularly troublesome to groups that want to view them as simply extensions and modernizations of existing institutional structures such as libraries within universities. All the early experience suggests that digital libraries, at least when they are done right, are *different*. They are potentially disruptive. The observations here underscore tensions linked to audience, economics, and control that were always, but only subtly, present in traditional library models; these are greatly amplified by digital libraries.

If we take the view that digital libraries are information systems that provide not only access to information but also analytic and collaborative tools (the view of digital libraries as tools of revolution), then we should note by that definition that there are perhaps no general-purpose digital libraries; certainly nobody has built one yet, and it may be a contradiction in terms. The more general a digital library, the more it resembles a traditional library (which is relatively general in purpose) that has simply employed information technology to provide access to its holdings in electronic form, instead of creating a customized set of information resources and services designed to support a specific kind of work by a specific community. We may well see many different digital libraries that serve specific communities and purposes. Libraries may well end up providing access to many digital collections in addition to providing technology-based access to a wide range of their existing collections.

In the print world, libraries have provided a tremendous coherence to the base of knowledge, although the services that they directly offer for organizing, manipulating, refining, and adding to this base are quite limited. Digital libraries are bringing customization by community—and perhaps even ultimately to the level of the individual—at the cost of a great loss of coherence. This loss is uncomfortable for traditional libraries for several reasons. They have a tradition of playing a key role in providing coherence and uniformity,[5] which have been valuable to their users and are particularly important in enabling interdisciplinary work and the reuse of materials by different disciplines across time. No institutional library can build the full spectrum of "active" community-specific digital libraries; if a library provides access to these, it will in essence fragment its user community into many communities that all use the externally provided systems that the library helps to finance. Its users become disintermediated; the role of the institutional library becomes less visible (we can see this happening already in some of the sciences). An institutional library that is confronting this possible future might prefer to emphasize the view of the digital library as a more passive digital extension of access to existing library resources.

Digital libraries are not place-based. They resist geography and geographically or institutionally based user communities. Instead, they emphasize the sharing and aggregation of information from multiple sources and multiple owners. Indeed, they tend to dismiss place in favor of intellectual and nature-of-work coherence. In the humanities, scholars are using the digital medium to circumvent the haphazard historical gerrymandering of knowledge into institutional collections. And in the sciences, digital databases belong to the entire scientific community and reflect the state of knowledge in that community. It doesn't matter which institution actually maintains them because they are viewed as "belonging to" and are underwritten by the scientific community and its funding agencies. Colaboratories, which can be thought of as extreme cases of digital libraries where collaborative work support dominates, are usually explicitly designed to transcend and, indeed, obliterate the arbitrary bounds of geography and institutional affiliation. But as we will see, the combination of intellectual property constraints and institution-based funding of content acquisition works at cross-purposes to these trends.

When they are free to do so, digital libraries are showing a disconcerting and exciting tendency to find their own user communities, which may be very different from the user communities envisioned or designed for by digital library developers. But they are seldom free to do so. We are also seeing the development of digital libraries in university settings that are more tightly bound to specific predefined user communities than are traditional libraries, due to large numbers of complex licensing agreements for much of the key content that comprises the digital library. Except where institutions underwrite access to expensive licensed content, these audience-restricted systems do not seem to be adding significant value for their users. They certainly don't foster new capabilities for collaborative information sharing and analysis in scholarly and professional work. Indeed, they can make it very hard (and illegal) to share citations or source materials among same-discipline colleagues from different organizations. Ironically, commercial systems that are paid for by the end-user remove this constraint: they open the "community" served by the digital library to anyone with the ability and willingness to pay. In this sense, commercial systems provide much more flexibility than complex institution-based license agreements in their ability to find and serve audiences.

Intellectual property and licensing issues are, I believe, a much-underestimated factor in defining audience and use of digital libraries. In the real world, they dominate; they determine who can participate and who cannot. For public libraries or statewide library consortia, they are perhaps *the* determining factor in establishing the audience and potential relevance of systems. They trump all other factors. Indeed, the licenses that can be negotiated and afforded *define* what is possible in developing digital libraries. Rightsholders have so much control that they can effectively determine where digital libraries can be established.

A Framework for Analysis: Control and Governance, Economics and Sustainability, and Audience

Control and governance, economics and sustainability, and audience are key factors that shape the digital library. How they align and support each other (or disconnect) are fundamental in understanding the evolution and character of various digital libraries.

In a simple, idealized world, a digital library finds an audience—perhaps the one that it was designed to serve or perhaps another. Everyone who wants to be part of this audience—all members of a community sharing specific work practices or a common interest in a collection of information—can participate. This audience funds and sustains the digital library; as long as the services offered by the digital library are sufficiently valuable to enough users, the digital library is sustainable.

The control and governance of the digital library is accountable to the users, either directly or indirectly (for example, through the ongoing choices that the users make in the marketplace). In most commercial digital libraries, the relationships among audience, economic support, and control tend to follow simple lines.

The world is seldom this simple, of course. Funding for the digital library may come from sources that are only indirectly, and nebulously, accountable to the user community or perhaps not accountable at all. The digital library may be funded to serve one user community but actually be used by a somewhat different one (though presumably with some overlap with the intended one). In this case, only the opinions of the intended user community may be considered relevant in guiding the digital library's development. The digital library may be restricted to members of a specific institutional community (and the scoping of this community may not mirror, in any reasonable way, the work practices or collaborative activities of the members of the audience). The digital library may be governed and controlled not by the user community but by a small subset of it. Or it may be governed by a group of intermediaries on behalf of the broader user community (and these intermediaries in turn may be accountable, more or less directly, to some, or all, or little, of the actual user community).

To the extent that the digital library deals with long-term strategic missions of an institutional or broader cultural nature, such as preservation, funding and governance may be far removed from the current day-to-day user community. All types of disjunctions between intended audience and actual use, governance, and economic support are possible. One need only consider digital libraries being deployed by academic or research libraries to observe many of these disconnects in action. Of course, many of them are not new to the digital library world; they are due to the basic structural characteristics of libraries within universities.

Thus, when we speak of user studies in these environments, we must always ask *which* users are the target of the research—the actual users of the system or the intended users?

Commercially Based Digital Libraries

When we think about commercial digital library systems, perhaps the most obvious (and oldest) examples are Lexis and Westlaw, which serve the legal community. These emphasize access to a very large, comprehensive, coherent collection of information to support professional work activities, along with a limited set of tools to facilitate these activities; they do not emphasize collaboration among users. Their heritage is traditional commercial information retrieval systems, such as Dialog, which never really achieved a critical mass of content in any specific subject area. Dialog also never provided work-based tools, except to its primary user community: the specially trained searchers (intermediaries) who did not themselves typically perform professional work in the disciplines of the users they served but did perform professional searcher activities like database selection and duplicate record elimination. Lexis and Westlaw for legal professionals and a few other commercial systems (Bloomberg for the financial community and some of the systems serving the chemical, biotechnology, and pharmaceutical industries come to mind) have moved beyond service to intermediaries and found an audience among working professionals. Interestingly, none of these have academia as the predominant part of their user base.

Some other much more interesting and subtle digital library services are emerging—subtle because one might not immediately recognize them as digital libraries. One of the most striking is the set of systems, services, and information resources developed to serve investors and stock traders during the stock market boom of the late 1990s. These represent a highly competitive economy of services that are federated in a low-tech, user-driven, relatively *ad hoc* fashion. They include portfolio analysis, quotes, analyst reviews, chat rooms, bulletin boards, newswires, and other facilities. Some are freely available, while others require registration, and still others charge subscription fees. They are used intensively by many thousands of people day after day, and they are, in fact, a venue for performing collaborative decision making and analysis though, unlike scientific publication, the ultimate outcome is individual investment decisions (mainly through online brokers). They integrate complex and rich social protocols. Also, this is a digital library (or a set of digital libraries) that is being constructed bottom-up by investors who choose from a wide range of component services being offered on various terms. These include not only commercial services (advertiser supported and for-fee) but also government services such as the Security and Exchange Commission's Electronic Data Gathering, Analysis, and Retrieval (EDGAR) database.[6]

Similar developments can be identified in other areas, such as collecting (linked to auction services like EBay) or consumer health (including such things as the location of clinical trials). Amazon.com is a fascinating example of a highly collaborative digital library environment devoted to the selection and purchase of books and, to a lesser extent, music and other materials.

All of these systems represent real changes in "work" practice by their users ("work" is really too narrow; what is at issue is decision-making and, more broadly, general behavior—consider a patient seeking information and therapeutic options, or a person shopping for a book). But they clearly move toward the more "active" end of the digital library functional continuum discussed earlier.

One of the most powerful, and attractive, characteristics of commercial offerings —or perhaps better termed "general public" offerings since they may appear free to the user due to advertiser support—is that they are open to anyone who wishes to use them (subject only to ability to pay). The audience is self-defining, and issues of audience and economics are not distorted by institutional sponsorship or affiliation. Governance by users is expressed through choice in the marketplace. In a marketplace characterized by many start-up companies unconcerned with protecting their own user base or product line but focused only on competing with those of other companies, we've seen a great deal of innovation and the emergence of systems that clearly are digital libraries but have little formal connection to the heritage of libraries as institutions or social structures.

Another striking thing about many of these commercial systems, however, is their lack of cultural values or social missions; they exist purely on the basis of their ability to attract users in the marketplace. They have no commitment to preservation and continuity of access beyond what their users will pay for (unlike, for example, research libraries). Other library or academic values (for example, privacy of patrons and quality control) are cast purely in marketplace terms. Even the acceptable level of civility in discussion is a purely commercial decision.

These complexes of commercial (or market-based public) services raise interesting questions about public libraries on the Net. Public libraries are having a particularly difficult time incorporating digital library services. To the extent that they want to offer commercial materials rather than just digital versions of the limited and typically specialized materials to which the library actually owns rights, they have often been unable in many cases to negotiate license agreements that extend beyond those physically present in library facilities. In essence, the marketplace in digital intellectual property has precluded them from establishing a meaningful presence on the network. Perhaps in these commercial or quasi-commercial systems, bulletin boards, mailing lists with archives, and Web sites we are seeing the beginnings of a series of focused "public digital libraries" on the Net, though without the coherence of the traditional public library collection. If they are one view into the future of public libraries, they show us a strange landscape that is largely disconnected from the output of traditional commercial publishing that has been such an important part of our cultural discourse. And traditional public libraries may well ultimately help to organize and preserve, evaluate, and offer access to these new systems as content that they can provide to their patrons "anytime, anywhere" (unlike the traditional printed materials from their collection, as these materials migrate to digital form).

One point is important to clarify. Several large commercial organizations have invested heavily in research in socially grounded engineering and sponsored outstanding work in this area. Xerox Palo Alto Research Center (PARC) is perhaps the best example of this with its pioneering ethnographic studies, but others include AT&T Labs (with the work of Bonnie Nardi) and, more recently, Microsoft. But these are not the organizations that are actually *building* commercial digital libraries of the sort under discussion here. Typically, the organizations building commercial digital libraries limit themselves to rather mundane usability testing activities.

Library-Based Digital Libraries in the Academic Setting

Until academic libraries began to connect to computer-communications networks in the 1980s, the library at a given institution of higher education usually enjoyed a relatively unchallenged monopoly in serving its institutional community. Except in a few cases where several universities existed in close proximity[7] and members at one university could conveniently visit the library at another university, students and faculty could reach only the local library established by their home institution.

In the 1980s and early 1990s, academic libraries conducted (not necessarily with clear and carefully considered management planning and sanction) an experiment in network-based access to information resources that offers powerful lessons for digital libraries. By the late 1980s, most major research libraries had replaced their card catalogs with online catalogs. These information-retrieval systems provided users with the ability to search the print holdings of various libraries. Users could use computer systems to *discover* what was available in print at their local libraries (and sometimes materials that might be obtained from other sources through interlibrary loan). The systems did not actually offer access to the content online (and indeed, even today, relatively little of this content is available online). Because these systems provided only access to bibliographic records, they could be made publicly available; there were no licensing issues. As these online catalog systems were deployed, many of them were connected to the Internet and, in effect, became globally accessible, most often as an unexpected and unplanned by-product of the library making its catalog available to the campus community via the campus network.

The audience for these systems was (or at least was intended to be) mainly the local campus community, and librarians "owned" the systems on behalf of that campus community. A number of strange, unexpected, and occasionally wonderful things happened. A competitive economy of online catalogs developed. People would use online catalogs at other institutions because they were better or more comprehensive than the local system and then take the results of searching a remote catalog and hand them off to the local library to actually obtain the materials. They would see features in other online catalogs on the Net and start asking hard questions about why they couldn't have the same features in their local catalogs.

At the University of California's Office of the President where I was working on the MELVYL information system (one of the early and more sophisticated online catalogs) in the 1980s and early 1990s, I heard many anecdotes supporting the prevalence of this kind of behavior, and our usage statistics indicated that 20 percent of use was coming from outside the University of California system. More than once, I visited other institutions, some half a world away from California, that trained people to use the University of California's system.

There were serious—though almost always unsuccessful—proposals at some universities to reduce the visibility of remote institutions' online catalogs or to restrict access to the local online catalog to members of the local institutional community, even though there were no licensing or other constraints that would require such access limitations. In essence, these proposals sought to dismantle (or at least downplay) the competitive economy of online catalogs that emerged so naturally in the network environment and to return to the days of the comfortable exclusive franchise that academic libraries held within their institutions.

Of course, proposals to limit access were not this forthright. Instead, the proponents of restricting access to institutional catalogs argued that they were attempting to protect library collections from hordes of external users who were not part of their funding base or primary user community but who would either show up in person to use the libraries or make heavy use of their collections through interlibrary loan requests.

This was actually an argument with considerable validity. Public and open access to online catalogs made the collections of good libraries much more subject to exploitation by other user communities whose primary libraries were of lesser quality. Another argument—that restricting access would protect response time and avoid the need to invest in additional computer capacity—was less persuasive, particularly as hardware costs continued to drop. Also less convincing was the argument that reducing the visibility of other remote online catalogs within a given institution would limit demands by its own primary patron community for user support and interlibrary loan. As time went on, it became clear that these systems, while an enormous advance over the card catalogs they replaced, were being limited in their potential functionality; the groups that guided development were tied to the functions of the traditional catalog and its role in making library holdings visible in some specific and neutral ways. Online catalogs, by and large, never incorporated result ranking, for example, though this was a well-developed technology in the information retrieval research community by 1990. It was in use in some large-scale systems outside the library community, such as WAIS (wide-area information server technology), by the early 1990s and was quickly implemented by the Web search engine developers as part of their earliest production systems later in the decade.

Even rather modest searching facilities that progressed beyond the card catalog (such as the ability to search by publication date or language of publication, which are powerful tools for certain kinds of catalog users) were relatively controversial. This was because the development of these systems was governed by librarians rather than by the audience of users; governance was intermediated. And the library community had (and still has) an enormous intellectual investment in the traditional catalog and the practices that surround it. Yet there was also enormous competitive pressure because the audience for the local system could see other competing systems (and could even use them rather than the local system). But the competition was among fundamentally similar online catalogs. The problem was that no institution was willing to take sufficiently radical steps to establish a standard that would move the competition to the next level and prevent library retrieval systems from becoming marginalized by a new generation of systems that broke conclusively with the tradition of the library catalog.

There is an interesting conclusion, or at least an update, to the story of the economy of online catalogs. Today the debate is largely over. People tend to use their local catalog, along with a handful of "megacatalogs" such as the Library of Congress catalog, the University of California's Melvyl, or the Online Computer Library Center's (OCLC) Worldcat. But they also use Amazon.com, which is really a digital library for book selection and acquisition, as another substitute for the local catalog. Amazon.com has produced a system that is much more effective for many kinds of book finding than current academic library online catalogs. Amazon has made a great effort to include enriched data such as book covers, tables of contents, and reviews that help users to make choices. Amazon has started to develop a sense of user community by incorporating user-written reviews and ratings and harnessing the power of that community to improve its system. Finally, Amazon has been more willing than research libraries to make compromises about user privacy to implement a variety of innovative popular and collaborative filtering tools. Library catalogs have indeed proven highly vulnerable (in terms of audience share) to competition from a new generation of systems that could never have been produced (philosophically, not technically) by the research libraries that controlled the development of the online catalog.

It is important to note that there was no lack of evidence about what users wanted. Literally endless articles going back to the early 1980s (and perhaps earlier) describe and propose many of the kinds of features that ultimately became part of systems like Amazon.com but are still missing from academic library catalogs. Some of these articles are based primarily on theorizing by system developers and researchers; others are based on user community studies. But these proposals were almost always considered too radical, too risky, or too expensive by the libraries that controlled the development of their online catalogs.

The other great change that occurred in the 1990s was that library catalogs simply became much less important. For many users of academic libraries, journals rather than books constituted the key literature; the online catalog represented little improvement in service for these users. Print abstracting and indexing services—Medline, Inspec, Current Contents, and a host of others—migrated into licensed online databases accessible to the campus community and thus opened up the journal literature to the power and convenience of online searching.[8] Later in the 1990s, substantial amounts of primary content began to be available in electronic form, both from publishers (under license) and for free (accessible freely on the Web). Paper materials rapidly became much less interesting to many patrons, who were eager for the immediate gratification and convenience of networked access to content, much to the horror of some librarians as they watched convenience repeatedly trump quality. Some publishers moved relatively slowly to make their content available electronically and lost readership; libraries, which continued to invest large sums of money in acquiring these materials, even though they were available only in print, faced questions about their choices. Libraries never saw the Web and Web search engines as actual competition to their collections and the access systems supporting them until very late in the game. Indeed, I would argue that many libraries are still in denial about this. But the library's audience is clearly changing its behavior.

As libraries moved down the path toward constructing more extensive digital collections, new issues emerged that again changed the balance between libraries and their audiences and introduced a new factor—publishers as access providers. Libraries added proprietary, licensed, access-restricted content—abstracting and indexing databases and, later, full-text journals in various formats. In the early days of their implementation, abstracting and indexing databases were often mounted locally. They were similar to online catalogs technically, and the same software could frequently be extended to support abstracting and indexing files. For both technical and economic reasons,[9] it became clear, based on the experiences with some early prototypes such as the Elsevier-sponsored TULIP project, that journals in electronic form would have to be mounted centrally at publisher or aggregator Web sites rather than separately and locally at each research library. Much of the electronic content turned into collections of commercial services (access to publisher Web sites), and local systems became less and less significant and distinctive. At one level, the availability of large amounts of important licensed electronic material reestablished the local library "monopoly"; the services offered by each library were limited to that library's institutional community due to licensing issues. Intellectual property considerations (and economic considerations deriving from them) again structured and defined audience.

In this environment, libraries just chose which of a portfolio of commercial services they would license on behalf of their user communities, and these services looked the same from one institution to another. In fact, there is at least strong anecdotal evidence that many users of these publisher Web sites don't even understand that their local library is playing a major role both in negotiating and paying for licenses. The sites appear free to the user, and the user simply assumes that the library is no longer playing a role in the digital information world.[10] The users don't distinguish well between free information accessible through the Web and information licensed at great expense that is also accessible through the Web, except when they run into various annoyances and anomalies—such as when they can't share pointers to material on the publisher Web sites with colleagues at other institutions unless the other institution licenses the same content or when the "published" material licensed by the libraries doesn't show up when the user employs Web search engines. But the fundamental issue is that the library's intermediary role has been greatly reduced in the digital world: publishers and users are talking to each other as never before, and *publishers* now have access to usage data that was historically held by libraries. Indeed, the libraries have to beg the publishers to share it with them and are only sometimes successful.

We have recently seen the next step on this road with the emergence of companies, such as Questia, that have business plans to license digital libraries of materials directly to users for a fee, completely bypassing the role of the institutional library. When we think about user and usage studies in this new environment, it is not even clear what organizations will be sources of data or consumers of research results, whether needed data will even be available (either because of lack of instrumentation to collect it or lack of access to the data by researchers), or to what extent study results will be viewed as proprietary business information. Finally, who can effectively respond to what is learned in these studies? Take just one example: it is overwhelmingly clear that, at least among the sciences, one of the things that readers want badly is the ability to click on a citation in one paper and move from there to view the cited work (with a minimum of access control barriers). This has been known for some years; it is not a startling new revelation that has just recently emerged from studying users. Yet we are making only gradual progress. Turning this goal into reality involves a monumental set of technical, economic, and legal negotiations among the entire community of publishers, standards developers, and system designers. Most libraries are playing a relatively minor and indirect role in both the progress and the process.

Libraries actually followed two separate paths simultaneously in the move to digital content. They licensed commercial commodity information services on behalf of their users, as already discussed. But they also mounted efforts to digitize unique local special collections of manuscripts, old books, photographs, and other materials (which the holding libraries most commonly held intellectual property rights to or which were in the public domain and thus unencumbered by rights constraints), often with the aid of one-time, frequently opportunistic, grant funding. There were compelling reasons to do so: digitizing these materials opened them up to much more extensive use and exploitation by the scholarly community, both locally and worldwide. Many of these collections were fragile and deteriorating; digitizing also offered a way of protecting the collections as well as making them more accessible. But if special collections define the unique greatness of an institutional library, digitizing them was problematic because this shared them with the world. As scholars developed overlays that organized these special collections along intellectual lines, submerging the accidents of institutional ownership or stewardship, this problem became particularly acute, as even the identity of individual institutional special collections threatened to disappear.

Some libraries are getting nervous about the implications of digitizing their special collections and making them available worldwide. Some are talking about charging for access from outside the institutional community; others are talking about limiting access to other institutions that can offer similar high-quality digitized special collections reciprocally.

This is an area where socially informed system design may be feasible and have a high payoff because of the transformative implications of such digital collections (and the tools to use and manipulate them) on certain communities of scholars, particularly in the humanities. The digitization of special collections may also change the nature and scope of their attendant user communities by opening participation much more fully for scholars who are not at major research universities with great special collections and who are not able to find the travel funds to visit such collections.

But ultimately, for the vast majority of users today, the special collections of research libraries are not the most compelling part of library services or available content. Far more interesting free content is being offered in digital form on the Net by organizations other than the local university library. And just as in the period of the competitive economy of online catalogs, libraries are vulnerable to losing audience. The difference today is that, unlike the situation in 1990, the competition is not simply among academic libraries but among a broad range of organizations, with the offerings of libraries playing a relatively small role.

Over the period from roughly 1980 to 2000, we have seen kaleidoscopic rearrangements of the relationships between audience, control, and funding in the academic library world, with the library's role (and power) as an intermediary rising and falling repeatedly. These have been technology-enabled social, political, and economic processes, to be sure, but not processes driven primarily by socially grounded design and evaluation—at least not in large-scale trends. An interesting and important question is how such an approach might usefully be introduced into the current situation (other than as an adjunct to product design by the commercial or quasi-commercial content and service providers).

Prospects and Barriers in User-Centered Design of Academic Digital Libraries

This section examines the prospects for introducing user-centered design into academic digital libraries. It moves back and forth between what we have learned about institutional imperatives and audience considerations, about user needs and user-based engineering, and considers all of this against a backdrop of systems that are increasingly defined and controlled by commercial providers.

Competition for audience in an open economy of services is a good thing for users. It encourages experimentation and innovation. It tends to disseminate innovation among systems. To the extent that many systems from many sources enter into the economy, it provides a hospitable environment for radically new and different systems to try to find an audience and to test and validate their approaches. For institutions, it is more dangerous because it puts them at risk of losing audience and sometimes creates dilemmas about whether to sacrifice fundamental principles to facilitate innovation and to effectively compete with offerings from players in the public marketplace.

There is such a competitive economy operating today but not among systems so much as among different kinds of content—a different kind of competition than the one that occurred in the 1980s and early 1990s. Few libraries, as we have seen, are implementing distinctive systems. Rather, they are licensing access to content that is often exclusively mounted on a single service or perhaps available from the original publisher and one or two aggregators. It is difficult to build innovative systems (for example, those that move across the spectrum of digital library functionality toward collaboratories) using published content from traditional sources. The content simply is not readily available to be integrated into such an environment.

There are both licensing (business) problems and technical problems with such content. The audience-defining power of intellectual property control means that there will be little competition within traditionally published materials—neither among libraries that supply access to it nor among commercial services that incorporate it. Instead, the competition may be between traditional literature from traditional publishers in electronic form and nontraditional content (or traditional content that isn't disseminated through traditional publishers) embedded in highly interactive collaboratory type systems. All of this will occur with a subsidiary competitive economy of services offering access to at least some of the nontraditional content (which itself may be part of the appeal of the nontraditional content).

Of course, there will be linkages (connections) among the two worlds. You can follow links from investing Web sites to *Wall Street Journal* articles, or you can move back and forth between Medline, protein sequences, and articles in the biomedical literature. But it is suspected that more of the linkages will be from the new to the old rather than from the old to the new; the new genres and environments will incorporate the relevant bits of the old by reference.

In the United States, we have thousands of colleges but only at most a few hundred research libraries (usually embedded in research institutions). Much of the effort thus far has been concerned with extending these research libraries to include extensive digital collections and with studying how user communities at these relatively elite research institutions will use digital information and digital environments. These institutions have significant financial, human, and technical resources to do local development if they choose or to integrate and manage access to a wide range of external systems.

The mass market—a large number of the postsecondary academic institutions—does not have these resources. They will be served (probably with considerable economic benefit in terms of costs and benefits compared with the print world) by purely commercial products, probably licensed from a small number of vendors. By subscribing to at most a handful of aggregator services, these institutions will obtain digital libraries that have substantially more material than was available to their user communities in print form. Such aggregators will be less important, and less successful, in the research universities where faculty and graduate students are part of the communities that the individual publishers will connect with directly and where extremely timely dissemination of materials, supplementary data files to support research, and interactive facilities for discussion will be essential. Members of research university communities will be heavily involved in the new digital library environments that will compete with the published literature. But for the typical smaller academic library, aggregation will be an essential strategy for controlling the costs of license negotiation and user support.

There will not be many aggregators; becoming an effective aggregator will require technical expertise, time, and substantial capital. Aggregators will shape their product offerings in response to their perceptions of marketplace demand and marketplace ability to pay for, deploy, and support their systems. Little system design will occur at these academic institutions, and hence socially grounded design will play only a minimal role. They will simply be choosing among a limited number of products in the marketplace. The point here is that the vendors may do usability studies to refine their products but aren't, by and large, much interested in the social implications or transformative impacts on community practices that their products may offer.

In all institutions—whether major top-tier research universities or small liberal arts colleges—the library will continue to develop the digital library as an extension of the academic library and will continue to worry about losing audience to commercial marketplace digital library systems that reach their audiences directly and thus represent competition. It is useful to honestly recognize some of the constraints and historical biases placed on libraries that will influence their actions and, in reviewing these, to think about where user-centered design will and will not really be feasible or useful.

Libraries will slowly redirect a large historical investment in print collections into investment in new digital content. They will have a natural bias toward the products of the existing publishing system with which they have a long relationship (though they may try to redirect investments to different players, such as scholarly societies rather than commercial publishers) *within* this publishing system. They understand how to value the products of this system, and the system produces products that are amenable to consistent and systematic acquisition. The new genres of digital content that are developing outside the publishing system are more problematic, and libraries will be slower to accommodate them. Libraries will insist on information discovery systems that span print and electronic information and that honor their past and ongoing investments in content. And just as they have a large investment in content, they also have a large historical investment in descriptive practices such as cataloging and the skills to create and manage such descriptive apparatus as the online catalog.

All of this history, tradition, and investment is a barrier to user-centered design, and it is important to recognize that, on the time scale of libraries, user-centered design is a relatively new idea. The principles of cataloging and print collection development were not necessarily informed by our current ideas about user-centered design and socially grounded systems engineering; nonetheless, this history and tradition is a powerful force that must be recognized. And I do not want to suggest that the problem is simply one of outdated and irrelevant tradition; rather, it is a problem of how to most effectively meld the results of employing these traditional collection and descriptive practices over the past couple of centuries with the best of current user-centered practice for organizing and providing access to electronic information. Our heritage, like it or not, is still mainly in printed works.

Librarians will take the lead in selection and evaluation of digital library systems and digital collections that extend the traditional library, and yet they are ultimately intermediaries. A danger that has caused endless problems with any number of system procurements is that librarians, as well as faculty, students, and staff, will use the systems. But the needs and preferences of librarians, who are driving the selection process, often do not accurately mirror the needs and preferences of the larger institutional community. Librarians do consult faculty but find it difficult to engage a wide range of faculty in system design and evaluation questions. But students are typically the heaviest users of library systems. We need much more research into student needs and behaviors, and we need to consider how to give these research findings sufficient political weight in decision making. This is particularly important as we deploy commercial digital library systems in the broad range of academic institutions, where undergraduates—perhaps the most disenfranchised, ignored, and difficult to reach sector of the user community—will likely be the dominant users.

Library-based digital libraries have stayed in the passive role, in part due to political mandate for generality—to serve the whole campus community, all the disciplines, with all the differences in behavior and work practices. These differences can be smoothed over by providing a lowest common denominator service of information access rather than trying to facilitate use, analysis, and even creation of new information. In fields where funding exists for discipline-based, active and interactive digital libraries oriented around work practice, these are being created largely outside of the library in individual communities of scientists and scholars.

Libraries may tend to ignore these systems and their implications. They may be reluctant to commit funds to help underwrite their operation (deepening the conflict for funding between the user communities that want to ensure the financial sustainability of their community systems and the libraries who are more comfortable underwriting the acquisition and thus the publication of traditional literature). A key research area for socially grounded engineering as digital libraries scale up will be how to ensure that discipline-based, specialized, "active" systems can coexist with the library traditions of consistency, coherence, reuse of materials across disciplines, and preservation of the heritage of all disciplines for the long term.

It may be necessary to reexamine the principle that virtually all information is centrally funded, on an institutional basis, through the library. Some funding may need to move back to the users so that they can use it to pay for public marketplace digital library services. The balances between the good of the individual and the good of the institution and between short-term information needs of individuals and longer-term information needs of today and tomorrow's institutional community as a whole (and even society more broadly) will be difficult to negotiate. This will also need to be part of the socially grounded engineering of the future digital library environment, if we can move away from highly polarized political positions to a more open-minded and scientific inquiry and analysis.

Finally, we must recognize that institutions of higher education (and the libraries that are part of them) both cooperate and compete. An important and complex debate is emerging about what it means to be a great university in a world of distance education and what it means to be a great library in a world where unique special collections are globally accessible and where many information resources are highly aggregated and licensed by large numbers of libraries identically. To understand how digital libraries operate as social constructs, we need to understand the contradictions between the potential for universal accessibility that the network can offer and the exclusivity that has been an important part of defining institutional uniqueness and competitive advantage.

The intermediary role of the library, a conservative institution with strong and long-standing traditions, in shaping library-based digital libraries means that the evolution of this class of systems is likely to be conservative. *Conservative* doesn't mean "bad"; it means "conservative." It also tends to be strongly infused with academic values such as privacy, intellectual freedom, and the importance of preservation of the intellectual record; these values have served society in good stead. Intermediation by librarians in design, governance, and control means that, where the traditional library worked well in terms of serving social, cultural, and institutional values and goals, digital libraries, as extensions of that library, should also work well. But such intermediation creates barriers for digital libraries as transformative and revolutionary ways of managing information and providing an environment in which digital libraries can be actively engaged to produce and disseminate new knowledge. Efforts to integrate user-centered design and socially grounded engineering of digital libraries must recognize these realities if they are to be relevant within the program of extending and ultimately transforming the academic and research library.

Postscript: The Focus on the User and The Ultimate Heresy

Much of user-centered design or socially grounded engineering is based on user reactions or user community behavior as it relates to a specific system, since inquiry into user behavior is often explored (and funded) as part of the development of specific systems. These systems set a context. Most of these would like to be the user's primary source for all forms of information—to be the exclusive provider (or, realistically, near-exclusive, since they always admit that their users do obtain at least some information about some things from other sources, such as personal use of mass media and colleagues). I don't know of anyone who fulfills all of his or her information needs (much less his or her professional or broader personal needs) by interacting with any single system.

Because this discussion focuses heavily on who supports and who benefits from user studies, I cannot resist underscoring the tremendous importance of continuing to support work on the individual user, workgroup, or intellectual community of practice "in the abstract" as a free-standing entity that chooses to employ (or not employ) a range of systems and sources (rather than in relation to a specific system). This is a tremendously hard problem, and most of the work that I'm familiar with in this area has been fairly superficial—along the lines of determining "information sources that people typically consult in making a decision about X (75 percent asked friends or colleagues, 50 percent consulted published works, etc.)." Such results don't help much with refining the engineering of information systems and deciding what functions to incorporate or exclude in a given information system. As so many things are moving into the networked information environment, the character and ease of use of different sources—everything from interactions with colleagues to looking at newspapers to the new capability to harvest the opinions of huge numbers of anonymous strangers—are changing. We need to understand how people, in this new networked environment, form opinions, track developments, and make decisions, as well as what information and resources are helpful in these processes.

I suspect that we know very little here. We don't even know what questions we should be asking. We need to better understand factors like convenience and desperation. We have already seen that many students, particularly when writing term papers the night before they are due, will immediately limit their consultation to materials that they can get right away on the Web. Better sources (even printed works) might be available, but, if they cannot be consulted in time, they are irrelevant. Here convenience or access dominates. On the other hand, we can also see many examples of people who will at least tolerate inconvenient and horrible user interfaces. These might include the critically ill person searching for a clinical trial that may offer hope, the investor seeking to make a profit, and the airline reservation clerk doing his or her job. Libraries tend to have one set of biases about how important convenience is or should be; the consumer experience with the Web offers another. Similarly, people have well-developed ways of establishing belief, confidence, and assessing quality; I suspect these are often quite different from the approaches that libraries incorporate. To give just one more example, it is unclear how important an information resource's position on the continuum from repository to collaboratory really is to different people for different purposes at different times.

Ultimately, we are going to have to develop information retrieval and management systems that actually empower the user in his or her day-to-day activities, and those systems are going to require an understanding of which systems the user employs and why and when each system is utilized. It will require a holistic view of user behavior, both as an individual and as a member of multiple workgroups and communities (and an understanding of how these workgroups and communities overlap and relate to each other). It is going to lead to technology that is very close and personal to the user and that helps the user to mediate with continually evolving information sources (both fixed and new), costs, and quality. Another key issue will be recognizing interesting new services, not just choosing among a fixed portfolio of existing services.

This maneuvering is likely to be unpopular with institutions that oversee the development of specific systems, since, in some sense, it limits the power of the providers of every system that the user employs in the course of the day (each of which would like to claim pride of role as *the* comprehensive information system). At some level, the problem is integrating all information systems used by individuals, eliminating duplication, and thereby adding coherence and intelligence, close to the user, that integrates information from multiple systems in smart ways. We must not overlook these ultimate user-centric design mandates in our quest for a socially grounded design of information retrieval systems.

We need to think through how to ensure sources of funding, support, and data to perform these user-centered studies and also how to make them as useful as possible for the institutions that build, control, and sustain the systems that populate the individual's information universe. We can reduce but not resolve the tension between empowering users and advancing institutional goals that, in reality, include only the empowering of individuals within specific institutional contexts.

Acknowledgments

Ann Bishop not only waited patiently and endlessly for this chapter but produced extensive comments on an early draft that were so thoughtful and insightful (and humbling) that I have been able to do them only very partial justice here. They should become a paper in their own right. But I must take the blame for all of the opinions expressed in the chapter, despite her best efforts to temper or refine them. Cecilia Preston was also kind enough to offer some very helpful comments on several drafts of the paper.

Notes

1. It is worth noting that the National Library of Medicine has a long history of high-quality user and usage studies. In some ways, its efforts could serve as a model for socially grounded

development. But the National Library of Medicine also has benefited from unusually clearly defined goals and constituencies that have facilitated this approach.

2. ERIC has recently embarked on a major evaluation program whose details are still being defined but will undoubtedly include impact and user studies.

3. I am grateful to Ann Bishop for suggesting this term, which I understand comes from the work of Bill and Susan Anderson at Xerox (Anderson & Anderson 1995), though I am not sure I am using it in quite the same sense that they use it. But I think it conveys the right sense. There is actually a tremendous void involved in exploring the nuances of socially grounded engineering, but that is largely irrelevant to the points being explored in this discussion. My point is really to distinguish between socially grounded engineering practices (design and development practices that focus on user needs, opinions, and behaviors and on the impact of the system on user communities) and nonsocially grounded design and development practices.

4. Ironically, the best libraries are constantly trying to ensure their accountability to their users and seeking user participation in governance. But users are busy, and their primary professional responsibilities and intellectual interests usually lie elsewhere. As long as things are going reasonably well, they are often content to provide occasional advice and let the library lead developments. It is often only when things go terribly wrong that they become energetically engaged in their governance roles.

5. This is not to say that a typical research library provides uniform service to all disciplines it serves; there may be great variability in the depth of the collection from discipline to discipline. But the type of service is relatively constant across disciplines with very few exceptions, perhaps the most notable being those disciplines that make heavy use of images and slide collections.

6. The story of the EDGAR database, which contains corporate filings required by the U.S. Securities and Exchange Commission (SEC) is a fascinating case study in social and political dynamics that has never, to the best of my knowledge, been examined in detail. Until the mid-1990s, this database was purchased by commercial information providers, repackaged, and marketed to the "institutional" financial and investing community at very high prices, even though it was public data. Carl Malamud, working with a group of researchers at New York University, obtained a National Science Foundation grant to pay for a license to the data and to place it on the Internet for public access, where it found wide use by individual investors and played an important role in leveling the investment playing field. When the grant ran out, so much demand for the database, and reliance on it, by the general investing public had been created that the SEC was virtually shamed into making it publicly available for free on the Net. We have seen somewhat similar situations where the political process persuaded the federal government to open other databases such as patents.

7. There were a handful of other unusual cases. At universities in Washington, D.C., students and faculty had easy access to the Library of Congress and the National Libraries of Medicine and Agriculture, and a few urban public libraries were actually great research libraries in their own right, such as the New York Public Library.

8. Online access to the contents of journals continued to be through mechanisms and databases that were separate from the ones used to find books, just as in the print world. The story of how this decision was made in the print world was later reaffirmed as the transition

to electronic resources occurred, and reflected or failed to reflect user needs, preferences, or work practices would be a fascinating study that, as far as I know, has never really been carried out.

9. In particular, dealing with a large number of constantly changing and poorly standardized input streams from many publishers was an overwhelming prospect for libraries, both operationally and from the point of view of software development and maintenance. In addition, few of the library automation systems in use could accommodate journals in electronic form (including all the problems of images, mathematical notation, and the like). Unlike abstracting and indexing databases, which were accommodated relatively easily, making journals available electronically would have represented a significant software development effort.

10. Further, although a detailed discussion would take us too far afield, among those faculty members who actually do understand the ownership, licensing, and economic structures surrounding the published literature, there is a major war developing about whether (and how) to restructure the system of scholarly publishing to make this material publicly available. Debates about e-print and preprint archives and the establishment of low-cost noncommercial journals to compete with commercial journals—such as the activities of the Association of Research Library's Scholarly Publishing and Academic Resources Coalition (SPARC) program—are all part of this.

References

Anderson, W. L., and S. L. Anderson. 1995. Socially Grounded Engineering For Digital Libraries. Paper presented at the 37th Allerton Institute, Monticello, Ill., October 1995. ⟨http://www.lis.uiuc.edu/gslis/allerton/95/anderson.html⟩.

Borgman, C. L. 1999. What Are Digital Libraries? Competing Visions. *Information Processing and Management* (special issue on Progress toward Digital Libraries), 35(3), 227–243.

Levy, D. M. 2000. Digital Libraries and the Problem of Purpose. *D-Lib Magazine*, 6(1). ⟨http://www.dlib/org/january200/01levy.html⟩.

Part III

9

Information and Institutional Change: The Case of Digital Libraries

Philip E. Agre

Introduction

The term *digital library* (DL) has been used to refer to both a machine and an institution (Borgman 1999). Computer scientists are willing to build a database and call it a library. But society will evaluate digital libraries in terms of the ways that they fit, or fail to fit, into the institutional world around them (Kling and Elliott 1994; cf. Berg 1999). Institutions, for present purposes, are the enduring categories of social roles, legal systems, linguistic forms, technical standards, and so on through which human relationships are conducted (Avgerou 2002, Dutton 1999, Gandy 1993, Kling 1996, Mansell and Steinmueller 2000; for theoretical background see Commons 1924; Goodin 1996; March and Olsen 1989; North 1990; Powell and DiMaggio 1991). An institutional field is an enduring ensemble of institutional categories; examples include particular historical forms of the market, the political system, or the university.

In this chapter, I propose to explore certain ways in which the fit between technology and institutions might be conceptualized and evaluated. I cannot survey all technical issues or all institutional issues. Instead, I focus on the boundary between technology and institutions. Investigation of this boundary requires considerable preparatory analysis, and it will be necessary to identify and transcend several intellectual traps. In particular, evaluation of the interaction between technology and institutions must commence with substantive ideas on both sides. If we derive our institutional ideas from the metaphors already embedded in the technology, then the process may be tautological from the start. Every technology is embedded in the social world in complicated ways, and this is particularly true for digital libraries, which are intertwined with the cognitive processes of a complex society.

Unless our conceptualization of society stands on an equal footing with our conceptualization of the technology it uses, our analysis will inevitably be overwhelmed by myths.

Digital libraries are distinctive in another way. The library world, like any institutional field, maintains a distinct identity. But the library world also articulates with other institutional fields—that is, it interacts with them in relatively stable and structured ways. In fact, the library world (along with some others, such as the legal system, higher education, the legislature, and journalism) articulates with virtually every other institution in society. The central institutional tension of the library is its need to maintain relatively uniform practices despite the great diversity of the social worlds whose members it serves (Agre 1995).

Although it has never been easy, librarians have historically been able to maintain this uniformity of practices because of the limited variety of physical media (books, films, records, magazines) that various social institutions have produced. Networked computing, however, permits libraries to be articulated much more elaborately with the institutions they serve.

Will the "common coin" of digital representation enable libraries to maintain a manageable uniformity in their practices despite the diversity of their articulations with other institutions? Or will the immense flexibility of digital computation unleash unmanageable pressures for heterogeneity? My approach in this chapter is analytical, not empirical, and I will proceed as follows.

I will begin by arguing for the *priority of analysis*—the idea that sociological conceptualization of user communities and institutions is logically prior to the design and evaluation of technical systems. When the priority of analysis is not respected, an intellectual vacuum is created, and various patterned cultural myths flow in to colonize our thinking.

The next section of this chapter considers the case of scholarship, with particular attention to the institutional conditions that are necessary for the construction of healthy scholarly communities. In both cases, conventional theories mislead us by dividing the world into extremes when the reality, both descriptively and normatively, falls in the middle.

I then consider the case of the public sphere. The health of a democratic society is founded, at least partly, in the pervasive processes of collective cognition that tie innumerable overlapping subcommunities. The conditions of these processes can be usefully compared and contrasted with those already analyzed in the world of scholarship.

The chapter builds on these cases by considering more abstractly the embedding of digital libraries in their institutional environments. I sketch several potentially useful themes, including convergence, specialization, standards dynamics, organizational boundaries, and genres.

I conclude by summarizing some of the positive contributions by which digital libraries might be evaluated.

Priority of Analysis

The design of computer systems begins with concepts that describe the people, places, and things that the computer is supposed to represent, the attributes they can possess, the actions they can take, and the actions that can be taken toward them. The concepts that become embodied in computers are part of intellectual history. They come from somewhere, and indeed the usefulness of the computer will be demonstrated largely in the accuracy with which the users' concepts can be used to explain what the computer does. When designing a computer to predict the weather, for example, most of the relevant concepts are derived from meteorology. Because the concepts of meteorology are already stable and codified, the design process has a clear starting point.

Designing a digital library, on the other hand, involves comprehending social phenomena of great complexity. No single discipline will provide all of the necessary concepts. Instead, several distinct levels of analysis must be employed. One level of analysis pertains to the physical and cognitive mechanics of work; on this level the necessary concepts derive from ergonomics and human-computer interaction. Another level of analysis pertains to the principled organization of information and the search habits of individual library users. On this level, the necessary concepts can be obtained in reasonably stable and codified form from the tradition of library and information science (Borgman 2000).

In this chapter, I am concerned principally with an even higher level of analysis, the embedding of a digital library in the larger social world, for which the necessary concepts derive from social theory. The design of technical systems and institutions has not usually been informed by concepts from social theory, however, and so in this section I consider the role of social theory in design.

A central challenge for social theory is the great complexity of social phenomena. No single concept explains everything. The social theorist is therefore necessarily engaged in traffic control, working consciously with the relationships among a large number of concepts. For example, useful concepts are found on several different scales or levels, and I have already informally sketched some of the levels that are relevant to the analysis of digital libraries. Each level of analysis is equally important, and analyses on the different levels will regularly inform one another. The concepts themselves differ from the concepts of science—that is, they cannot be defined in mathematical terms. Their purpose, rather, is to help describe particular examples of social practice. These descriptions are necessarily intricate and do not necessarily lead to simple generalizations. The purpose of theoretical work, for example, in this chapter, is to clarify concepts and their relationships. Whether the concepts are useful in the analysis of particular empirical situations is a different question, and each project—the theoretical and the empirical—provides important guidance to the other.

Concepts play at least three roles in design: (1) they are employed in studying the task and the context of use; (2) they are inscribed directly into the software and into the categories and policies of the institution; and (3) they define the criteria by which the technical and institutional systems are evaluated.

It follows that the analysis of concepts should precede design or at least that conceptual analysis is necessary for design to make any progress. This is the *priority of analysis.* Although the point may seem like common sense, few design projects make any explicit provision for this kind of conceptual work before making irreversible design commitments. The traditional methods of systems design do employ the word *analysis* but in a narrower sense, in which the concepts are assumed to be given in advance. This is a dangerous assumption without analysis; designers must necessarily employ whatever concepts they find lying around. These concepts might be incomplete or incoherent, and they might distort the practices or omit large parts of them. Concepts that derive from the millenarian ideologies of computerization movements (Kling and Iacono 1988) are likely to be misleading as well. A design process that does not analyze its concepts forecloses much of the design space before it even begins, and it risks catastrophe if its concepts are broken. And because the same potentially problematic concepts are used in evaluating the system that is designed, fundamental design flaws will not necessarily be detected.

The design of digital libraries requires conceptual analysis because of the great complexity of a library's relationship to its institutional context. Advanced computing and broadband networking will enable digital libraries to become highly integrated with the institutions in which they are used, but little is known about the forms that this integration might take. A digital library does not require its users to extract themselves from their ongoing patterns of activity. To the contrary, the library can conform itself to those patterns of activity in numerous ways. Thoughtful design will require substantive ideas about those patterns of activity and about what it might mean for a digital library—or anything else—to "fit" within them.

Conceptual analysis faces other challenges as well. Information technology often requires designers to revisit and clarify old concepts so that design thinking does not fall into simple dichotomies. What, for example, is a library? A concept of "library" that is too fully rooted in past historical forms will make innovation impossible, but a superficial concept of "library" that draws out only a few aspects of those past historical forms (for example, the library as a big container of documents) will pass over phenomena whose absence in a newly designed system may be fatal.

The middle ground between the maximal and simplistic conceptions of "library" is enormous and is not easily mapped. In attempting to map that ground, it helps to have two kinds of concepts. *Bridging concepts* are concepts that enable designers to move back and forth between the technical and institutional sides of their work. An example of a bridging concept is *inscription*—the process by which social discourses are translated into the workings of software (Agre 1998b; cf. Latour 1993). And *meso-level concepts* are concepts that describe medium-sized social phenomena, such as institutions and social networks, thus avoiding the sterile opposition between macro and micro that frustrates many applications of social theory (e.g., O'Neil 1998, p. 10).

Much of the skill of conceptual analysis consists of looking for common traps that can confine a project's concepts within the bounds of unnecessary assumptions. Here, for example, are several conceptual traps that may afflict the unwary designer of digital libraries:

• *The trap of presupposing standardization*   Fantasies about computers in popular culture often assume an implausibly high level of interoperability among systems that have arisen independently of one another. This is certain to be a substantial issue as digital libraries are integrated with the systems of their diverse users. The effort that goes into technical implementation in a narrow sense may be slight in comparison to the effort of consensus-building around standards.

• *The trap of deriving political consequences straight from the technology*   Authors such as George Gilder (1992, pp. 48–50, 126) have predicted that the decentralized nature of networked information technology would lead to a decentralization of power in society. But this consequence hardly follows. Computer networks are just as capable of projecting the instruments of control into far-flung locations. Likewise, librarians know well that uniform technical standards for access to digital libraries do not imply equal access in any effective social sense.

• *The trap of automation*   The word *automation* often slips back and forth between two distinct senses. In one sense, the word simply refers to any use of technology. But in another sense, it refers to a particular way of designing and using technology, whereby the workings of a machine are modeled on the activities involved in a particular job and the purpose of the machine is to replace the human effort that the job involves. When the word does slip silently between these two meanings, the design process can be led to presuppose the narrower sense of the term rather than consciously choosing it. It is sometimes both practical and beneficial to replace a human job with a machine on a one-for-one basis, but the possibilities of technology are vastly larger. In most cases, a new technology will lead to a renegotiation of the roles of people and machines, and this renegotiation should be part of the design process.

• *The trap of assuming rapid change*   The capacities of computer chips and fiber optic cables are growing rapidly, but it does not follow that social institutions will change as fast, that they *can* change that fast, or that they should. Institutions become intertwined in with information technologies in many ways (Kling and Iacono 1989). Technical standards, once entrenched in the installed base and practices of an institution, are exceedingly difficult to change. The institutions themselves, carriers of collective memory and skill, are usually slow to change as well and for good reason. A design process that assumes rapid change will become preoccupied with "keeping up" and with "not being left behind" and will therefore not perceive the need for sober analysis of concepts—including concepts about change itself.

• *The trap of all-or-nothing change*   Many highly developed discourses presuppose that computing will give rise to total discontinuous changes, either in society generally or institutions in particular (in higher education, for example, see Dolence and Norris 1995). The world is thus divided, in Manichean fashion, into revolutionaries and reactionaries, those who embrace change and those who resist it. This kind of opposition is understandable in the absence of analysis exactly because the changes are new. Society has not needed concepts to describe the relationship between what is changing and what is not. Describing that relationship is a central role of conceptual analysis in the design process. Visions of discontinuous change can also arise from an overly simple understanding of the relationship between technology and sociology. Designers often associate technology with the future and sociology with the present and past. Because technology is supposed to change things, conceptually sophisticated investigation of the social world as it exists can seem irrelevant to design. But this understanding of design is simplistic and even dangerous; it amounts to a willful blindness to the context in which the designed systems will be used.

• *The trap of command-and-control computing*   The main tradition of computer system design arose in military and industrial contexts in which the designers were closely allied with authorities who possess great power to direct the activities of the users. As a result, computers have long been associated with rationalization and hierarchy. The rise of the Internet, with its decentralized control structures, has shaken this association somewhat, but many legacies of the command-and-control era remain. For example, database design still assumes that individuals will be assigned unique identifiers, despite the serious privacy problems that this practice can raise. A new generation of privacy-enhancing technologies (Agre and Rotenberg 1997) has not been integrated into day-to-day design practice.

• *The trap of inventing a new world*   Several of these traps can combine to persuade designers that they can use technology to impose entirely new patterns of activity on their users. After all, the purpose of design is innovation, and true technical innovation is impossible unless the users change their habits. While it does make sense to speak of institutional design (Goodin 1996), new institutional structures usually cannot be imposed through technology. Existing patterns of activity are usually shaped by many factors beyond technology. Designers can consciously choose to amplify an existing force in society, but they probably cannot create new forces.

• *The trap of blaming "resistance"*   Some technologies are rapidly adopted, and others are not adopted at all. The difficulty of predicting adoption can frustrate designers, and the language of "resistance" provides a simple explanation of the problem. But a responsible designer will try to distinguish between resistance that is irrational and resistance that arises from a poorly designed system.

• *The trap of assuming away intermediaries*   Networked information technology is frequently created to eliminate the need for intermediaries—those individuals and organizations that facilitate connections between buyers and sellers, citizens and government, people and information, and so on. After all, if the network can connect the parties directly, what is the purpose of the intermediary? This argument depends on an ambiguity in words like *connect*. A computer network can transport data between point *a* and point *b*, and it can make information available in a standard format at many points *a* so that computers located at various points *b* can search it. But intermediaries can serve many other purposes, and most of the successful new businesses on the Internet are, in fact, intermediaries (Sarkar, Butler, and Steinfield 1995; Shapiro 1999). A more suitable term—and a less constraining concept—is *reintermediation* (Halper 1998; Negroponte 1997).

• *The trap of technology- and economics-driven scenarios*   Institutions must be described using vocabularies from several disciplines, and great havoc can result when one discipline's language is employed to the exclusion of all others. As this list should already have made clear, design is too often technology-driven. Given a hammer, one looks at things as if they were nails. Nonetheless, economics-driven design is equally hazardous. Economics is a powerful mode of analysis, but economic theories simplify and idealize the world. The mainstream neoclassical theories, for example, almost entirely ignore information and institutions (Casson 1997; Hodgson 1988). Economic theories have also tended to homogenize things by treating them as a uniform array of resources to be allocated, and they have likewise tended to oversimplify the web of human relationships within which economic exchange takes place (Granovetter 1992). When economic analyses are turned into institutional prescriptions, these simplifications can become serious blind spots. Economic analysis is an increasingly important component of the design process, but it is useful only as part of a dialogue.

• *The trap of designing for a limited range of cases*   Much of the design process necessarily takes place far from the places where the resulting systems will be used. As a result, designers must depend on their own imaginations. Designers whose imaginations are shaped by experience with one setting or one type of user risk designing systems that discriminate against other settings or users (cf. Friedman and Nissenbaum 1996).

• *The trap of presupposing transparency*   Experts usually forget what it is like to be a beginner, and designers usually cannot imagine what it is like to confront their systems anew. A generation of user-interface design has developed great solicitude for the situation of the beginning user (Shneiderman 1998), but this work has been primarily ergonomic and cognitive in orientation. Digital libraries exemplify a new generation of systems that cross institutional boundaries (Friedman 1989), and little is known about the challenges that such systems present to the beginner. Designers can too easily assume that the user possesses the whole tacit world view of the designers' own community, and analysis will be required to understand the ways in which differences in world view can affect the assumptions that users bring to a system.

These, then, are twelve challenges for conceptual analysis as part of the design process. A good design will seem deceptively simple, precisely because these issues do not arise. A good set of analytical concepts will avoid these traps as well, and it will be useful to keep them in mind when evaluating the substantive discussions of the remainder of the chapter.

Scholarly Community

The scholarly community is deeply intertwined with the library. Much of the library's contents is produced by scholars, and the structure of these materials reflects the institutional structure of scholarship. Much of a scholar's professional persona lives on the library's shelves. The success of scholarship depends on the health of scholarly communities, and digital libraries will participate in changes that can affect the health of these communities for better or worse. A central theme, then, will be the internal workings of these communities and their links to the rest of the world.

Scholars need a space apart from the world. What is this space? It is not physically localized and indeed stretches around the earth in the "invisible colleges" into which scholars form themselves (Crane 1972). Scholars' space is understood in many other ways—time to think, a private workspace, access to books and journals, freedom from political pressure, the opportunity to try ideas that might not succeed, and so on. In particular, scholars need a space for the self-organizing mechanisms of their community. While popular authors and journalists make their living selling their writings in a straightforward market system, scholars cannot use market mechanisms to govern their work because their task is to produce public goods—ideas and discoveries that are difficult to buy and sell (Hallgren and McAdams 1997). This is the purpose of peer review and the informal assignment of credit to innovators (Latour and Woolgar 1986). These mechanisms obligate scholars to monitor one another's careers—for example, by reading journals and through professional meetings and rumor networks—and scholars thereby have a powerful incentive to adopt new information and communication technologies.

Scholars' space is also a container for conflict. In his astonishing sociological history of philosophy, Randall Collins (1998) has demonstrated that the intellectual health of philosophy has depended crucially on robust debate between scholarly movements or schools. When the institutional conditions are present to support scholarly work, Collins argues, philosophical schools strategically split and merge so that only a few schools compete in the intellectual "attention space." The debates among these schools keep them honest, and the need to respond to opposing schools' arguments is the motive force that moves philosophical inquiry forward. When the institutional conditions of orderly debate fail, for example, through economic collapse or political controls, philosophical inquiry becomes rigid or fragmented.

New technologies are quite capable of affecting the system of incentives that makes these mechanisms work. If scholars can advance in their careers by leading coherent intellectual movements, then technology can make it easier to organize such movements by maintaining communications among their members.

The Defense Department's Advanced Research Projects Agency (ARPA), which invented the Internet, has also pioneered methods for using the Internet to operate research communities. Indeed, rather than leaving the creation of such communities to the career agendas of individual researchers, ARPA has largely internalized the process, using its own research-funding procedures to organize technical communities (for example, the Image Understanding community) that are defined in relation to ARPA's own needs. Technology can also make it easier for scholars to move from one community to another, staying long enough to apply their skills to particular problems and then moving along to another community when a given line of work yields diminishing marginal returns.

These factors may have contributed to the tremendous growth of interdisciplinary research during the 1980s and 1990s—not just the crossing of disciplinary boundaries but the continual creation of new communities with new, permeable boundaries. Assuming that Collins's argument applies to disciplines besides philosophy, what does it say about the current situation? The increased fluidity of scholarly communities may be a sign of fragmentation. Or perhaps the attention space can now accommodate a larger number of schools. Collins's argument also points to the need for a loose coupling between individual research communities and the rest of the world. On the one hand, a community must be sufficiently coherent to define a common language and a shared set of problems, methods, and goals. Otherwise, it could not contain its own internal conflicts or develop any depth of learning. On the other hand, each community must be accountable to the arguments and objections of other communities.

If this accountability becomes too rigid, then research will devolve into a purely political struggle. But if it becomes too loose, then misguided research communities will be able to reproduce themselves indefinitely. This is fundamentally a question of institutional design, but technology can either hinder or facilitate any institutional mechanism.

A loose analogy might be drawn between scholarly communities and ecosystems. Island ecologies permit evolution to go in new directions without being constrained by the competitors in other ecologies. Likewise, the intellectual communities of different countries have historically been somewhat isolated from one another. Scholars have always corresponded and traveled, not least because they can advance professionally by bringing new ideas home with them. But communications and travel have always been laborious and expensive, and the interactions among national scholarly communities have always been limited as a result. Language has been a barrier as well. Nonetheless, new technologies decrease the costs of scholarly interaction, so that now it is necessary to determine the optimum level of coupling among different intellectual ecosystems. The dangers of excessive technologically facilitated homogeneity can be seen in the computer industry, which went through a long series of manias during the 1990s (virtual reality, agents, network computers, and so on). Institutional changes will be required if these wasteful storms of intellectual fashion begin to disrupt scholarship.

Technology might contribute to constructive changes in the institutions of scholarship through the invention of new genres (Agre 1998a). As a thought experiment, consider the problem of professional mobility—the possibility of advancing in a career, either by building stature within a particular research community or by moving to another one. Either kind of move requires an individual to master a complex landscape of scholars and their work. Where can maps of this landscape be found? One map is available in traditional library catalogs, but that map does not reflect many of the most important features of the territory (Agre 1995). Other maps are available in the narratives by which authors of scholarly literature give credit to the authors who have gone before them. And yet others are available in survey articles.

A digital library might make these narratives more systematically available. Graduate students who are defining an area of dissertation research could be obliged to produce an extensive survey, as a structured hyperlinked document conforming to a particular XML document type, of the literature in that area. The scholarly community could use the Internet to organize a peer-review and publication system for these documents, and these mechanisms would help students to develop their professional voices. Once deposited in the digital library, the world's entire collection of literature surveys could then be searched by anyone wishing for an introduction to a given literature. Reverse links, from works to the surveys that mention them, would make available several narratives of a given work's place in intellectual history.

Whether this proposal is feasible, of course, depends on much more than technology. It would require a great deal of consensus building, and it would require individual research programs to surrender some degree of control over the progress of their students' careers. A digital library can be designed to support these kinds of technical mechanisms, and it can be evaluated in terms of the support it offers to the laborious process of building consensus around them. Increased professional mobility may have disadvantages as well.

Students today are strongly bound to their dissertation advisors both by cognitive limits (it is hard to learn any other intellectual system than that of one's teacher) and by the mechanisms of professional evaluation (only one's teacher is in a position to write the evaluations on which one's advancement depends). But if these bonds are loosened, then it might become impossible to build a stable intellectual community.

These few ideas hardly exhaust the range of institutional links between a research community and the rest of the world. Other links include those between research and teaching, between scholarship and government, and between theoretical work and applications. A more careful treatment would consider these links systematically, revealing how each works in the present day and inquiring how a digital library might facilitate or disrupt the existing dynamics (cf. Lamb 1995). These considerations return in the next section.

Public Sphere

In addition to their role in supporting research, digital libraries can also be evaluated in terms of the contribution they make to the health of democracy. This would seem obvious enough. Democracy is supposed to be a matter of rational deliberation, and a digital library ought to support the activities of research, reflection, and communication that rational deliberation requires. But much depends on particular concepts of democracy and of the cognitive processes that support it. Liberal political theory, for example, locates the practices of democracy in the individual—individual people who gather information, debate one another, and express their choices through aggregating mechanisms like voting. To the extent that library science conceptualizes library patrons as individuals, it embodies a liberal theory of politics. Given the epistemic and cognitive limitations of isolated individuals, however, library or no, such a theory cannot explain how citizens can effectively deliberate on matters that involve far-flung facts and affect the community as a whole. At the opposite extreme, authors like V. N. Volosinov (1973) interpret both politics and cognition entirely in collective terms, leaving no analytical space for the individual. This theory is no better, and (strikingly), for the same reason: collective cognition, to be effective, requires a substantial division of labor (cf. Hutchins 1996; Weick and Roberts 1993). Somewhere between these positions, communitarian authors imagine individual cognition and action to be constituted, to a large extent, by the norms and language of the community but do not imagine that the community completely determines the individual's choices (Etzioni 1995).

This is progress, but it provides no real theory of interests and conflict. What is needed, therefore, is a substantive account of the cognitive basis of social movements and other social groups (Melucci 1996) and of the ways that technologies and institutions can either support or disrupt this collective cognition.

One starting point might be the rough analogy between social groups and scholarly communities. Each type of community needs an autonomous space, loosely coupled to the spaces of other communities. Each provides its individual members with relatively safe opportunities to develop their public voices. But the analogy stops there. Scholars need to be accountable in material terms for the coherence and utility of their ideas so that the institutions of scholarship can allocate their resources in a productive way. Otherwise scholars would be paid to talk nonsense. No outside regulation is required to prevent a social movement in a democracy from talking nonsense since the movement's ideas must be coherent enough to organize effective action and appealing enough to form the ideological basis for coalitions with other groups. Incoherent ideas can be exposed by other movements that compete to recruit the same social groups to their own coalition.

The principal question, therefore, concerns the conditions under which different groups are able to organize themselves cognitively: unequal access to the means of collective cognition can lead to material inequalities of other sorts. Herein lies one of the central political claims for the Internet: online discussion groups provide cognitive infrastructure for a vast range of constituencies (Agre 1998a), and digital libraries seem certain to do the same. Hubertus Buchstein (1997, p. 251) observes that "viewed in terms of contemporary democratic theory, the positive qualities attributed to the Internet strikingly resemble the Habermasian unrestricted public sphere." The public sphere is not singular but multiple, and "[t]echnologies of communication … make possible a highly differentiated network of public spheres…. The boundaries are porous; each public sphere is open to the other public spheres" (Habermas 1987, pp. 359–360; cf. Fraser 1992). Several authors have even spoken of new communications technologies as providing the conditions for a collective intelligence, whether in organizations (Fisher and Fisher 1998; Smith 1994), on a societal level (Hayek 1948, pp. 50–54), or globally (Levy 1997; Rossman 1992; Wells 1938). But just as obviously, technology does not provide all the necessary conditions. Interest-group politics notoriously suffer from free-ridership (Olson 1965): group members who do not participate in developing an intelligent group consensus will nonetheless benefit from it.

Collective cognition requires a shared identity, social skills, and morale, each of which has conditions of its own. What is more, technology has also raised the stakes by facilitating the explosion of "information-driven politics" (Greider 1992, p. 46) that has been accelerating since its origins in the open-government movements of the 1970s.

The purpose of "think tanks" is precisely to generate the steady stream of convenient facts, persuasive phrases, and finely tuned ideologies that assemble coalitions around the agendas of their paying supporters. These organizations expose the great complexity buried beneath simple concepts such as the "marketplace of ideas" (e.g., Baker 1989; Ingber 1984). Ideas are public goods, and I have already mentioned the role played by scholarly communities in alleviating the economic pathologies that public goods raise. But the marketplace of ideas is strange in another way: ideas in the public sphere are useful to me not because I "buy" them but because other people do. And that is the role of the think tank—selling one group's ideas to others. Every group has an interest in influencing the thinking of every other group, for example, through the public relations practice of providing "information subsidies" to the media (Gandy 1982), and so it can be extremely difficult for a social group to conduct its collective cognition autonomously (Habermas 1987).

The problem is partially one of scale: a social group whose members are few in number but command great resources can organize its institutions of collective cognition more easily, other things being equal, than a group whose members are more numerous and less wealthy. Larger groups are easier to infiltrate and thus provide easier targets of surveillance, and several public relations firms now routinely monitor public Internet discussions, among other popular communications channels, on behalf of their clients (see, for example, ⟨http://www.ewatch.com⟩).

The problem of autonomy arises on the most basic level when provocateurs set about disrupting a community's cognitive institutions, and some online communities have developed sophisticated methods for maintaining their boundaries in the face of such attacks (Phillips 1996). New technologies can support the development of autonomous processes of collective cognition if they provide social groups with the tools to minimize or at least equalize these dangers.

Finally, digital libraries bear on the relationship between the professions and the rest of society. This has been a crucial issue for democracy since the days of Walter Lippmann (1922) and John Dewey (1927). Although opinions differed on the extent of formal political power that should be invested in the experts, the elite consensus of that era was that, nonetheless, democracy should concede a great deal of cognitive authority to professions and their expertise (Schudson 1998, pp. 211–219). Subsequent experience, however, has made clear that democracy requires an irreducible creative tension between professional and popular voices. Digital libraries will presumably continue to facilitate the production and authorization of professional knowledge, but they also may enable nonprofessionals to appropriate this knowledge in their own ways (cf. Blau 1999, pp. 125–127). Once again, the conditions are largely institutional now that it is technically possible to make professional knowledge accessible to the public; new incentives might be useful to encourage professionals to make professional knowledge accessible in a fully effective sense. Digital libraries should also be evaluated for their capacity to support forms of collective cognition that differ from those of the traditional organized professions.

Institutional Embedding

The previous sections of this chapter have sketched a few of the ways in which a digital library might fit, or else fail to fit, into the institutional world around it. The discussion is necessarily schematic, and it will not be possible to offer any meaningful generalizations until digital libraries are being used on a large scale.

Nonetheless, some general patterns can be anticipated. Most fundamentally, the design of digital libraries will require a dynamic approach—neither ignoring the institutional context nor trying to legislate it but participating in the dialectical interaction between technology and institutions.

Institutional processes shape technologies, and the technologies that result are then appropriated by the institutions' members. Experience with these appropriations helps to shape new generations of technology, which are appropriated in turn. These appropriations are famously unpredictable, but they can in fact be predicted to a certain degree. Given an analysis of the existing forces in a given institutional field, one can safely say that those forces will shape the community's understandings of the technology and its potential uses.

The dialectical interaction between institutions and technology does not happen in isolation but is increasingly mediated by the global dynamics of technical standards (David and Shurmer 1996; Kahin and Abbate 1995). In the 1970s, much software was produced by organizations for their own use, but in the 1990s, the inherent economics of software created tremendous forces away from these dedicated applications and toward packaged software so that the enormous software development costs could be spread across many different customers. As a result, few organizations determine their own fate.

Even a whole institutional field, such as libraries or schools, can be held hostage to global standards that emerge and develop a critical mass of users in other sectors. It is easy to speak of the design of digital libraries as if designers can freely choose their own directions, but in practice, digital libraries emerge through negotiations in a tremendous variety of standards coalitions. Some of these coalitions are specific to libraries as an institutional field, but most are not. It follows that digital libraries can be designed intelligently only if their stakeholders join these negotiations (Oddy 1997, p. 83).

Questions of institutional fit also arise in other design contexts, of course, and the substance of these standards negotiations will often pertain, explicitly or not, to fundamental ideas about institutions and the social relationships that they define. Despite their esoteric reputation, standards can very easily embody substantive commitments that shape and constrain people's activities (Reidenberg 1998), and they can bias a playing field toward some players and away from others (Mansell 1995).

Technology and institutions interact especially in regard to issues of centralization and decentralization. The library community has already gone a long way toward eliminating duplicate efforts by pooling catalog records, and that experience can serve as a template for future issues of digital library governance. Centralization is also fundamental to the establishment of compatibility standards inasmuch as standards require consensus that must usually be coordinated through some central body. In many cases, centralized power is required to create the incentives for compatibility, but compatibility then creates the conditions for power to be decentralized.

An institution field can easily become "stuck" with an overly centralized concentration of power, but it can just as easily become stuck at the opposite extreme when sufficient consensus cannot be established to adopt and implement new standards. These governance challenges are great enough when standards change slowly and in isolation from one another, but they become crucial when a large number of standards are being developed and adopted simultaneously, as they are right now. In the worst case, the direction of digital library standards setting could be captured by a single interest—for example, a software vendor who can leverage a standard operating system or a coalition of intellectual property owners who can leverage their contractual control over digital library content. The potential for monopoly rent extraction in that scenario is enormous, and so libraries will have to learn how to maintain their boundaries against such effects.

Digital libraries also face strong centrifugal forces. I have already mentioned the great diversity of institutional fields with which libraries interact, and each of these institutional fields is likely to have developed its standards and practices in relative isolation from the others. The technologies and policies of a digital library can be deeply integrated with any one of those neighboring fields or with a few, but it will be hard to integrate with many of them. If the design of digital libraries is biased by the needs of a small number of powerful user groups (experts, for example, as opposed to lay persons), then they might discriminate against others. Or they may simply be pulled to pieces, with different digital libraries heading in different directions without being interoperable with one another.

Managing these tensions will be a great institutional challenge. Digital libraries may also become the terrain on which diverse institutions negotiate a common set of standards that facilitate activity in each area without artificial constraint.

Conclusion

This chapter has sketched some of the institutional problems with which the development of digital libraries must contend. It has also made clear that librarians are far from being automated into nonexistence by new technology and retain a considerable role in ensuring that libraries continue to encourage these values (cf. Nardi and O'Day 1996). This role is centrally one of design, not the command-and-control style of design from which computers first emerged but a participatory style in which the well-being of social institutions and their participants cannot be separated from the construction of technical systems. This new style of design thus involves leadership skills of a high order. But it also involves analytical skills, and social theory has a role to play in the practical work of designing digital libraries that can be truly useful in a complicated world.

Among the contributions of social theory has been a clear sense in which a library, even when it is digital, is still a place—the place where a scholarly community or a social movement can conduct its collective cognition with a reasonable degree of autonomy. We still know little about the construction of such places, but perhaps we can renew our appreciation of the need for them.

Acknowledgments

I appreciate helpful comments by Nancy Van House and Jeremy Roschelle.

References

Agre, P. E. 1995. Institutional Circuitry: Thinking about the Forms and Uses of Information. *Information Technology and Libraries*, 14(4), 225–230.

Agre, P. E. 1998a. Designing Genres for New Media. In S. Jones, ed., *CyberSociety 2.0: Revisiting Computer-Mediated Communication and Community*. Thousand Oaks, CA: Sage.

Agre, P. E. 1998b. The Internet and Public Discourse. *First Monday*, 3(3). ⟨http://www.firstmonday.dk/issues/issue3_3/agre/index.html⟩.

Agre, P. E., and M. Rotenberg, eds. 1997. *Technology and Privacy: The New Landscape*. Cambridge, MA: MIT Press.

Avgerou, C. 2002. *Information Systems and Global Diversity*. Oxford: Oxford University Press.

Baker, C. E. 1989. *Human Liberty and Freedom of Speech*. New York: Oxford University Press.

Berg, M. 1999. Accumulating and Coordinating: Occasions for Information Technologies in Medical Work. *Computer Supported Cooperative Work*, 8(4), 373–401.

Blau, A. 1999. Floods Don't Build Bridges: Rich Networks, Poor Citizens and the Role of Public Libraries. In S. Criddle, L. Dempsey, and R. Heseltine, eds., *Information Landscapes for a Learning Society*. London: Library Association.

Borgman, C. L. 1999. What Are Digital Libraries? Competing Visions. *Information Processing and Management* (special issue on Progress toward Digital Libraries), 35(3), 227–243.

Borgman, C. L. 2000. *From Gutenberg to the Global Information Infrastructure Access to Information in the Networked World.* Cambridge, MA: MIT Press.

Buchstein, H. 1997. Bytes That Bite: The Internet and Deliberative Democracy. *Constellations*, 4(2), 248–263.

Casson, M. 1997. Institutional Economics and Business History: A Way Forward? *Business History*, 39(4), 151–171.

Collins, R. 1998. *The Sociology of Philosophies: A Global Theory of Intellectual Change.* Cambridge, MA: Harvard University Press.

Commons, J. R. 1924. *Legal Foundations of Capitalism.* New York: Macmillan.

Crane, D. 1972. *Invisible Colleges: Diffusion of Knowledge in Scientific Communities.* Chicago: University of Chicago Press.

David, P. A., and M. Shurmer. 1996. Formal Standards-Setting for Global Telecommunications and Information Services. *Telecommunications Policy*, 20(10), 789–815.

Dewey, J. 1927. *The Public and Its Problems.* New York: Holt.

Dolence, M. G., and D. M. Norris. 1995. *Transforming Higher Education: A Vision for Learning in the Twenty-first Century.* Ann Arbor, MI: Society for College and University Planning.

Dutton, W. H. 1999. *Society on the Line: Information Politics in the Digital Age.* Oxford: Oxford University Press.

Etzioni, A. 1995. *New Communitarian Thinking: Persons, Virtues, Institutions, and Communities.* Charlottesville: University Press of Virginia.

Fisher, K., and M. D. Fisher. 1998. *The Distributed Mind: Achieving High Performance through the Collective Intelligence of Knowledge Work Teams.* New York: AMACOM.

Fraser, N. 1992. Rethinking the Public Sphere: A Contribution to the Critique of Actually Existing Democracy. In C. Calhoun, ed., *Habermas and the Public Sphere* (pp. 109–143). Cambridge, MA: MIT Press.

Friedman, A. L. 1989. *Computer Systems Development History, Organization and Implementation.* Chichester, UK: Wiley.

Friedman, B., and H. Nissenbaum. 1996. Bias in Computer Systems. *ACM Transactions on Information Systems*, 14(3), 330–347.

Gandy, O. H., Jr. 1982. *Beyond Agenda Setting: Information Subsidies and Public Policy.* Norwood, NJ: Ablex.

Gandy, O. H., Jr. 1993. *The Panoptic Sort: A Political Economy of Personal Information.* Boulder: Westview Press.

Gilder, G. 1992. *Life after Television.* New York: Norton.

Goodin, R. E., ed. 1996. *The Theory of Institutional Design.* Cambridge, MA: Cambridge University Press.

Granovetter, M. 1992. Economic Action and Social Structure: The Problem of Embeddedness. In M. Granovetter and R. Swedberg, eds., *The Sociology of Economic Life* (pp. 53–81). Boulder, CO: Westview.

Greider, W. 1992. *Who Will Tell the People? The Betrayal of American Democracy*. New York: Simon and Schuster.

Habermas, J. 1987. *The Philosophical Discourse of Modernity: Twelve Lectures*. Translated by Frederick Lawrence. Cambridge, MA: MIT Press.

Hallgren, M. M., and A. K. McAdams. 1997. The Economic Efficiency of Internet Public Goods. In L. W. McKnight and J. P. Bailey, eds., *Internet Economics* (pp. 455–478). Cambridge, MA: MIT Press.

Halper, M. 1998. Middlemania. *Business 2.0*, 3(7), 45–60.

Hayek, F. A. 1948. *Individualism and Economic Order*. Chicago: University of Chicago Press.

Hodgson, G. M. 1988. *Economics and Institutions: A Manifesto for a Modern Institutional Economics*. Cambridge: Polity Press.

Hutchins, E. 1996. *Cognition in the Wild*. Cambridge, MA: MIT Press.

Ingber, S. 1984. The Marketplace of Ideas: A Legitimizing Myth. *Duke Law Journal*, 1984(1), 1–91.

Kahin, B., and J. Abbate, eds. 1995. *Standards Policy for Information Infrastructure*. Cambridge, MA: MIT Press.

Kling, R., ed. 1996. *Computerization and Controversy: Value Conflicts and Social Choices*, 2d ed. San Diego: Academic Press.

Kling, R., and M. S. Elliott. 1994. Digital Library Design for Usability. In J. L. Schnase, J. J. Leggett, R. K. Futura, and T. Metcalfe, eds., *Proceedings of Digital Libraries '94: The First Annual Conference on the Theory and Practice of Digital Libraries (College Station, TX, June 1994)* (pp. 146–155). St. Louis, MO: Hassan.

Kling, R., and S. Iacono. 1988. The Mobilization of Support for Computerization: The Role of Computerization Movements. *Social Problems*, 35(3), 226–243.

Kling, R., and S. Iacono. 1989. The Institutional Character of Computerized Information Systems. *Office Technology and People*, 5(1), 7–28.

Lamb, R. 1995. Using Online Information Resources: Reaching for the *.*'s. In F. M. Shipman, R. K. Futura, and D. M. Levy, eds., *Proceedings of Digital Libraries '95: The Second Annual Conference on the Theory and Practice of Digital Libraries (Austin, TX, June 1995)* (pp. 137–146). College Station: Texas A&M University Press.

Latour, B. 1993. *We Have Never Been Modern*, trans. C. Porter. Cambridge, MA: Harvard University Press.

Latour, B., and S. Woolgar. 1986. *Laboratory Life: The Construction of Scientific Facts*. Princeton: Princeton University Press. (Originally published in 1979)

Levy, P. 1997. *Collective Intelligence: Mankind's Emerging World in Cyberspace*. Translated by R. Bononno. New York: Plenum Press.

Lippmann, W. 1922. *Public Opinion*. New York: Macmillan.

Mansell, R. 1995. Standards, Industrial Policy and Innovation. In R. Hawkins, R. Mansell, and J. Skea, eds., *Standards, Innovation and Competitiveness: The Politics and Economics of Standards in Natural and Technical Environments* (pp. 228–235). Aldershot, UK: Elgar.

Mansell, R., and W. E. Steinmueller. 2000. *Mobilizing the Information Society: Strategies for Growth and Opportunity*. Oxford: Oxford University Press.

March, J. G., and J. P. Olsen. 1989. *Rediscovering Institutions: The Organizational Basis of Politics*. New York: Free Press.

Melucci, A. 1996. *Challenging Codes: Collective Action in the Information Age*. Cambridge: Cambridge University Press.

Nardi, B., and V. L. O'Day. 1996. Intelligent Agents: What We Learned at the Library. *Libri*, 46(2), 59–88.

Negroponte, N. 1997. Negroponte, Reintermediated. *Wired*, 5(9), 208.

North, D. C. 1990. *Institutions, Institutional Change, and Economic Performance*. Cambridge: Cambridge University Press.

Oddy, P. 1997. *Future Libraries, Future Catalogues*. London: Library Association.

Olson, M., Jr. 1965. *The Logic of Collective Action: Public Goods and the Theory of Groups*. Cambridge, MA: Harvard University Press.

O'Neil, P. H. 1998. *Revolution from Within: The Hungarian Socialist Workers' Party and the Collapse of Communism*. Northampton, MA: Elgar.

Orlikowski, W. J. 1993. Learning from Notes: Organizational Issues in Groupware Implementation. *Information Society*, 9(3), 237–250.

Phillips, D. J. 1996. Defending the Boundaries: Identifying and Countering Threats in a Usenet Newsgroup. *Information Society*, 12(1), 39–62.

Powell, W. W., and P. J. DiMaggio, eds. 1991. *The New Institutionalism in Organizational Analysis*. Chicago: University of Chicago Press.

Reidenberg, J. 1998. Lex Informatica: The Formulation of Information Policy Rules through Technology. *Texas Law Review*, 76, 553–593.

Rossman, P. 1992. *The Emerging Worldwide Electronic University: Information Age Global Higher Education*. Westport, CT: Greenwood Press.

Sarkar, M. B., B. Butler, and C. Steinfield. 1995. Intermediaries and Cybermediaries: A Continuing Role for Mediating Players in the Electronic Marketplace. *Journal of Computer-Mediated Communication*, 1(3). ⟨http://www.ascusc.org/jcmc/vol1/issue3/sarkar.html⟩.

Schudson, M. 1998. *The Good Citizen: A History of American Civic Life*. New York: Free Press.

Shapiro, A. L. 1999. *The Control Revolution: How the Internet Is Putting Individuals in Charge and Changing the World We Know*. New York: Public Affairs.

Shneiderman, B. 1998. *Designing the User Interface: Strategies for Effective Human-Computer Interaction* (3rd ed.). Reading, MA: Addison Wesley.

Smith, J. B. 1994. *Collective Intelligence in Computer-Based Collaboration*. Hillsdale, NJ: Erlbaum.

Toulmin, S. 1972. *Human Understanding: The Collective Use and Evolution of Concepts*. Princeton, NJ: Princeton University Press.

Volosinov, V. N. 1973. *Marxism and the Philosophy of Language*. Translated by L. Matejka and I. R. Titunik. Cambridge, MA: Harvard University Press.

Weick, K. E., and K. H. Roberts. 1993. Collective Mind in Organizations: Heedful Interrelating on Flight Decks. *Administrative Science Quarterly*, 38, 357–381.

Wells, H. G. 1938. *World Brain*. Garden City, NY: Doubleday, Doran.

10

Transparency beyond the Individual Level of Scale: Convergence between Information Artifacts and Communities of Practice

Susan Leigh Star, Geoffrey C. Bowker, and Laura J. Neumann

Introduction

Interviewer: How do you keep up with your field?

Respondent (engineering professor of forty years): ... [T]he field that I work in—I have been in it for so long and I have trained a lot of the people in it. I know practically everybody in the world who is working in it. I am also on the editorial board of a number of journals, so in a lot of cases I see things before they are even in print. In my case, I probably make very good usage of both of those things. I am in correspondence with a lot of people, and they tell me what they are doing, just because I know them.... And I go to meetings, to quite a few. This is where I meet my colleagues, and we talk. I probably am almost never in the situation where I am having to do a search in an area where I don't know anything about the field. I may never.

Problems that undergraduate students have with the library system (from a focus group with undergraduates)

- Journals are scattered all over campus.
- The article I need is always ripped out.
- I don't know how to phrase things just right, so I don't get what I want.
- The professors won't let us write papers from the abstracts alone.
- I'd take local, irrelevant articles over remote, relevant ones.

Not all the work that has made the ICD more applicable has been done internally through modifications to the list. Indeed, one background factor that has had a great impact has been the convergence of international bureaucracy. What we mean by this is that throughout this century in general people have become more and more used to being counted and classified. (Bowker and Star 1994, p. 202)

As information systems are used by more people and permeate more of our working and leisure lives, scholars studying the human side of computing must scale up various concepts traditionally seen as individual or psychological. In a general sense, this problem has been troubling information scientists for years. For example, in discussing how the field of human-computer interaction (HCI) has changed,

Jonathan Grudin (1990) notes that it has gone from individual (screen) to groups (groupware) and now logically must seek to extend beyond the group into the wider social sphere. When large groups of people are using a widely distributed system, old notions of "one person, one terminal" as the basis for design are inadequate. Liam Bannon (1990) made a similar point in discussing his "pilgrim's progress" away from the individualized, cognitivist notions of HCI. Rob Kling and Walt Scacchi's (1982) classic notion of the "web of computing" sought to expand beyond the single-user terminal and formal program stereotype to locate design in a larger context of usability, the workplace, and its social networks. John King and Susan Leigh Star (1990) apply the scale-up problem to decision support systems (DSSs), noting that the scale-up from decision support to group decision support system (GDSS) to organizational decision support system (ODSS) is complex.

The model is no longer about one person or a small team optimizing decisions in a controlled environment. At the organizational level, deciding includes issues of social justice, multiple interpretations, and adjudication of conflict across social boundaries. Within the field of computer-supported cooperative work (CSCW) and the design of large-scale information systems such as digital libraries (DLs), such social scaling questions are critical. Steve Griffin, program director for the National Science Foundation (NSF) Digital Libraries Initiative (DLI), in announcing a second round of large digital library grants, noted: "The focus is shifting very much from the technology to the content, the users, and the usability of the content" (Kiernan 1998, p. 427). What will be the proper unit of analysis for design of such digital libraries—group, organization, community, network? What does a community interface or system even look like? With the widespread use of the World Wide Web and the growth of digital libraries, these problems become more urgent.

In this discussion, we focus on just one aspect of the social side of scaling up in information systems—how it affects the concept of *transparency*. Transparency for one user means that he or she does not have to be bothered with the underlying machinery or software. In this sense, an automobile is transparent when a driver sits down, turns the key, and drives off, all without the foggiest notion of how internal combustion works.[1] As we move up from a single user to a community, transparency becomes more complex. At larger scales it means that we must have an idea about these issues:

· For whom and when and where is a particular tool transparent?

· What happens when degrees of transparency are different for various subgroups of users?

• How does something become invisibly usable at larger levels of scale, and what differences are required in process and design content?

• How are newcomers taught to make the tool, interface, or retrieval system transparent for themselves?

Transparency and ease of use for groups are products of a shifting alignment of information resources and social practices. It is never universally or permanently achieved but rather requires staying on top of changing politics, knowledge, material resources, and infrastructure.

One impetus for clarifying a scaled-up idea of transparency comes from changes in the information landscape itself. In the current wave of building and research on digital libraries and information infrastructure, we find places and possibilities where new kinds of arrangements are being forged for information organization and access. Pockets that have been semiautonomous until now are getting interconnected, including widespread federations of information repositories. One such example is the Illinois Digital Library Initiative. One of the goals of this project is to create seamless interconnection of federated repositories. Also, traditional models of information retrieval are being turned completely upside down; they can no longer presume an individual user, an information intermediary such as a librarian, and a repository with discrete documents that are then returned to the workplace or home. Information documents and multimedia are being fractured and distributed in different ways across the work- and playscape. Thus a fuller understanding of people, their activities, and the scaling of transparency is especially crucial for a workable digital library but applies as well to other complex and large-scale information systems (Bishop and Star 1996).

Transparency Results from the Convergence of Communities of Practice and Information Artifacts

Transparency at larger levels of scale is partly achieved through the convergence of knowledge and resources across groups of users. A *community of practice*[2] is a group of people joined by conventions, language, practices, and technologies (Lave and Wenger 1992). It may or may not be contained in a single spatial territory; in the modern information world, it often is not. It contains strong ties that are not covered by the terms *family*, *formal organization*, or *voluntary association*. For instance, the community of practice of chief financial officers (CFOs) is distributed across many organizations; each organization has only one CFO, but members of the community of CFOs have much in common through their jobs. Other examples of communities of practice are stamp collectors, rock climbers, activity theorists, and socialist feminists, all of which have discussion groups on the Web. Communities of practice may divide into subgroups or have intersecting and partially overlapping concerns; the degree to which they engage members or are voluntary is variable (Star 1994). An *information artifact* means any of a wide array of tools, systems, interfaces, and devices for storing, tracking, displaying, and retrieving information, whether paper, electronic, or other material.

Communities of practice and information artifacts converge when use and practice fit design and access. While we speak of them separately here, in practice they are difficult to pull apart. In a sense, information artifacts and communities of practice are fitted to each other and even come to define one another over time. That is to say, the sharing of information resources and tools is a dimension of any coherent social organization—whether the community of homeless people in Los Angeles sharing survival knowledge via street gossip or the community of high-energy physicists sharing electronic preprints via the Los Alamos archive.

Any given community of practice generates many interlinked information artifacts, which may be loosely coupled and sometimes indirectly relevant to the purpose of the community. People without houses log onto the Internet, and physicists indulge in street gossip at conferences—as well as engage in a whole set of other information practices (Palmer 1996). Put briefly, information artifacts undergird communities of practice, and communities of practice generate and depend on these same information resources. *Convergence* is a term for this process of mutual constitution.

Convergence has been well described in the work of Elfreda Chatman (1985, 1991, 1992) in her coining of the useful holistic term *information world*. An information world is the collection of information resources employed by an individual, organization, institution, or other group to solve problems, learn, play, and work. Information worlds may be as formal as libraries and databases, but they also may be informal, including the opinions of family and friends on health matters or books and films. Such resources are also material in the sense that they are enabled and constrained by time, transportation, and material cultural practices. Information artifacts and communities of practice form information worlds in Chatman's sense.

Through three examples, we will explore how convergence occurs. First, we discuss how convergence is experienced as a result of membership by academic researchers. Second, we examine a study of nursing classification wherein nursing researchers are making their classifications of nursing activities converge with professional clinical practice. Finally, we describe the convergence that subtends a global well-established information collection and analysis system administered by the World Health Organization—the *International Classification of Diseases* (ICD). This convergence goes beyond a single occupation, nationality, or locale; in fact, it consists in the end of a great number of intersecting bureaucracies and information worlds (Bowker 1994).

Through each case, we examine the operation of key feedback processes that underwrite the fitting-together of social worlds and information artifacts. We then discuss how the concept of convergence can be of use in developing the integration of sociological and informational analysis toward which much recent work in library and information science has pointed.

Convergence and Membership

Interviewer: How do you find information or references?
Respondent (professor of twelve years):   Often a paper comes across my desk. It is an archaeological dig here ... simply because I will review articles or journal submissions, and something seems interesting, and I will file it. Or something seems interesting in one of the publications I subscribe to.

Sometimes people walk into a library or other information system and can put their fingers on things. They seem to sit at the metaphorical center of the social web and can tweak a vast complex and mature system of social networks to "get stuff." The system is quite transparent for these people. At the other end of the spectrum are the people who see the information system as confusing, chaotic, insurmountable, and unusable. They try to follow given directions and miss by a mile and a half. Much professional socialization concerns moving from this lost state into the state of obviousness or naturalness. Analysis at this individual level asks how people adapt to and work around formal structures when they don't fit, how they put together and maintain networks that "feed" their work, and how they understand the interaction between vernacular and formal systems of information and classification. For example, professors often point to articles they review or items colleagues send them as more important information resources than the library.

Convergence for the apprentice professional is a trajectory—the shaping of individuals so that they see themselves as having the set of information needs that can be met by their new social world's information resources. As individuals move along this trajectory, they learn a field that they can "put their mark" on and are changed by their socialization so that their original "information needs" are not the same as their final ones. For individuals, this feedback process, especially the ways in which their own information needs change, is frequently invisible. People often see themselves at the end of an academic question as asking the same "big dumb question" (Linde 1997) that they asked going in, without recognizing that the constitution of the question has thoroughly changed through its adaptation to their relevant information and social worlds.

Thirty-eight research scientists and students were interviewed for a large information infrastructure-building project over a three-year period. The goal of these individual interviews and focus groups was to learn more about the information-finding behaviors of research scientists and students, including their use of information systems. In line with established research findings in user studies of information retrieval, most of the senior-level respondents did not often use formalized information systems (Garvey and Griffith 1980; Pinelli 1991; Taylor 1991). These researchers acquired the information they needed through an entire suite of other information-seeking activities, such that they rarely found themselves going out and searching for anything. It was often difficult for these people to describe how they had access to items that they needed; these were simply ready-at-hand, part of their immediate transparent environment.

The means for making this information readily available is related to a person's location at a particular point in the academic trajectory. It involves insider membership status in an academic community—having been through the apprenticeship process. Full participating members have access to needed resources, yet membership must be achieved before that access is granted. When asked to name the most fruitful resource in his information world, a professor of fourteen years noted, "the conferences, talking to colleagues, conference papers and proceedings—those would probably be most valuable. But then that means you have already mastered a particular area. In a way, it is a chicken or the egg question."

In this process of becoming a member, one becomes attached to the installed base and extant infrastructure of academia. For example, membership in academia requires conference attendance, discussions with colleagues, and membership in various professional organizations. Professors publish in certain key journals and track career paths through key institutions. In our interviews with digital library users, professors had an already taken-for-granted map of the field, including acceptable methods and vocabulary. As one professor of twenty years put it, "I don't do much research because I have to constantly keep up. I have a pretty good idea of what is out there." Interviewees were supported in this by a strong infrastructure: library services copy and send on needed articles; online archives send automatic reminders of new work in the field. Professors reviewed and wrote articles and often became journal editors. Further along the career trajectory, they participated in professional governing bodies or perhaps joined a funding agency that shaped future research trends.

In the process of establishing this kind of senior membership, channels were established along which information flowed. The workings of the profession into which academics enter facilitate their participation in a community of exchange of information on both formal and informal levels. In most ways, knowing many members of a research community and talking to them frequently are more important than reading professional journals (Crane 1972; Garvey and Griffith 1980; Wolek and Griffith 1980). Many respondents emphasized that the idea is to know what will appear in the journals *before* they are released.

Throughout their careers as students and in the process of becoming professionals, researchers focus their interests, which are represented in collegial networks, in the piles and files of papers kept, in research groups, and in journal publications. One researcher (a professor of nine years) noted that her most important resources began with "subscribing to a lot of journals myself, and I am a member societies, so I subscribe to a lot of stuff through those, and I get their conference proceedings. And I also buy a lot of books.... So first of all I get a lot of things just by myself." Thus people set up the means, artifacts, and processes that were necessary for social and information convergence. The further a professional travels in the career line and into the specialized reaches of an academic specialty, the more intense the convergence becomes. Bruno Latour (1987, p. 152) notes the somewhat counterintuitive aspect of this: "There is a direct relationship between the size of the outside recruitment of resources and the amount of work that can be done on the inside.... an isolated specialist is a contradiction in terms. Either you are isolated and very quickly stop being a specialist, or you remain a specialist, but this means you are not isolated."

This reinforcement of the knowledge webs of many researchers can also be their Achilles' heel. One researcher pointed out that she did indeed rely almost entirely on her network of contacts and subscriptions, and if those resources had holes in them, then she would probably never know it. Reliance can become total (Swanson 1986). In this way, too tight a convergence becomes overdetermination; it becomes a rut that cannot be altered. Overly tight coupling between a social world and its information artifacts can be seen as a powerful force that closes off other possibilities of finding information or using the imagination because they are not part of the routine. Marcia J. Bates (1979) describes how these overtrodden paths of thinking can be redirected specifically for finding information. Donald T. Campbell's (1969) essay on the "fish-scale" approach to interdisciplinary knowledge also intuits the value of partial overlaps and a looser coupling between social worlds and information.

For the individual member, while convergence can produce transparency and ease of use, the opposite may also occur. This frequently is the experience of newcomers to a social world or of those who do not meet membership criteria for the formal information systems. This results in fracture and frustration—a plight often felt by undergraduates who do not have a clear academic home, may not have a "major" field of study, and do not know what the pertinent journals are, who the main people are, or even what language to use to learn these things. As we were told in focus groups about undergraduates' information needs, "I just need five references. Any five references. I want to get those as quickly and as painlessly as possible." It does not matter who wrote the article, when, or why. This situation is typical of many undergraduates' relationship to the field of study. Those markers and the community gestalt that define quality, relevance, and usefulness are functionally absent and thus irrelevant. If, and as, students become members of a particular community through participation, these community standards and needs become their own and eventually transparent.

As Lave and Wenger (1992) describe, "learning the ropes" of a new community requires *legitimate peripheral participation*. We learn through gradually being made a fully participating member of a community, not through information stripped of context. The ideal academic system is structured to provide that space for legitimate peripheral participation—to cause a convergence between the student's developing academic, social, and artifactual worlds and available information artifacts. In practice, of course, as Lave and Wenger (1992) and Penelope Eckert (1989) point out, the ideal is often not realized. Learning may be treated as merely information acquisition, as if students were not juggling multiple complex memberships. The trajectory to convergence, when traveled, is rarely smooth.

We now turn to a second example of convergence from the point of view of a community of nurses. In this example, we see the same feedback process occurring through which the nursing profession, by attempting to integrate into the professional world of medicine, is shaping and being shaped by the information world it is becoming part of.

Community and Convergence: Crafting and Aligning Information Systems with Community Practices

I attended the American Nursing Association (ANA) Database Steering Committee in June, and a recommendation coming out of that meeting (and I'm assuming making its way through the ANA channels) is that ANA needs to put forth a major lobbying effort toward nursing representation at the National Committee on Health and Vital Statistics (NCHVS). The ANA has spent too many efforts on Current Procedural Terminology (CPT) [Used for encoding billable procedures] when it is clear (at least to me) that NCHVS is where the lobbying efforts should be focused so that nursing procedures can eventually be included in the procedures section. Nursing has not been able to crack the inner circle there. (participant in a discussion on the Nursing Interventions Classification Listserv, November 18, 1996)

In this section, we examine how a community of practice attempts to create convergence between information systems and the classification of work and to embody that convergence in accounting systems, research protocols, and clinical practice and knowledge.

In the Nursing Interventions Classification (NIC), nursing administrators are attempting to produce a standard list of everything that nurses do in the course of their working day—from changing a bedpan to giving an injection to telling a joke (McCloskey and Bulechek 1996). Why do they need a classification system? NIC's developers argue that nursing has traditionally been so idiosyncratic that no portable body of knowledge has been developed (cf. Turnbull 1997 and Latour 1987, on transporting local knowledge).[3]

Nursing researchers argue that they should be represented in (but are currently absent from) integrated hospital information systems. They need this for two reasons: to scientifically demonstrate the value of nursing within hospitals so that it can receive proper funding and recognition and to develop nursing knowledge: "Documentation of care with standardized classifications allows for the integration of research and practice" (McCloskey 1996, p. 14). Each of these goals—institutional and scientific—aims to describe the way things are; each in practice also involves the operation of a principle of convergence. Achieving this involves fundamental changes in accounting for nursing practice. In this example, the information needs of the nurses are fitted to the professional worlds they are trying to integrate with, at the same time as these worlds are themselves modifying information systems to represent nursing. At stake is the creation of an information world where the practices of nurses can be transparently represented in both hospital accounting schemes and in comparative nursing research.

NIC brings many threads of nursing practice—practitioners, specialists, educators, and students—together into one legitimated form of representation. It will no longer be enough, argue nursing informaticians, for nurses to act locally within the context of their home institutions. To help build a body of nursing knowledge, they will have to act as other nurses act in other institutions. If NIC wins the day, nursing practice will be modified both to address and record NIC interventions. This is further supported by the inclusion of NIC in nursing classroom curricula. Thus the information tool, the classification system, will change both practice in the field and training: the users will be disciplined to both act and represent their actions in NIC form.

The language of NIC will be shaped to fit with the language of other sciences (Kritek 1988, p. 25):

The most frequently focused upon language is that developed by medicine and the numerous natural or physical sciences that contribute to medicine's scientific knowledge base. This language is focused upon most frequently by nurses, because of medicine's long history of dominance in nursing, its relative specificity about physiological phenomena, its apparent measurability, and its familiarity. It is also valued for its social status as a scientific language, used by a politically powerful professional group, physicians.

This has meant that nurses have, in building their information infrastructure, emulated the successful research apparatus of other disciplines. Through the classification, nursing work is deliberately made subject to scientific testing. The informational goal is to create manipulable data by ensuring comparability between nursing interventions carried out at different local sites. The institutional goal is to make nursing "look and feel" like other sciences, especially medicine, to tap the same set of accountability and accounting information practices that the other sciences have so successfully developed. They are working to make their classification align with medical informatics and medical standardized languages by making alliances and integrating themselves in other formal medical language systems—such as the Unified Medical Language System (UMLS) and the National Library of Medicine's Medical Subject Headings (MeSH).

The nurses who are promoting NIC clearly see that their classification must become compatible with the rest of medical informatics at this higher level. They argue that NIC must also be accepted by doctors and other health-care professionals. They have in front of them the specter of a failed classification of nursing diagnoses—North American Nursing Diagnosis Association (NANDA): "When other professionals hear us use NANDA terms, they about laugh us out of the hall because some of them [NANDA terms] don't seem to be directly related to what we are talking about at all!" (personal communication with NIC developer 1993). NANDA, however, was developed without the feedback from the wider setting that characterizes convergence.

Nursing interventions need to be compatible with both medical and pharmaceutical interventions (McCormick 1988, p. 176):

Without a unified language system that is integrated into other health care standard initiatives, the nursing profession will not be able to use the standard language developed in collaborative health care initiatives. Nor will the unified language be capable of being integrated with other developments in clinical practice, reimbursement formulae, or case simulation models without a collaborative health care initiative.

In general, "It is not enough for nurses to accurately record what they do—they must record it in a language acceptable to physicians and administrators, and productive of a new, higher status." (Berg and Bowker 1997, p. 529)

The feedback process among information needs, classification schemes, and the wider professional communities with which nursing seeks to align, produces a trajectory whose end result is a nursing informatics that can both integrate with wider medical informatics and a nursing practice that will be sensibly altered to permit this integration. The convergence here is between information tracking systems and the community of practice of nurses—a new kind of nursing information world, more broadly defined and strategically situated.

As in the case of the individual academic discussed above, then, professional communities also find themselves struggling with transparency. Here the issue is not so much one of membership but one of aligning information systems and practice in such a way that the emerging information world is strategically situated among other information worlds.

The fledgling academic must achieve transparency through learning and professional socialization. The professional community of practice as a whole here achieves its transparency through a series of translations and negotiations with its neighbors. In neither case is the picture entirely deterministic: both the individual and the community have the ability to reconfigure the world.

A curious forgetting of the learning and negotiating process often happens over time for both individual members and communities. The achievement of transparency becomes naturalized: this is just how the world is. For the mature researcher, it is easy to forget the barriers and blockages that the newcomer faces; as communities of practice converge with wider-scale information systems, the categories come to seem entirely natural rather than negotiated (Becker 1994; Bowker and Star 1994). In each case, the operation of the feedback process of convergence has underwritten the mutual fitting of a changed individual or community and a new information world.

In the following section, we examine what convergence looks like when communities of practice are stitched together and overlap in yet larger information systems. The *International Classification of Diseases*, tenth revision (ICD-10) was first developed over 100 years ago and today is used by such various groups as epidemiologists in national health institutions and individual health practitioners in small villages around the world.

Convergence and Infrastructure: Linking Multiple Communities of Practice with Wider-Scale Information Systems

Alain Desrosières (1988) has shown beautifully how census breakdowns of the populations of Germany, France, and England have remained closely tied to the history of work trade unions and government intervention in those countries. We suggest that as the ICD [*International Classification of Diseases*] "naturally" becomes more universally applicable, this is partly the result of the hidden spread of western values through the application of our own bureaucratic techniques. These techniques appear rational and general to us but when looked at in detail prove highly contingent. (Bowker and Star 1994, p. 202)

We now sketch how multiple overlapping information worlds are generated and supported through wider-scale convergence. At this overview level, we are concerned with the development of large-scale information systems that are shared by many communities of practice—for example, census data shared by economists, policy analysts, municipalities, and government agencies or biodiversity databases shared by environmentalists, biologists, community groups, and so forth. The same principle of mutual feedback between the information artifacts and the communities of practice is applied here. That is, the wider collection of information worlds changes in the same way as individual membership or professional practice changes in the process of developing an information world. Transparency at this level begins to merge with the very notion of infrastructure—something that is readily available across a wide scale of operation. This too is never a pure case, given the complexity of transparencies in such heterogeneous mergers (Star and Ruhleder 1996 discuss this at some length).

Convergence here fits together the representation of knowledge and the phenomena that are being represented through highly distributed knowledge systems. Consider, for example, job classifications in Europe: there are large differences, as Alain Desrosières (1988) points out, between *professional* in England and *cadre* in France.

Each classification system has been incorporated into a number of instruments of government such as the census and into a series of organizational decision—for example, the organization of trade unions. Each has spawned its own particular varieties of statistical knowledge: epidemiologists track "stress," for instance, by occupational category; sociologists structure their surveys using those same categories; government policy makers then adopt the findings to target specific groups found by sociologists.

In England and France, two different information worlds have historically reified two incommensurable subsets of the population, and these subsets have become ever more real and identifiable subsets as a result of this convergence. Is stress differently distributed in England and France? The question becomes virtually inseparable from the structuring of the information artifacts as well as from the practices of the different social worlds employing them. When they converge within nations, two different transparencies may emerge between countries (Desrosières 1988, 1993). If they are standardized across national boundaries, the transparency will extend further.

To explore the relationship between knowledge and infrastructure further, we will turn our attention briefly to the *International Classification of Diseases*, tenth revision (ICD-10) (WHO 1992). The ICD is one of a series of international classification systems that were developed in the late nineteenth century. These ranged from classifications of labor to freight packages to ears. The shape of ears, originally thought to be linked to types and degrees of criminality, became enshrined in the classification of people in the collection of vital statistics and immigration. Each scheme involved the creation of new international governing bodies together with the development of an information world in which those bodies could work. And many of the new classifications became strongly entrenched in information gathering practices. Bertillon's classification of ears, for example, is still the basis for the "three-quarter" pose displayed on U.S. immigration photographs (Tort 1989).

In the case of the ICD, the series of overlapping communities that needed access to the same information to coordinate their own professional activities included—but was by no means restricted to—the following:

• *Epidemiologists* who needed comparable data on diseases in different countries to discern patterns in outbreaks and then deduce causes;

• *Public health officials* who needed to be able to patrol the boundaries of the nation and keep out infectious immigrants as well as monitor health conditions at home;

- *Health insurance company employees*   who needed to file data on diseases using doctors for information-collection purposes and who needed to use a classification system that doctors would understand;
- *Census officials*   who needed to create comparable international data sets to serve the needs of the various professions that drew on the census.

No one of these communities had priority in the development of the ICD. To the contrary, the final classification as it has developed has entrenched flaws from every perspective. Let us consider two examples. First is the number of diseases that are coded. To get doctors or, in outlying areas, lay personnel to fill in death certificates using the code, there must be a restricted number of disease categories. This feature of medical practice is a blight for epidemiologists for whom a nuance in a particular classification might well make a large difference in the context of a specific inquiry. This case, however, displays the same principle of convergence that we observed above. Over the past 100 years, doctors have been trained in medical school in the use of the ICD throughout their practice. They use it in accounting forms, personal histories, and death certificates. They have modified their practices so that they can support the use of a developing ICD, which has grown consistently in scope and size and so become more epidemiologically useful.

Our second example is the use of the ICD in developing countries. People in developing countries frequently complain that the ICD is not relevant to their interests. In countries where infants usually die before they can be "accurately" diagnosed, policy makers and leaders question the purpose of propagating intricate diagnostic tracking systems (Bowker and Star 1994). They argue that the ICD does not represent their own diseases. Against this view runs Michel Foucault's (1991) observation that the ability of governments to count people, things, and diseases is a feature of recent "development." We have a relatively impoverished vocabulary for talking about natural accidents, and the ICD is richest in its description of ways of dying in developed countries at this moment in history. Other accidents and diseases can be described, but they cannot be described as well. Differentiating insect and snakebites, for example, is important for those living in the rural tropics. However, while arthropods, centipedes, and chiggers are singled out under "bites" in the ICD index, snakes are divided only into venomous and nonvenomous, as are spiders.[4]

This pragmatic classification makes sense to some extent: making fine distinctions between types of accident makes sense if those distinctions might make a difference in practice to some agency—medical or other. At the same time, those agencies have traditionally been more accountable to Western allopathic medicine and to the industrial world than to traditional systems.

As the ICD gains a representational foothold in developing countries, so also does Western medical practice, which itself is tied to the ICD. Thus, the ICD becomes the most apt descriptor of disease. The application of representational practice onto this single classification scheme is tied directly to the spread of allopathic medicine— and, where allopathic medicine is dominant, the ICD provides the best description of diseases.[5] A continuing problem for the World Health Organization (WHO) comes in trying to integrate traditional systems of medicine. For example, acupuncture has no category of disease per se but models imbalances of forces in the body. It is difficult to map the one onto the other in any simple sense without violating someone's ontology. The ICD comes as a package. It would be a mistake to see it as "just" a classification system that can be applied as is to developing countries: it marks the convergence of many communities of practice and an extended system of information artifacts, including death certificates, software, and the epidemiological record keeping of the WHO.

Thus the development of the ICD can be seen as the mutual interadjustment of multiple information artifacts and multiple communities of practice. Given the complexity of the processes involved, the resultant large-scale information infrastructure is not seamless—nor, we are careful to note, "right." However, it operates as efficiently and reliably as it does because of the continual feedback process of convergence. In the case of large-scale infrastructure, attachment to the installed base, affiliation with multiple communities, and compatibility with other systems of knowledge are necessary for success (Star and Ruhleder 1996). The ICD's success, as seen by its almost universal adaptation and use, is based as much on the general Westernization, industrialization, and bureaucratization of the world as on the classification itself (Bowker and Star 1994). The ICD is a central point in an international system of interlocking classifications and standardized medical languages as well as a key structuring agent in medical linguistics generally (Berg and Bowker 1997).

The Ontological Status of Convergence

We have looked briefly at three examples of convergence from the perspective of membership and information retrieval, a professional community of practice and its codification and accounting procedures, and a large-scale information infrastructure involving multiple communities and information artifacts. We have witnessed some similar processes in each case, mediated by developing information infrastructure: the world comes to look as if the convergent description of it is accurate and "natural." The degree of transparency achieved in each case depends on location, strategy, and politics. As the system scales up, transparency becomes more subject to contention arising from the heterogeneity of the participating social worlds; at the same time, once achieved, it acquires considerable inertia and even coercive power.[6]

The co-creation aspect of convergence is central to our argument here: while transparency may seem wholly natural, it is never inevitable or universal. In this we follow Yates and Orlikowski's use of Giddens's structuration model: structure and agency are dialectical co-creations (Orlikowski 1991; Orlikowski and Yates 1994; Yates 1989, 1997). Lynda Davies and Geoff Mitchell (1994) use a similar approach.

Mary Douglas (1986, p. 45) states that "How a system of knowledge gets off the ground is the same as the problem of how any collective good is created." Social institutions, she explains (Douglas 1986, p. 48) are about reducing entropy, and

the incipient institution needs some stabilizing principle to stop its premature demise. That stabilizing principle is the naturalization of social classifications. There needs to be an analogy by which the formal structure of a crucial set of social relations is found in the physical world, or in the supernatural world, or in eternity, anywhere, so long as it is not seen as a socially contrived arrangement. When the analogy is applied back and forth from one set of social relations to another, and from these back to nature, its recurring formal structure becomes easily recognized and endowed with self-validating truth.

Douglas is at her strongest when she unflinchingly argues that basic classificatory judgments are social in the sense that they are created, maintained, and policed by institutions: "Nothing else but institutions can define sameness. Similarity is an institution" (Douglas 1986, p. 55). The relevance of her position in our discussion of convergence is clear. She is adopting the position—close to our own—that the development of a single information world in any particular setting is not about the enlightened discovery of the truth about the world but is about the consolidation of social institutions and information systems. For us, this consolidation explains some people's transparent use of information infrastructure and others' inability or lack of desire to do so.

We disagree with how Douglas reifies "tine social" and sets it apart from technology or information systems.[7] On the one hand, she defines society as apart from the existence of things-in-the-world, but then, for a causal argument, she has trouble setting "society" as a determinant on one side of an equation and "the world" or "knowledge" on the other. There has never been room for co-construction in this form of functionalist argument. A related weakness is that, as with many functionalist positions, she makes stability too much the norm.

For our needs as information scientists, what needs empirical explanation is both information convergence and the lack of convergence. Neither can be assumed. As sociologist Everett Hughes (1971, p. 26) reminds us, we need always to keep in mind that "it might have been otherwise." Douglas's position implies that we all belong to one professional world, one social class, one ethnicity, and so forth—and that the alignment of these memberships into a coherent social institution acts as unproblematic guarantor for information convergence and thus for transparency. Those for whom this knowledge is not transparent must be deviant.[8]

In spite of this criticism, our understanding of convergence is enhanced by Douglas's work. It reminds us that convergence is a result of the consolidation of institutions and technologies. The information science literature also lends studies of the multiple paths along which information converges—such as colleague networks, personal collections, and community practices (cf. Aloni 1985; Bishop and Star 1996; Garvey and Griffith 1980; Lancaster 1995; Pinelli 1991). Network analyses have shown the importance of interconnections (Haythornthwaite 1996; Scott 1992; Wasserman 1994). More work is also being done to investigate how computer networking affects social networks. Barry Wellman (1997) notes that computer networks can be used to support a wide variety of social relations—dense and bounded to sparse and unbounded. According to Wellman, computer networks are indeed being used to support all of these relationships. The actor network perspective (Callon 1989; Callon, Law, and Rip 1986; Latour 1987, 1993) has demonstrated the importance of multiple translation networks across different communities of practice to the success of any scientific endeavor. To these we add that convergence is a process in which status, cultural and community practices, resources, experience, and information infrastructure work *together* to produce transparency within a wider information world. Convergence is fully situated and is not universal or exclusive. In times of change, it also has a certain fragility. It is a situation, for the most part, of privilege.

From a conservative point of view, such "obviousness" or naturalness would mean successful integration with the normative structure of social life. This occurs only when convergence is taken as natural or unproblematic. We do not hold this view: people belong to many communities of practice and participate in many information worlds. Success is relative and partial. For any given information world, the degree of convergence is not a matter of "proper" socialization or internalization of the correct norms. Rather, people balance conflicting requirements from different information worlds; they may have good reasons (such as more important memberships) for undertaking pathways to membership or convergence (Becker 1967), or they may be excluded on other grounds (such as gender, age, or race). So the situation of transparent, obvious, easy information retrieval on a routine basis for work and play is in some ways a special case.

Because communities of practice and information artifacts often do *not* converge, it is important to conceptualize failures of transparency as well as the successful formation of information worlds. The term *anomie* seems suggestive of those situations where there is a failure of convergence. *Anomie* in common parlance means something like "at loose ends." The *Oxford English Dictionary* lists a range of definitions beginning with "a disregard of divine law" through the nineteenth- and twentieth-century sociological terms meaning "an absence of accepted social standards or values." Most social scientists associate the term with Émile Durkheim (1951), who used the concept to speak of the ways in which an individual's actions are matched, or integrated, with a system of social norms and practices. When such a map for behavior is present, he argued, the person is less likely to experience the angst of normlessness, less likely to internalize a situation with no structure, and therefore less likely to kill themselves.[9]

While we disagree with aspects of this explanation, the word *anomie* does capture some sense of what happens with a profound mismatch between community of practice and information artifacts. The dispirited undergraduates we interviewed about library usage were experiencing anomie in the commonsense meaning—at loose ends, with no road map.

Durkheim also formally defined *anomie* as a mismatch, not simply as the absence of norms. Thus, a society with much rigidity and little individual discretion could also produce a kind of anomie—a mismatch between individual circumstances and larger social mores. Thus, fatalistic suicide arises when a person is too rule-governed. Information anomie may likewise arise in both kinds of settings. It may occur when there is an absence of social standards, guidance, or values that are accepted as useful by those successfully working with the system. It may also arise when the system is overly rigid, allowing for no tailoring or customizing to individual needs in a place of practice (Trigg and Bødker 1994).

Terms like *anomie* can carry much undesirable freight, such as the notion that there is one right user interface and only deviants will fail at transparent usage with that perfect recipe. Similarly, the notion that there are "good users" and "bad users" is widespread. We feel it is more productive to understand the nature of mismatches and a lack of convergence than to blame users. At higher levels of scale, this becomes a question for social justice. Lack of convergence may mean the exclusion or invisibility of whole social worlds or classes of people—institutionalized opacity rather than transparency (Star and Strauss 1998).

Related Work in Library and Information Science

Library and information science (LIS) research has dealt with most of the components of the contents of the information worlds to which we refer above. Our task in social informatics is to describe how these components work together and to analyze longer-term interaction of information resources and communities of practice. In this section, we describe how our model of convergence can be used to situate the social informatics dimension of LIS research.

Many important LIS studies are centered on libraries, journals, and other formal information systems. They are grounded in a model of a "user" first conceiving of an information need and then going out and getting that information (Bates 1989; Lancaster 1995). The role of the information scientist or intermediary is between those two points—interpreting the information need, creating and utilizing document surrogates, using or building a system to retrieve an item that will satisfy the user's need for information. While bringing attention to the often overlooked role of the user, this research usually questions the circumstances influencing the utility and effectiveness of different forms of document and system use, representation, and retrieval under different circumstances (Bates 1989; Covi 1996).

The view that these studies take is often segmented along different axes, such as exploring the use of specific document forms or the use of single systems among particular communities (e.g., Aloni 1985; Baym 1995; Bizot, Smith, and Hill 1991; Crane 1972; Lancaster 1995; Pinelli 1991; Warden 1981). Lisa M. Covi and Rob Kling (1996) critique the more segmented view as a "closed rational, or bounded database perspective." Specifically, they define this as a perspective that reviews the use of a system as if that system were not situated in other resources online and offline, in community norms, or in other social practices. They call for more use of the open and naturalistic inquiry. This integrated naturalistic view as a departure for LIS research is less common although growing in interest and importance in the LIS community (Bishop and Star 1996; see the papers in Bowker and Star 1999).

We seek in the model presented in this chapter to include and add to the scope and perspective of current understanding of information use. Indeed, they are complementary and supportive to the many studies that have demonstrated the importance of colleague interaction for information exchange (Crane 1972; Garvey and Griffith 1980; Lancaster 1995; Wolek and Griffith 1980). Others point to the significance of personal collections and resources (Chatman 1991, 1992; Soper 1976). Also, the importance of physical proximity in use of resources has been well documented for some time (Gould and Pearce 1991; Hertz and Rubenstein 1954; Palmer 1981; Pinelli 1991; Rosenberg 1966; Waples 1932). All of these point to the importance of, and describe the significance of, various elements in information worlds, which we are then drawing together.

Further complementary research is found in the strong tradition of studies of communication patterns examining the social networks of individuals. This particular thread of social network analysis "is an approach and set of techniques used to study the exchange of resources among actors" (Haythornthwaite 1996, p. 1). It traces exchanges between actors and finds patterns of relationships of exchange. These relationships are synthesized into networks. These networks can be examined for patterns of exchange of information. Relationships between "nodes" are measured and evaluated for several properties, including strength, intensity, direction, and content (Scott 1992; Wasserman 1994). Bradford W. Hesse, Lee S. Sproull, Sara B. Kiesler, John B. Walsh, and Barry Wellman have all indicated how computer and social networks interpenetrate in creating these relationships (Finholt and Sproull 1990; Hesse, Sproull, Kiesler, and Walsh 1993; Sproull and Kiesler 1991; Wellman 1997). The development of the strength of these ties points to the communicative processes associated with convergence. Further research into the similarities and differences between social networks and communities of practice would further specify these links.

The advent of wide-scale networked information systems has opened many new discussions about communities and information. The possibility of information acquisition, retrieval, and tailoring at the community level (including finding communities on the Internet and designing for communities rather than individuals) is an exciting one. Its success depends in part on the multiple factors discussed here. Not surprisingly, we know the most about the individual level of information retrieval, access, and use; we know some about the community or occupational level, particularly in the areas of scientific communication; and we are just beginning to learn about infrastructural convergence as a problem in design and use. Ole

Hanseth and Eric Monteiro (1996), Hanseth (1996), and Hanseth, Monteiro and Morten Hatling (1996) have begun investigations into information infrastructure and standardization that address many of these issues. They argue in particular that the tensions between standardization and flexibility in the creation of information infrastructure are a key design factor and that analysis of institutions supporting the creation of standards and standardized classifications is a key to successful design in the networked computing environment. This point receives support in Bowker and Star (1994, 1996, 1999) and Star and Karen Ruhleder (1996).

One challenge to LIS research, and social informatics in particular, is to tie these many factors together into a holistic and dynamic model. As other researchers (e.g., Bates 1989; Covi 1996; Covi and Kling 1996) have pointed out, the search-and-retrieval model is too simplistic, rational, and mechanical to describe the dynamics of how people, groups, and infrastructure come together as resources. They are calling for study of the use of digital libraries in terms of open natural systems and a social-world perspective, looking at the use of the system in terms of a bounded world and rational actors and also looking at the physical and temporal workplace, the professional field, and the occupational niche of the user (Covi 1996; Covi and Kling 1996). To this we are adding the idea that these variables are crucial in understanding the use and usability of digital library systems or other information systems and also in understanding the amount of convergence or the lack of convergence that a person, a community, or multiple communities experience. We are concerned with how the contents of information worlds or individual units therein interact dynamically and ecologically.

In addition, search and retrieval are not compartmentalized tasks but rather products of social processes and relationships. How individuals or communities are linked is the result of the position of each. Information is not a static product that is "out there" waiting to be found but is rather a construct of particular situations at specific points in time. Convergence is a process by which things come together through social process such as socialization into a profession or the formation of a community with the co-construction of information artifacts at different levels of scale.

Conclusion

We know that no universal scheme can make information easily available to everyone. Some systems of information organization are widely adopted, such as that of the Library of Congress or the telephone book. However, as Sanford Berman (1981) illustrates for the Library of Congress Subject Headings, these same headings may carry with them systematic exclusions of points of view of ethnic minorities; they overemphasize Western religions and do not well represent common names for everyday objects. Similar points hold for other attempts to organize and classify the world of knowledge. Current attempts to catalog and classify the chaos of the Internet are just beginning to grapple with universal organizational schemes for cataloging or indexing. They must link together "islands of indexing," working with the powerful new capabilities of search engines and limitless hypertext.

This state of affairs is an opportunity for researchers in social informatics. We have in this chapter seen it as an opportunity to understand something about the relationship among people, history, information, and material and social order. While there is no socially perfect transparency and never will be, there is sometimes relative transparency at many levels of scale. In our work studying designers and users of information, we see "clear and present differences" between people who are very comfortable within some world of information (who are "plugged in" to a range of sources and are able to find all the information they need easily) and those who are intimidated, confused, and limited by their relationships to that same information. The differences are subtle. To some degree, they are hidden by models of individual user needs and concepts such as levels of expertise or idealized transparency unmodified by time or place.

This has given us some insights into those experiencing information anomie. Differences between insiders and outsiders are neither trivial nor easily understood. Sometimes differences in career stage explain relative ease and familiarity: an old hand will likely move more easily through a world of information than a novice, as we have noted, but might not if that world itself is shifting and changing rapidly and training situations favor students over professors. Sometimes there are insider/outsider differences of the sort noted by Berman—systematic exclusions of members of a certain group or differences in access to materials and infrastructure that make something like a digital library a current impossibility (Berman 1984). Sometimes other kinds of memberships are more salient (Eckert 1989).

The greatest issue for the creation of information systems is that many of these processes are invisible to traditional requirements analysis. They can be seen only through the developmental or naturalistic analysis of information worlds.

Convergence can be a tricky thing to analyze. It often masquerades as naturalness, but it is the result of a particular configuration of relationships between information and social order. Convergence is not interoperability but rather a layering of solutions and conventions, memberships and standards. Usability is an emergent property that is not currently addressed in this fashion. The reason that some things are obvious to some people and not to others has to do with the connectedness that some people and not others experience and the memberships that some have and not others.

The challenge in understanding and recognizing these differences is to design to them. This does not mean creating "simple" versus "complex" interfaces to systems but rather recognizing the fundamentally different processes of searching and use germane to different information worlds and levels of scale. Recognizing differences in needs implies building systems that actually do different things. The new relationships and points of access being built and developed now benefit only a small group of people at a particular place in their careers. Systems are now being linked with hopes of interoperability without taking into account the trajectories of information on a systemic or infrastructural level. More than ever, interdisciplinary teams for design and use are needed to address the multifaceted nature of information—a social informatics.

Acknowledgments

This chapter was supported in part by the National Science Foundation, Defense Advanced Research Projects Agency, and National Aeronautics, and Space Administration Digital Library Initiative (DLI) under contract number NSF 93-141 DLI and by the National Science Foundation with a grant for research on classification and infrastructure, contract number 9514744. Our thanks to people who read drafts of this work and discussed the ideas herein, especially Ann Bishop and the DLI Social Science Team and members of the Graduate School of Library and Information Science's proseminar. Thanks to Robert Alun Jones for reading the section on Durkheim.

Notes

1. Total transparency is, of course, an ideal type, nowhere to be found in the real world (everyone's car stalls now and then or has idiosyncracies requiring a deeper involvement with a tool).

2. After some debate, we chose the term *community of practice*, which has some currency in information science circles, rather than *social worlds*, although we believe that they have the same meaning. *Social world* is a term used in sociology and first coined by Anselm Strauss (1978b). It is cognate with the notion of *reference groups*.

3. Julius Roth (1963) makes a similar point about tuberculosis classification. Until a common classification of patient status had been introduced, patients could not carry a single

description of their condition from one institution to another. It was almost impossible to share information between institutions or to build a community or shared language for comparison purposes. See also Star and Bowker (1997) for an analysis of TB classification.

4. Under External Causes of Morbidity and Mortality (WHO 1992, chap. 20, the largest chapter in the ICD-10), contact with venomous snakes and lizards is X20. A list of eight snakes and one Gila monster is included, but these are not broken down in the actual coding. So rural inhabitants cannot distinguish the density of sidewinders versus rattlesnakes, as they probably would want to do for safety purposes.

5. Where social movements against Western medicine have become strong, the ICD has been weakened correspondingly.

6. This finding is akin to the discussion of how scientists handle anomalies in Star's *Regions of the Mind: Brain Research and the Quest for Scientific Certainty* (1989), to Latour's analysis of scientific networks in *Science in Action: How to follow Scientists and Engineers through Society* (1987), and at the level of large-scale control to Foucault's original sense of discipline.

7. She falls prey to Latour's (1993) arguments against all great divides that engender determinism.

8. Illustrative of this point is functionalist sociologist Robert Merton's (1973) categorization of scientific discovery (novelty) as a form of deviance!

9. In *Suicide* (1897), Durkheim accounts for the lower suicide rate among Catholics (compared with Protestants) by noting the stricter, more binding, and more clearly specified sets of norms and practices affecting Catholics. Durkheim's work has been widely criticized on empirical and methodological grounds.

References

Allison, P. D., and S. Long. 1990. Departmental Effects on Scientific Productivity. *American Sociological Review*, 55(4), 469–479.

Aloni, M. 1985. Patterns of Information Transfer among Engineers and Applied Scientists in Complex Organizations. *Scientometrics*, 8, 279–300.

Bannon, L. 1990. A Pilgrim's Progress: From Cognitive Science to Cooperative Design. *AI and Society*, 4(4), 259–275.

Bates, M. J. 1977. Factors Affecting Subject Catalog Search Success. *Journal of the American Society for Information Science*, 28(3), 161–169.

Bates, M. J. 1979. Idea Tactics. *Journal of the American Society for Information Science*, 30(5), 280–289.

Bates, M. J. 1989. The Design of Browsing and Berrypicking Techniques for the On-line Search Interface. *Online Review*, 13(5), 407–424.

Baym, N. (1995). From Practice to Culture on Usenet. In S. L. Star, ed., *The Cultures of Computing* (pp. 29–52). Sociological Review Monograph Series. London: Basil Blackwell.

Becker, H. 1967. Whose Side Are We On? *Social Problems*, 14(3), 239–247.

Becker, H. 1982. *Art Worlds*. Berkeley: University of California Press.

Becker, H. 1994. Foi por acaso: Conceptualizing Coincidence. *Sociological Quarterly*, 35(2), 183–194.

Berg, M. 1997. *Rationalizing Medical Work–Decision Support Techniques and Medical Problems*. Cambridge, MA: MIT Press.

Berg, M., and G. C. Bowker. 1997. The Multiple Bodies of the Medical Record: Towards a Sociology of an Artifact. *Sociological Quarterly*, 38(3), 513–537.

Berman, S. 1981. *The Joy of Cataloging: Essays, Letters, Reviews, and Other Explosions*. New York: Oryx Press.

Berman, S., ed. 1984. *Subject Cataloging: Critiques and Innovations*. New York: Haworth Press.

Bishop, A. P., and S. L. Star. 1996. Social Informatics of Digital Library Use and Infrastructure. In M. E. Williams, ed., *Annual Review of Information Science and Technology* (vol. 31, pp. 301–401). Medford, NJ: Information Today.

Bizot, E., N. Smith, and T. Hill. 1991. Use of Electronic Mail in a Research and Development Organization. *Advances in the Implementation and Impact of Computer Systems*, 1, 65–92.

Bowker, G. C. 1994. Information Mythology and Infrastructure. In L. Bud-Frierman, ed., *Information Acumen: The Understanding and Use of Knowledge in Modern Business* (pp. 231–247). London: Routledge.

Bowker, G. C., and S. L. Star. 1994. Knowledge and Infrastructure in International Information Management: Problems of Classification and Coding. In L. Bud-Frierman, ed., *Information Acumen: The Understanding and Use of Knowledge in Modern Business* (pp. 187–216). London: Routledge.

Bowker, G. C., and S. L. Star. 1996. How Things (Actor-Net)work: Classification, Magic and the Ubiquity of Standards. *Philosophia*, 25(3–4): 195–220.

Bowker, G. C., and S. L. Star. 1999. *Sorting Things Out: Classification, Infrastructure and Boundary Objects*. Cambridge, MA: MIT Press.

Brown, J. S., and P. Duguid. 1994. Borderline Issues: Social and Material Aspects of Design. *Human-Computer Interaction*, 9(1), 3–36.

Callon, M., ed. 1989. *La Science et Ses Reseaux: Genes et Circulation des Faits Scientifiques*. Paris: La Decouverte, Conseil de l'Europe: UNESCO.

Callon, M., J. Law, and A. Rip, eds. 1986. *Mapping the Dynamics of Science and Technology: Sociology of Science in the Real World*. Basingstoke, UK: Macmillan Press.

Campbell, D. 1969. Ethnocentrism of Disciplines and the Fish-Scale Model of Omniscience. In M. Sherif and C. Sherif, eds., *Interdisciplinary Relationships in the Social Sciences*. Chicago: Aldine.

Chatman, E. A. 1985. Information, Mass Media Use and the Working Poor. *Library and Information Science Research*, 7, 97–113.

Chatman, E. A. 1991. Channels to a Larger Social World: Older Women Staying in Contact with the Great Society. *Library and Information Science Research*, 13(3), 281–300.

Chatman, E. A. 1992. *The Information World of Retired Women*. Westport, CT: Greenwood Press.

Covi, L. 1996. Social Worlds of Knowledge-Work: How Researchers Appropriate Digital Libraries for Scholarly Communication. In G. Whitney, ed., *ASIS Mid-Year Meeting Proceedings: The Digital Revolution: Assessing the Impact on Business, Education, and Social Structures* (pp. 84–100). Medford, NJ: Learned Information.

Covi, L. M., and R. Kling. 1996. Organizational Dimensions of Effective Digital Library Use: Closed Rational and Open Natural Systems Models. *Journal of the American Society for Information Science*, 47(9), 672–689.

Crane, D. 1972. *Invisible Colleges: Diffusion of Knowledge in Scientific Communities*. Chicago: University of Chicago Press.

Davies, L., and G. Mitchell. 1994. The Dual Nature of the Impact of IT on Organizational Transformations. In R. Baskerville, O. Ngwenyama, S. Smithson, and J. DeGross, eds., *Transforming Organizations with Information Technology* (pp. 243–261). Amsterdam: North Holland.

Desrosières, A. 1988. *Les Categories Socio-Professionnelles*. Paris: La Découverte.

Desrosières, A. 1993. *La Politique des Grands Nombres: Histoire de la Raison Statistique*. Paris: Editions La Decouverte.

Douglas, M. 1986. *How Institutions Think*. Syracuse: Syracuse University Press.

Durkheim, É. 1951. *Suicide: A Study in Sociology*. New York: Free Press. (Originally published in 1897)

Eckert, P. 1989. *Jocks and Burnouts: Social Categories and Identity in the High School*. New York: Columbia University Teachers College.

Finholt, T., and L. Sproull. 1990. Electronic Groups at Work. *Organization Science*, 1(1), 41–64.

Foucault, M. 1991. Governmentality. In G. Burchell, C. Gordon, and P. Miller, *The Foucault Effect: Studies in Governmentality* (pp. 87–104). Chicago: University of Chicago Press.

Garvey, W., and B. Griffith. 1980. *Scientific Communication: Its Role in the Conduct of Research and Creation of Knowledge*. Key Papers in Information Science. White Plains, NY: Knowledge and Industry.

Gould, C., and K. Pearce. 1991. *Information Needs in the Sciences: An Assessment*. Program for Research Information Management. Mountain View, CA: Research Libraries Group.

Grudin, J. 1990. The Computer Reaches Out: The Historical Continuity of Interface Design. In J. Chew and J. Whiteside, eds., *Human Factors in Computing Systems: CHI '90 Conference Proceedings (April 1–5, 1990, Seattle, WA)* (pp. 261–268). New York: ACM Press.

Hanseth, O. 1996. Information Technology as Infrastructure. Ph.D. dissertation, Gøteborg University, Sweden, Department of Informatics. Report 10, Gøteborg Gothenburg Studies in Informatics.

Hanseth, O., and E. Monteiro. 1996. Inscribing Behavior in Information Infrastructure Standards. In B. Dahlbom, F. Ljungberg, U. Nulden, K. Simon, C. Sørensen, and J. Stage, eds., *Proceedings of the Nineteenth Information Systems Research Seminar in Scandinavia (IRIS): The Future (Gøteborg, Sweden, August 1996)* (pp. 293–332). Gøteborg: Department of Informatics, Gøteborg University.

Hanseth, O., E. Monteiro, and M. Hatling. 1996. Developing Information Infrastructure: The Tension between Standardization and Flexibility. *Science, Technology, and Human Values*, 21(4), 407–426.

Haythornthwaite, C. 1996. Social Network Analysis: An Approach and Technique for the Study of Information Exchange. *Library and Information Science Research*, 18, 323–342.

Hertz, D., and A. Rubenstein. 1954. *Team Research*. Boston: Eastern Technical.

Hesse, B. W., L. Sproull, S. Kiesler, and J. Walsh. 1993. Returns to Science: Computer Networks in Oceanography. *Communications of the Association for Computing Machinery*, 36(8), 90–101.

Hughes, E. 1971. *The Sociological Eye: Selected Papers*. Chicago: Aldine-Atherton.

Kiernan, V. 1998. U.S. to Offer up to $50 Million in Grants to Support Research on Digital Libraries. *Chronicle of Higher Education*, 44(25), 427.

King, J., and S. L. Star. 1990. Conceptual Foundations for the Development of Organizational Decision Support Systems. In *Proceedings of the Twenty-third Hawaiian International Conference on Systems Sciences* (vol. 3, pp. 143–151). Washington, DC: EKE Press.

Kling, R., and W. Scacchi. 1982. The Web of Computing: Computing Technology as Social Organization. *Advances in Computers*, 21, 3–78.

Kritek, P. B. 1988. Conceptual Considerations, Decision Criteria and Guidelines for the Nursing Minimum Data Set from a Practice Perspective. In H. H. Werely and N. M. Lang, eds., *Identification of the Nursing Minimum Data Set* (pp. 22–33). New York: Springer.

Lancaster, F. W. 1995. Needs, Demands, and Motivations in the Use of Sources of Information. *Journal of Information, Communication, and Library Science*, 1(3), 13–19.

Latour, B. 1987. *Science in Action: How to Follow Scientists and Engineers through Society*. Cambridge, MA: Harvard University Press.

Latour, B. 1993. *We Have Never Been Modern*. Cambridge, MA: Harvard University Press.

Lave, J., and E. Wenger. 1992. *Situated Learning: Legitimate Peripheral Participation*. Cambridge: Cambridge University Press.

Lévi-Strauss, C. 1973. *From Honey to Ashes*. New York: Harper and Row.

Linde, C. 1997. Personal communication.

McCloskey, J., and G. Bulechek, eds. 1996. *Nursing Interventions Classification (NIC)* (2nd ed.). St. Louis, MO: Mosby.

McCloskey, J. M. 1996. Standardizing Nursing Language for Computerization. In M. Mills, C. Romano, B. Heller, eds., *Information Management: In Nursing and Health Care*. West Dundee, IL: S-N.

McCormick, K. A. 1988. A Unified Nursing Language System. In M. J. Ball, K. J. Hannah, U. G. Jelger, and H. Peterson, eds., *Nursing Informatics: Where Caring and Technology Meet*. New York: Springer.

Merton, R. 1973. *The Sociology of Science: Theoretical and Empirical Investigations*. Chicago: University of Chicago Press.

Orlikowski, W. 1991. Integrated Information Environment or Matrix of Control? The Contradictory Implications of Information Technology. *Accounting, Management, and Information Technology*, 1(1), 9–42.

Orlikowski, W., and J. Yates. 1994. Genre Repertoire: The Structuring of Communicative Practices in Organizations. *Administrative Science Quarterly*, 39(4), 541–574.

Palmer, C. 1996. Practices and Conditions of Boundary Crossing Research Work: A Study of Scientists at an Interdisciplinary Institute. Ph.D. dissertation, University of Illinois, Urbana-Champaign, Graduate School of Library and Information Science.

Palmer, E. S. 1981. The Effect of Distance on Public Library Use: A Literature Study. *Library Research*, 3(4), 315–354.

Pinelli, T. 1991. The Information-Seeking Habits and Practices of Engineers. *Science and Technologies Libraries*, 11(3), 5–25.

Ricoeur, P. 1988. *Time and Narrative* (vol. 3). Chicago: University of Chicago Press.

Rosenberg, V. 1966. The Application of Psychometric Techniques to Determine the Attitudes of Individuals toward Information Seeking. Master's thesis, Lehigh University.

Roth, J. A. 1963. *Timetables: Structuring the Passage of Time in Hospital Treatment and Other Careers*. Indianapolis: Bobbs-Merrill.

Scott, J. 1992. *Social Network Analysis: A Handbook*. London: Sage.

Soper, M. E. 1976. Characteristics and Use of Personal Collections. *Library Quarterly*, 46(4), 397–415.

Sproull, L., and S. Kiesler. 1991. *Connections: New Ways of Working in the Networked Organization*. Cambridge, MA: MIT Press.

Star, S. L. 1989. *Regions of the Mind: Brain Research and the Quest for Scientific Certainty*. Stanford, CA: Stanford University Press.

Star, S. L. 1994. Misplaced Concretism and Concrete Situations: Feminism, Method, and Information Technology. Working paper in the Gender, Nature, Culture Network Working Paper Series. Denmark: Odense University.

Star, S. L., and G. C. Bowker. 1997. Of Lungs and Lungers: The Classified Story of Tuberculosis. In A. L. Strauss and J. Corbin, eds., *Grounded Theory in Practice* (pp. 197–227). Thousand Oaks, CA: Sage.

Star, S. L., and G. C. Bowker, eds. 1998. How Classifications Work: Problems and Challenges in an Electronic Age. *Library Trends*, 47(2).

Star, S. L., and K. Ruhleder. 1996. The Ecology of Infrastructure: Problems in the Implementation of Large-Scale Information Systems. *Information Systems Research*, 7(1), 111–134.

Star, S. L., and A. Strauss. 1999. Layers of Silence, Arenas of Voice: The Dialogues between Visible and Invisible Work. *Journal of Computer-Supported Cooperative Work*, 8(1–2), 9–30.

Strauss, A. L. 1978a. *Negotiations: Varieties, Contexts, Processes, and Social Order*. San Francisco: Jossey-Bass.

Strauss, A. 1978b. A Social World Perspective. *Studies in Symbolic Interaction*, 1, 119–128.

Swanson, D. R. 1986. Undiscovered Public Knowledge. *Library Quarterly*, 56(2), 103–118.

Taylor, R. S. 1991. Information Use Environments. In B. Dervin and M. Voight, eds., *Progress in Communication Science*. Norwood, NJ: Ablex.

Tort, P. 1989. *La Raison Classificatoire. Les Complexes Discursifs: Quinze Etudes*. Paris: Aubier.

Trigg, R., and S. Bødker. 1994. From Implementation to Design: Tailoring and the Emergence of Systematization in CSCW. In *CSCW '94: Proceedings of ACM 1994 Conference on Computer-Supported Cooperative Work* (pp. 45–54). New York: ACM Press.

Turnbull, D. 1997. On with the Motley: The Contingent Assemblage of Knowledge Spaces. Ph.D. dissertation, University of Melbourne.

Waples, D. 1932. The Relation of Subject Interests to Actual Reading. *Library Quarterly*, 2, 42–70.

Warden, C. 1981. User Evaluation of a Corporate Library Online Search Service. *Special Libraries*, 72(2), 113–116.

Wasserman, S. 1994. *Social Network Analysis: Methods and Applications*. Cambridge: Cambridge University Press.

Weech, T., and H. Goldhor. 1984. Reference Clientele and the Reference Transaction in Five Illinois Public Libraries. *Library and Information Science Research*, 6(1), 21–42.

Wellman, B. 1997. An Electronic Group Is Virtually a Social Network. In S. Kiesler, ed., *Culture of the Internet*. Mahwah, NJ: Erlbaum.

Wolek, F., and B. Griffith. 1980. *Policy and Informal Communication in Applied Science and Technology*. Key Papers in Information Science. White Plains, NY: Knowledge Industry.

World Health Organization (WHO). 1992. *International Statistical Classification of Diseases and Related Health Problems (ISC-10)* (10th rev. ed., 3 vols). Geneva: World Health Organization.

Yates, J. 1989. *Control through Communication: The Rise of System in American Management*. Baltimore, MD: Johns Hopkins University Press.

Yates, J. 1997. Using Giddens' Structuration Theory to Inform Business History. *Business and Economic History*, 261, 159–183.

11

Digital Libraries and Collaborative Knowledge Construction

Nancy A. Van House

Effective digital library (DL) design is not simply a matter of converting existing information practices and artifacts to a digital world. Digital libraries support cognitive or knowledge work. Designing effective DLs, then, requires understanding knowledge work and the way that it is not only supported but potentially changed by digital libraries. We must examine the work, its tools and practices, the people who do the work, the institutions that support it, and the interaction of all these with the DL. Understanding all of this might be too much to aspire to in a complex and shifting world, but we can make inquiries and reach some partial understanding that will help the continuing enterprise of knowing.

Although much discussion has focused on user-centered design of information systems in general and digital libraries in particular, a theoretical or conceptual base for such work has been lacking. In this chapter, I investigate the uses of irreductionist approaches to social theory with an emphasis on situated action, science studies and science and technology studies, and actor network theory as bases for understanding knowledge work and then DLs.

This chapter emphasizes three critical characteristics of knowledge work: it is situated, distributed, and social. That it is situated means that knowledge work is performed by specific people under specific conditions for specific purposes. It is distributed because it entails cooperation among people who don't know each another and those who do, across space and time. In fact, some current approaches to learning (Lave and Wenger 1991) argue that the community and not the individual "knows." Finally, it is social because we work and learn together and decide what and whom to believe and rely on in the community. Much of what we claim to know comes not from our own direct experience but from what others tell us is so, including our knowledge of whom to believe.

This focus on the nature of knowledge work applies to DLs in two ways. First, documents and other information artifacts, publishing, and libraries are critical to knowledge work. The digital library, as a new entry into this array of artifacts and institutions, is affected by and potentially changes existing processes and relations. Second, DL design, construction, and maintenance are also a form of knowledge work involving designers, implementers, content providers, and users.

This analysis is rooted in an empirical study of the University of California–Berkeley Digital Library Project, but it is not specific to any one DL. It addresses concerns raised by participants and nonparticipants in this DL about actual and potential changes in information production and use made possible by DL technology, not about the specific design and content of the Berkeley DL. Furthermore, I contend that these findings apply to many areas where information technology is changing the processes of knowledge work. I argue that these findings reflect not just issues specific to one DL or to DLs alone but to the social and material practices of knowledge work and the changes brought about by the electronic distribution of information. The DL challenges existing practices of knowledge work, the boundaries of knowledge communities, and the practices of trust and credibility, all of which are central to the creation and use of knowledge.

The first part of the chapter reports on conversations with actual and potential users, contributors, and creators of the UC Berkeley DL and discusses some of their concerns and problems. The second part begins the development of an analytical perspective from which to better understand these processes. It is rooted in situated action, science and technology studies, and actor-network theory (ANT). The final section considers the implications of this analysis for digital libraries.

In this chapter, I argue that the DL is not simply a new technology or organizational form but a change in the social and material bases of knowledge work and the relations among people who use and produce information artifacts and knowledge. I further argue that it crosses boundaries—from private to public and across work communities—in ways that are becoming increasingly common with networked information. Finally, I argue that the DL and its potential interaction with knowledge work, communities, and practices highlights critical issues of trust and credibility in the networked world. I then examine the implications for DL design and evaluation and for other forms of networked information.

The UC Berkeley Digital Library

This chapter draws on research conducted as part of the UC Berkeley Digital Library Project (⟨http://elib.cs.berkeley.edu⟩) (UCB DL), part of the Digital Libraries Initiative funded by the National Science Foundation and others. This research was conducted during Phase I, which developed a work-centered DL supporting environmental planning. Its premise was that workgroups require sophisticated digital library services and collaborative support to effectively utilize massive distributed repositories of multimedia information and that the work and the workgroup need to be understood so that DL services can be built to support them. During Phase II, the focus of the project shifts beyond the content from Phase I. This chapter refers primarily to the project as it existed under Phase I.[1] This is only one of many digital library initiatives being undertaken at UC Berkeley and within the UC system.[2]

The UCB DL developed tools for image retrieval, new document models, Web-based geographic information systems (GIS), distributed search, and natural-language processing for distributed search. A major component of the project was and remains a substantial and diverse test bed of textual documents, photo images, datasets, and various kinds of georeferenced data. As of November 2002, it held 185,000 photo images, over 2 million records in geographical and biological data sets, and 2,900 scanned documents comprising 300,000 pages of text for a total of over one terabyte of data.

Much of the discussion here refers to one major part of the test bed, called Calflora,[3] a linked collection of botanical datasets containing descriptive information, photos, and diversity mapping for over 8,000 vascular plants in California and nearly 675,000 records of plant observations from fifteen state, federal, and private databases. These records came in a range of formats from a variety of places. The size of this database of observations and its coordination with the other parts of the test bed make this DL significant for the botanical and environmental planning communities as both a resource and a model.

Interviews

The premise of the user needs assessment and evaluation component of the UCB DL is that we must understand the work before we design the DLs that will support that work (Schiff, Van House, and Butler 1997; Van House 1995; Van House, Butler, Ogle, and Schiff 1996; Van House, Butler, and Schiff 1998). We investigated the practices of environmental planning, particularly those of information use and production; the information artifacts used and produced; and the possible effects of a Web-based digital library. More generally, our goal has been to better understand distributed collaborative cognitive work, the role of information and information practices and artifacts, and the potential effects of digital information.

This chapter reports on a particular aspect of this research—issues related to the sharing and use of digital data that arose during interviews with two different communities.[4] The purpose of these interviews was to learn about environmental planning and its use and production of information and information artifacts. The interviews were about the potential of DLs in general, including the UCB DL.

One set of interviews was with people engaged in work related to water planning in California. Participants included employees of various state, local, and federal agencies; private consultants; members of the public; and people from a range of professions, including engineering, hydrology, biology, computer science, and planning. In earlier papers drawn from these interviews, we described the social world of water planning (Schiff, Van House, and Butler 1997; Van House 1995; Van House, Butler, and Schiff 1998).

The second set of interviews was with people whose work was associated with the botanical datasets included in the UCB DL. We interviewed people in state and federal agencies, in nonprofit research institutions, and within the University of California. We discussed with them a range of issues related to current efforts to make biological data available at UC and elsewhere.

These water and botanical groups share some important characteristics. Neither is homogeneous with clear boundaries. Both fields consist of multiple scientific disciplines engaged in coordinated work and data-sharing but with different knowledge bases, interests, methods, and terminology. Both include people working in state, federal, and local governments and nonprofit and commercial organizations doing research, planning, and applied work. Both have at least some involvement in decisions about land and water use that are often highly political[5] with potentially great impact on the environment and economy (such as housing development, agriculture, timber harvesting, and recreation).

Both rely heavily, at least for some purposes, on large datasets of records of observations and specimens and on summaries and analyses using these records. The records come in many forms and formats, have a variety of sources, and are collected over a long time by people with varied backgrounds, training, and interests. For example, to retroactively establish a baseline to determine changes in an ecological community, researchers may search out and correlate data about a particular geographic area collected over many years by different observers for a variety of purposes. Fine-grained data, such as the location of individual plants in a particular area, is often available only from local, sometimes nonprofessional, sources, while higher levels of government are often the source of lower-resolution data, such as aerial photos or statewide resource surveys.

In both fields, computing and telecommunications enable access to large datasets that were previously difficult to share. These linkages have accelerated recently with developments in digital libraries and the Internet.

Interview Findings

This section reviews issues raised in the interviews about the willingness of individuals, organizations, and knowledge communities to contribute to and use information from the DL and to participate in the DL's creation and maintenance. The interviews revealed concerns about DLs and their relationship to the situated, distributed, and social nature of knowledge. The following section presents an analytical perspective on knowledge work that can aid in understanding the issues raised in the interviews and the DL's role in the shifting processes of knowledge creation and trust.

DL Contributions   Identifying and acquiring content for a DL like the UCB DL—which does not begin with a preexisting collection or even a specific user community and includes both unpublished and published information—requires that some information producers agree to have their content included. It also requires that information be indexed, synthesized, and correlated. For example, the Hrusa California Plant Synonyms, a comprehensive table that tracks changes in California plant names created by a single researcher and made available to the DL, is indispensable for coordinating multiple databases of botanical specimens and observations.

In the interviews, we found a generally high level of interest in sharing data. In particular, local environmental and government organizations, which often have difficulty disseminating their work, saw the DL as a way to reach a wider audience.[6] But we also found many concerns, especially about access to unpublished data.

Previously, much of the data for which respondents were responsible was available to others mainly in summary form. Raw data were available directly from the data owners (on site or remotely) or within a professional community (for example, through closed file transfer protocol). These social and technical barriers limited the data to people within a professional community or to those who had dealt directly with the data owner, who could screen requests and explain the data and their limits. Now such data can be made available to anyone. However, the possibility of open access raised several concerns among respondents. I will focus on three of these concerns—fear of possible misuse of data, especially by people outside the professional community; the burden of making data available to others; and the hazards of making visible previously invisible work.

The first concern is "inappropriate use" of the data. Data available via the Internet may be used by people with interests, training, or expertise different from the data's producers. A widespread concern in water planning interviews, in particular, was that uninformed users would download and use a snapshot of a changing dataset, misinterpret the data, combine datasets inappropriately, run unsuitable analyses, or otherwise misuse the data. For example, fish populations are measured by netting a sample. Projecting from the netted sample to a population is complex. The time of day and year, weather conditions, type of net, depth, and many other conditions affect the catch, and various models may be used to project a population from a sample.

The next two sets of concerns relate to what one respondent called "productizing" work. Sharing data requires work—cleaning the data and creating documentation and metadata. One respondent said, "The Web ... has made a lot of our scientists realize that they are [information] providers and take much more seriously the notion that they have obligations." Those obligations may not be welcome to either the researcher or the institution. Creating and maintaining data are often considered preliminary to the "real" work of research. Making data available to others takes time, effort, and sometimes different skills. It may be inconsistent with the researchers' or organization's priorities. And it may primarily serve the needs of people in other organizations.

"Productizing" work may make previously invisible work visible (Star and Strauss 1999). Sometimes this is desirable—for example, in giving workers credit for workload and skill. But it may make a person accountable in new ways. For respondents, making available the data underlying their analyses may allow a critic to redo or refute their work—as one said, to "use our data against us." For example, we learned that water supply projections are derived from a large number of measures and data points collected over space and time. The process of incorporating the data into models to generate forecasts is not mechanical. Forecasters exercise considerable judgment about which observations are accurate, how to massage the data, and which forecasting models and assumptions to use. Opening up the datasets to scrutiny opens up the entire process of generating supply projections.

Two findings are worth emphasizing here. One is the potential problems in crossing boundaries (Marshall, this volume chapter 3; Star 1989; Star and Griesemer 1989)—that is, making data available to and from people in other professional or practice communities and making public what was private. The distributed nature of knowledge work means that people are continually sharing work, whether with people they know or impersonally through publication. But this sharing generally takes place within social, organizational, and professional boundaries, helping to ensure appropriate understanding and use and reciprocity. The DL has the potential of decontextualizing information and making it more readily available outside of the social world within and for which it was produced. The other finding that I want to stress is the need for at least some DLs to actively solicit people's participation to win their trust and to align with their interests.

Using the DL   Now we turn to some factors that affect people's willingness and ability to use DLs. The literature contains many discussions of factors influencing people's adoption of new technology. A DL is not simply a technological artifact, however; it is an information resource. It is not exactly analogous to a traditional library.

Three possible differences between DLs and physical libraries are particularly relevant to this discussion. First, traditional libraries are closely tied to the publishing system. The publishing system provides review and quality control, giving users some imperfect but indicative assurances about the sources, credibility, and appropriate uses of information. However, some DLs may provide access to previously unpublished information, placing them outside these established institutional safeguards.

Second, traditional libraries are established institutions, and librarianship is a respected profession with selection policies and procedures, standards of performance, and a code of objectivity. Some DLs are affiliated with traditional libraries and staffed by professional librarians, but many are not.

Third, a DL contains content, technology, and functionality, all of which users must be willing to trust if they are to rely on the DL as a source of information and possibly as an aid in analysis and to incorporate the work of others, carried by the

DL, into their own. The UCB DL provides innovative functionality, including the ability to manipulate and recombine data in a variety of ways. For example, it includes a GIS function—the ability to select, map, and overlay georeferenced data from unrelated sources.

Using published sources, data collected by others, or even technology designed by others is an act of trust. The distributed nature of knowledge work makes it a collective good. Much of what we claim to know is rooted in what we have been told by others, including what we have learned about who or what to trust and how to evaluate the trustworthiness of sources. Even in science, which purports to ground truth claims in direct experience, knowledge is grounded in the community's prior knowledge, its instruments and methods of observation and argument, and its assessment of competing knowledge claims: "We rely on others.... [T]he relations in which we have and hold knowledge has a moral character, and the word I use to indicate that moral relation is *trust*" (Shapin 1994, p. xxv).

Users of any library must assess the quality and credibility of its contents. This is an assessment partly of the library as an institution or collection (a medical school library will have different health information from a public library) and partly of the actual information resources. Users often take for granted the trustworthiness of a library based on their knowledge of the institution. A university library, for example, is assumed to adhere to collection development standards and procedures that, to some degree, warrant its contents.

Assessing the credibility of information is not easy. For example, a steady trickle of e-mail arrives questioning or "correcting" identifications of the UCB DL's flower images. No doubt some of the original identifications were erroneous. But assessing the correctness even of something as factual as plant identifications is complex. Some species can be positively identified only under a microscope. Other differences may be due to changes in plant taxonomy over time—names change, the boundaries between species and subspecies move, new species or subspecies are defined.

How does a user assess the credibility of information? A forest ecologist described her criteria for plant identifications. A record of a sighting is less credible than a specimen in hand. A report of a species in an unexpected place is less plausible than one from a place where it is common. She also considers the expertise of the person making the identification. She knows that some people are experts in a geographical area; she'll trust their identification of species common to that area but not necessarily rare species. Others are experts in a species, and she will trust them on that species and subspecies wherever they may occur. In other words, she judges the professional skills of data providers—directly or indirectly—to determine the credibility of the data.

We found that knowledgeable users consider not just the data providers' expertise but also their interests. The naive ideal of objectivity in data collection and analysis quickly broke down in politically charged discussions about such issues as water planning with people from, for example, resource-extraction industries and environmentalists. Many choices are made in collecting, analyzing, reporting, and interpreting data that reflect particular groups' interests and expectations. People who are knowledgeable about (and suspicious of) information providers' interests factor those in when assessing credibility.

How might a DL assist users in assessing data quality? First, a DL generally exercises informed judgment in the selection process. Collection-level assessments are made to determine which botanical data should be included in the UCB DL. Some in the user community have argued, however, that every image or record should be reviewed by an expert. However, this would require considerable skilled labor, and the UCB DL does not do this.

Second, a DL can provide informed users with information about the source. The UCB DL links each botanical observation to its source and to whatever metadata are provided. This shifts the burden of assessment to the user, which may not help the inexpert user but may be invaluable to the expert (like our forest ecologist).

DLs provide not only content but functionality. This includes search tools—that is, the DL's interface may act as an aid or a barrier between the user and the collection. The UCB DL also performs operations on the data, such as geographic information system (GIS) overlays, where graphical representations of multiple datasets are laid over the same geographical substrate. GIS data can be overlaid, so the level of resolution (even the choice of colors) affects the interpretation of the results. Denis Wood (1992) shows how maps, which appear to be simple representations of a territory, reflect interests by the choices of what is represented and how. These issues about trusting a DL's functionality were raised not by users but by UCB DL designers with expertise in the applications area, probably because they knew the most about the possibilities chosen and those missing.

In understanding knowledge work, we need to be concerned with how people decide whether to trust others' work and to incorporate it into their own and how the DL supports or undermines these processes. By removing information from its original setting, the DL may strip it of the contextual indicators needed to interpret, assess, and use it—the sources, methods, and terminology (Van House, Butler, and

Schiff 1998; Watson-Verran and Turnbull 1995) of people and practices that mark the work of a community.

Creating and Operating the DL   DLs require contributors and users but also many different kinds of designers, builders, and operators. The UCB DL, like many, is both a research project and an active test bed. Many DLs are collaborative projects among researchers, user communities, librarians, and technologists. Participation by at least some of these people may be voluntary.

Two major sets of interests to be reconciled in a DL like the UCB DL are (1) computer science researchers and staff and (2) content area specialists who are both information providers and users. The DL may be something different to each. Computer scientists, for example, tend to emphasize technological innovation, whereas users are more interested in content and stable reliable functioning. An interviewee from a natural history museum said that he had seen problems on similar projects "because the computer scientists are interested in blazing issues in computer science because that's what's going to make them famous, and we on the content side are trying to shovel mounds of [data], and it doesn't flow down well" (personal communication). He went on to say, "This gets into the political agendas of what serves any participant in this project best and how we can devise a workplan for a project that is going to give everybody the right benefits out of it. What we need to end up with here in the museum is more stuff digitized and systems that enable us to manage, capture, and maintain it after the project has run its course. We have certain limitations in the tool sets that we will be able to cope with" (personal communication).

Such differences are not unique to the UCB DL. Judith Weedman (1998) studied the Sequoia project, a research project that developed computer-based tools for earth science research. She found differences between the earth science researchers, who wanted a reliable, working system, an incremental improvement over existing tools requiring minimal effort on their part, and computer science researchers, who wanted to be at the cutting edge of their field. These differences are likely to surface in the choices about development priorities, content, functionality, the speed with which changes are introduced, and attention to user friendliness.

Another source of tension is that participating disciplines may have different norms and styles for decision making. Many computer science projects and many Web-based projects are highly democratic and decentralized. Whoever is willing to contribute is welcome. A member of the DL's technical staff noted differences between computer science and the botanical community over how much effort (and delay) to put into making a "right" decision versus getting something up and running and dealing with problems later. It is dangerous and perhaps unfair to generalize about two research communities from these comments, but some key participants believed that such differences existed. It seems reasonable to expect differences between participants focused on the technology and those focused on content and functionality in their assessment of what is good enough to make publicly available. Problems with content and functionality are likely to be more distressing to members of the user community.

While the most visible differences may be over major decisions, another issue is articulation work (Star and Strauss 1999)—that is, tuning, adjusting, monitoring, and managing the consequences of the distributed nature of work. This is the interplay between formal and informal. Articulation work is almost always invisible (especially when it is done well), and because of this it is often overlooked in technological innovation. In traditional libraries, librarians do the articulation work, whether ensuring the integrity of the collection or helping users locate information when the tools fail. In research DLs, there may be judicious discussions about major policy decisions but then a lack of attention to articulation work.

In the UCB DL document intake process, for example, the work of creating metadata for documents was initially left to the student assistants doing the scanning. In time, this was seen to be problematic, and a form was created for document providers to report metadata. However, libraries know well the problems of inconsistency that come from having metadata assigned by multiple people.

In summation, then, DLs balance multiple interests. In some DLs, information providers must be willing to allow their work to be included and to be used, possibly by people from other knowledge communities whose understanding and practices may differ from their own. Participants must be willing to use (to trust) the DL's contents, which may include documents, data, and other representations that are not available through the more traditional channels of publishing and libraries and without the warrants that these institutions provide. The DL provides not simply contents but also functionality—minimally, to search and filter the DL's contents but in some cases to operate on, recombine, and otherwise manipulate the contents, often in ways that are not transparent to the users. Finally, the design, implementation, and operation of the DL often requires the work of different professional groups, such as computer scientists, librarians, and members of the targeted user communities, each with different interests and priorities for the DL.

DLs participate in, are affected by, and potentially change the relations among the people, artifacts, and practices of knowledge work, which is situated, distributed, and social. The DL itself, furthermore, is an instance of distributed work that requires balancing the interests and knowledge frameworks of multiple communities.

To design and evaluate DLs, then, it would be useful to have an analytical base for understanding issues of knowledge creation, use, and trust; of crossing of boundaries between public and private and across knowledge communities; and of cooperation in the creation of sociotechnical systems such as DLs.

Understanding Knowledge Work

Up to this point, this chapter has been concerned with findings from interviews with participants and potential participants in the UCB DL, with reference to this DL and to the more general possibilities of sharing digital data. This section presents a framework for understanding these findings in the context of knowledge work and DLs as sociotechnical systems. It draws on three related analytical approaches—situated action/practice theory that is based primarily on the work of Jean Lave and Etienne Wenger, science studies and science and technology studies, and actor-network theory. In the interests of space, we concentrate on the aspects of these approaches, individually and collectively, that are most relevant to the present discussion. The following section then applies this approach to the empirical issues described above.

All three of these approaches are to some degree *irreductionist* frameworks (Kaghan and Bowker 2000), according to which social order is not preexisting but is continually produced and reproduced through people's ongoing practical action, the concrete day-to-day activity and interaction. The emphasis is on the processes by which order is (re)constructed and on the practices and artifacts that carry and shape understanding and activity across time and space.

In the situated action approach of Lave and Wenger (Lave 1988, 1993; Lave and Wenger 1991; Wenger 1998), learning is a dimension of social practice. Theories of social practice emphasize the interdependence of the person and the world. Learning concerns the whole person acting in the world as a member of a sociocultural community: "Learning, thinking, and knowing are relations among people and activity in, with, and arising from the socially and culturally structured world" (Lave and

Wenger 1991, p. 50). Knowledge is not an inert substance transferred from teacher to student. It is a complex social and material phenomenon, a "nexus of relations between the mind at work and the world in which it works" (Lave 1988, p. 1).

Practice—concrete, daily activity—is critical to knowing. Information artifacts, including texts and images, are not simply reflections or carriers of knowledge. They shape and reflect practice and are instrumental in creating and re-creating knowledge as well as coordinating work across space and time.

Knowledge, practices, and artifacts are tightly bound up with the knowledge community or community of practice (Lave and Wenger 1991; Wenger 1998). People learn and work within groups that share understanding, practices, technology, artifacts, and language. Communities of practice include professions, workgroups, and disciplines, as well as less easily labeled groups.

The recognition of trustworthy sources is a necessary component of all systems of knowledge and knowledge communities (Shapin 1994). Our perceptions are located within "hierarchies of credibility" (Star 1995, citing Becker). An important task of knowledge communities is establishing these hierarchies—deciding what is known, what processes and principles are used to evaluate knowledge claims, who is entitled to participate in the discussion, and who should be believed.

In our society, science is considered the prototype of rational knowledge construction and validation. Science studies[7] and science and technology studies (STS) investigate the culture and practices of science and technology and the ways that what is known is decided, including how hierarchies of credibility are created and how participants attain their places in them. The approaches subsumed by science studies and STS vary, but they share a contention that scientific and technical knowledge is determined not entirely by nature but also by social influences. This does not mean that knowledge is simply a product of social agreement. But this approach challenges the traditional boundaries between nature and society, science and politics, and knowledge-creation practices in science and in other areas of knowledge.

STS is further concerned with how sociotechnical systems—systems of people, technology, and practices—are created and maintained. It takes the stability of such systems as something to be explained rather than taken as given.

Actor-network theory (ANT)[8] originated within science studies and STS with ethnographic studies of science laboratories and the processes by which reputations were built and resources garnered via the practical daily work of the laboratory (e.g., Latour and Woolgar 1991). ANT sees power and order as effects to be explained. ANT argues that action at a distance and mobilization of allies are critical to the development of scientific knowledge, establishment of credibility, and acquisition of resources and that inscriptions (including texts, images, graphics, and the like) play a key role. It has also been concerned with explaining the stabilization of sociotechnical systems, which it describes as temporary, changing heterogeneous networks of resources.

ANT has its shortcomings.[9] Our purpose here is to see how ANT may assist in understanding DLs as sociotechnical systems and as collections of inscriptions used to act at a distance. To better understand ANT and how it may apply to DLs and to our interview findings, we need to examine some of its elements.

The basic ontological unit of ANT[10] is the *actor-network*, which is "most simply defined as any collection of human, non-human, and hybrid human/non-human actors who jointly participate in some organized (and identifiable) collective activity in some fashion for some period of time" (Kaghan and Bowker 2001, p. 258). ANT is concerned with how these pieces are held together as organizations, social institutions, machines, and agents, at least for a time. Perhaps the most radical contribution of ANT is the inclusion of the nonhuman. According to ANT, networks are heterogeneous, composed not only of people but of machines, animals, texts, money, and other elements.

Translation is a key process by which actor-networks are created and stabilized, however temporarily. Actors' disparate interests get translated into a set of interests that coincide in the network. For example, in Michael Callon's (1986a) study of the development of electric vehicles, the producers of vehicle bodies and of fuel cells both had to see their interests as being served by the electric vehicle to be willing to research and produce the needed components. Without their involvement, the vehicle could not be built.

Black-boxing is a process of closing questions and debates. Participants agree to accept something as presented. Ideas, knowledge, and processes can all be black-boxed, at least temporarily and for some people. "Standard" research methods, for example, get black-boxed within research communities. What is black-boxed for some groups may not be for others: different research communities may have different standard methods.

Finally, "an *intermediary* is an actor (of any type [i.e. human or nonhuman]) that stands at a place in the network between two other actors and serves to translate between the actors in such a way that their interaction can be more effectively coordinated, controlled, or otherwise articulated" (Kaghan and Bowker 2001, p. 258). Because networks are never completely stabilized, translation and black-boxing are continual. This is the work of intermediaries.

Inscriptions are an important kind of intermediary in ANT. Knowledge, according to ANT, is not abstract and mental. It takes material form in inscriptions such as journal articles and patents, conference presentations, and skills embodied in scientists and technicians: "The actor-network approach is thus a theory of agency, a theory of knowledge, and a theory of machines" (Law 1992, p. 389). Inscriptions play an important role in scientific and other regimes of knowledge. Modern information technology, beginning with the printing press, has made possible the immutability and mobility of inscriptions, allowing people to send their work and arguments across space and time and to gather, compare, combine, contrast, summarize, and refute others' work.

Another kind of intermediary of interest in the present discussion is the boundary object. Susan Leigh Star and James Griesemer (Star 1989; Star and Griesemer 1989) noted that scientific work has always required information that can be used by multiple users and communities for a variety of purposes, retaining its integrity across space, time, and local contingencies without losing its specific meaning in a local setting. To explain how this works, they propose a model based on symbolic interactionism's "social worlds" theory. Instead of a single actor trying to funnel others' concerns into a narrow passage point, they argue for multiple translations, entrepreneurs from multiple intersecting communities of practice, each trying to map their interests to those of the other audiences in such a way as to ensure the centrality of their own interests. One place that these interests come together is in a boundary object.

They describe boundary objects as being plastic enough to adapt to local needs, with different specific identities in different communities, and robust enough to maintain a common identity across sites and be a locus of shared work. They identify several types of boundary objects, one of which is of particular interest for the present discussion—repositories, which they describe as "ordered piles of objects indexed in standardized fashion" (Star and Griesemer 1989, p. 408). Their example is the University of California, Berkeley, Museum of Vertebrate Zoology (MVZ), a collection of specimens of amphibians, birds, mammals, and reptiles, with extensive standardized metadata including such information as descriptions of the location and conditions under which specimens were collected and by whom. (The MVZ's specimen records, coincidentally, are now accessible through the UCB DL.) Star and

Griesemer's account of the creation of the MVZ shows how the interests of the director, major funder, collectors, trappers, and university administrators came together to create an impressive array of specimens of California flora and fauna with standardized descriptive information to be used by a variety of scientific disciplines for many purposes.

Situated learning has profound implications for the design of information systems. Our ideas about knowledge creation and use and information transfer are rooted in our assumptions about learning. The very phrase "information transfer" implies that information is a thing that can be transferred. If learning is rooted in practices, artifacts, and communities, then information systems have to be co-constituted with these as well. If knowledge is tightly bound up with a community of practice, then information systems have to be aligned with communities of practice. At the very least, we have to look more carefully at the social aspects of knowledge and the role of information artifacts such as documents in practice and in communities of practice. In particular, we have to consider how DLs both support the recognition of trustworthy sources and are evaluated as trustworthy themselves.

Science studies, STS, and ANT are relevant to the UC Berkeley DL and DLs in general in a number of ways. First, much of the content of the UCB DL is scientific; most of the interviewees do work that is to some degree based in science. The practices of scientific work and of credibility are relevant to interviewees' concerns about sharing of data.

Second, as we have said, science is often considered the prototypical system of knowledge. Furthermore, ethnographic studies of science have concluded that what scientists do is not fundamentally different from ordinary activity (Latour 1987; Latour and Woolgar 1991). I contend that much of what has been learned in science studies probably applies to other knowledge communities and to digital libraries and other digital information generally. In particular, the recognition of trustworthy sources is a universal issue.

Third, STS and ANT are concerned with the building and maintenance of sociotechnical systems; DLs are sociotechnical systems. The UC Berkeley DL consists of technology, texts, images, databases, functionality, users, contributors, funders, designers, builders, operators, and more. ANT may provide us with resources for understanding people's willingness (or unwillingness) to use and contribute to the DL and its inscriptions and to participate in the DL's creation.

Fourth, ANT addresses the role of inscriptions in translation and action at a distance, the establishment of credibility, order, and power (e.g., Latour 1986; Law 1986). DLs substantially affect the creation, circulation, and combination of inscriptions.

Finally, I contend that DLs are boundary objects. The UCB DL is an ordered indexed repository of publications, images, and records of specimens and observations, of catalogs, indexes, maps, and mapping functionality. It has different identities in different communities (such as a repository of documents and records or a test bed for technology development) but a common identity across sites (the UCB DL). And it is the locus of multiple simultaneous translations as the different communities seek to enroll others to help create and maintain the DL in ways that support their interests.

Understanding Digital Libraries and Knowledge Work

Participants are concerned about contributing content to a DL, using it, and participating in its creation and maintenance. This section examines these concerns in light of the chapter's discussion so far.

The DL is a locus of shared work of two kinds. First, building and maintaining a DL involve a variety of groups that contribute content, do the work of design and operation, and use it. Second, the content of a DL supports the shared work of information providers and users. Botanical databases, for example, are used across communities doing botanical and other kinds of environmental work. The DL is the "host," so to speak, of other boundary objects such as databases.

The power of the DL is in the integration of its heterogeneous elements—its nature as a sociotechnical system, its usefulness to multiple professional communities. This integration, however, creates difficulties and conflicts as each participating group places its own concerns at the center and seeks, in ANT terms, to enroll others—or at least participates only as long as its interests are met, as long is it is willing to be enrolled. ANT warns us that network stability must be explained and not assumed and that networks tend to come apart. Some of the tension that is revealed in the interviews—including differing priorities between technologists and content-area specialists and concerns about the effort required of data providers—is a reflection of this tendency toward instability. They reflect not problems to be solved once and for all but expected and ongoing tensions in the DL as a dynamic locus of multiple translations.

The flexibility created by DLs is a major reason for building them. But this same flexibility may destabilize the processes and standards of knowledge creation. In science and technology, as in other areas, knowledge work is performed within an interdependent network of previous work, methods, tools, and colleagues—assemblages of people, practices, tools, theories, standards, publications, social strategies, and the like that are dynamically mutually constituted in and by knowledge work (Watson-Verran and Turnbull 1995; Van House, Butler, and Schiff 1998). Inscriptions play a key role in the creation of knowledge and the processes by which knowledge communities decide what and whom to believe, interests are translated, and networks are stabilized. Hierarchies of credibility are created within communities of practice; there too inscriptions are most easily understood (see, for example, the discussion of genres in Agre 1998). But DLs enhance the ease with which inscriptions cross community boundaries making Latour's "immutable mobiles" even more mobile—perhaps too mobile, when respondents worry about their data being misused by people without the needed professional qualifications.

Crossing the boundaries of knowledge communities may call into question that which is taken for granted within communities, opening black boxes and challenging the established cognitive order, hence the concerns in the interviews about misuse of data or the inclusion of data from unfamiliar sources. Shared practices and understandings can no longer be assumed; the indicators of credibility may be questioned or missing; users may not know how to assess credibility; inscriptions may be used in ways not envisioned (or accepted) by their creators. The DL potentially destabilizes the processes and the hierarchies of credibility and power relations.

Furthermore, the DL can challenge inscriptions' immutability (Levy, chapter 2). The DL is not simply a repository of inscriptions, a passive container or conduit. It makes them more manipulable and combinable. It even makes new kinds of inscriptions possible. The UCB DL, for example, allows the user to take a published table and reorganize it. For example, a table of dams arranged alphabetically can be reordered by size. The concerns about data being misused, for example, include fears that the "black box" of an inscription such as water supply forecasts will be opened, the modeling replicated with different assumptions, and the forecasts challenged.

Of course, the originator never retains control over his or her inscriptions. But the DL allows them to move even more freely. It may enable more fluid recombinations of them, and it may even open them up in ways not possible before.

The DL opens some black boxes, but it may close others. For example, its analytical and search tools may be opaque to the user, making it difficult for the user to understand, evaluate, and trust what the DL is doing.

In sum, then, the DL is not simply a passive repository of inscriptions. By potentially changing the mobility and mutability of inscriptions, it may challenge the established order and destabilize the network of relations. It becomes a participant in the processes of knowledge creation and the establishment of trust and credibility.

Implications for Digital Library Design, Management, and Evaluation

In this chapter, I have discussed knowledge work—its social and material practices and their interaction with the digital library. I argue that electronic distribution of information potentially challenges existing practices of knowledge work, boundaries of knowledge communities, and practices of trust and credibility—all of which are central to the creation and use of knowledge and the stabilization of the cognitive order.

I have emphasized three characteristics of knowledge work and their implications for DLs. Because knowledge work is situated, DLs have to articulate with the contexts and practices of specific knowledge communities. Because it is distributed, DLs support the coordination of work across space and time. And because it is social, they are entangled in the processes by which communities collectively decide what they know and whom to believe.

This focus on knowledge work applies to DLs in two ways. First, digital libraries support users' knowledge work, so understanding knowledge work will help us to ensure that DLs are useful and used. Second, DL design, construction, and maintenance are also a form of knowledge work, so this understanding may help to ensure that DLs are sustainable (Lynch, chapter 8).

The DL is not simply a new technology or organizational form but is a change in the bases of knowledge work and its network of social and material relations. The DL may cross boundaries—from private to public and across knowledge communities—in ways that are becoming increasingly common with networked information. Finally, the DL and its potential interaction with knowledge work, communities, and practices highlight critical issues of trust and credibility in the networked world.

A digital library is a heterogeneous network of users, researchers, funders, operators, and other people; of documents, images, databases, thesauri, and other information artifacts; of practices and understandings; and of technology. It is a boundary object, both created and used by different communities for different purposes. It is the locus of multiple translations as various participants try to enroll others to ensure that the DL meets their needs. It is an active participant in the creation and circulation of documents, images, and other kinds of inscriptions.

This understanding of DLs has implications for DL design, management, and evaluation and for our understanding of knowledge work and communities. First, it suggests that the work of translation in DLs is ongoing. The problems and tensions identified in these interviews are indicative of the dynamic tension among the DL's participants. We cannot expect a DL to settle into an easy equilibrium of interests. We can expect conflicts, compromises, and jockeying for advantage among various stakeholder groups to be a continual part of DLs.

Second, it cautions us to be sensitive to the variety of communities' existing practices of knowledge creation and work and indicators of credibility. A successful DL has to fit with these practices. In particular, the DL has to articulate with participants' hierarchies of credibility and processes of establishing trustworthiness so that people will be willing to use and contribute to the DL, and these vary across communities of practice.

The world and the work that the DL serves are continually changing; so too must the DL. Its design needs to be deliberately fluid and dynamic to accommodate emergent work practices and ongoing enrollment and co-constitution.

There is currently much discussion about involving users in the design of technology, but how (and whether) it is done varies. This analysis suggests that high-order user involvement in DL design is critical. Only members of the knowledge community fully understand the complexity of their network of practices, tools, and participants, including their processes and criteria of credibility and how the DL might interact with them. However, the question of who is to be involved (that is, which communities) can become even more uncertain if the value and the threat of DLs are in their ability to cross the boundaries of communities of practice. Which communities are involved, and how are their differences reconciled?

Understanding the DL as a boundary object helps us understand some of the ongoing tension between generality and specificity in DL design. But it also helps us to see DLs as continuous with past boundary-spanning practices in science and other areas. A DL designed for specific communities and tasks can mesh with the methods, genres, understandings, and language of a specific community and support existing and emergent practices of trust and credibility. But a major value of DLs is their ability to cross communities and tasks. Furthermore, customization is costly. For whom will the DL be customized? Which group's interests will dictate design decisions?

The implications for DL evaluation are several. The DL needs to be evaluated not just for how well it performs its intended functions and meets targeted users' identified needs but for its interaction with work, practices, artifacts, and communities. Evaluation needs to be targeted to specific user communities and tasks, with the understanding that DLs are different things to different groups. This makes evaluation more complex and uncertain but also more realistically aligned with users' concerns and with the changing array of users and other stakeholders.

Evaluation needs to address not just service delivery but also the organizational issues involving the creation and maintenance of the DL, the stabilization of the DL as a heterogeneous network, and the ongoing enrollment and coordination of resources and participants. A DL without users is a failure, however good its technology and its contents.

Common to science studies, STS, and ANT is a reliance on ethnographic research methods—the study of activities in their natural settings, of what people actually do as well as their own accounts of their behavior. The appropriate methods of evaluating DLs are varied (Van House, Bishop, and Buttenfield, chapter 1), but ethnography is certainly central.

I argue that these findings reflect not just issues that are specific to one DL or to DLs alone but those that apply to the social and material practices of knowledge work and the changes brought about by the electronic distribution of information. DLs challenge existing practices of knowledge work and boundaries of knowledge communities, both of which support the interpretation and understanding of information and the assessment of its accuracy. DLs raise questions about much that we take for granted in knowledge work. Taking these questions seriously will help us to better understand these communities and their work, their relationship to artifacts and technology, and the way that we can design better DLs.

Two areas to which this analysis might be extended are knowledge management and the growing problem of trustworthiness of Internet-based resources. Knowledge management is concerned with making better use of organizational knowledge resources. It often tries to decontextualize knowledge and information artifacts— "repurposing" them, to use one of the uglier neologisms. But potential users need to be able to understand the information's context to assess its credibility and appropriateness for their needs. Similarly, the Internet has made a vast amount of information of uncertain provenance accessible to people who may lack the expertise to evaluate it, medical information being a prominent example. Both are areas where information is crossing boundaries in ways that challenge its interpretation and assessment and where attention to existing practices of information activity and cognitive order, and especially of trust and credibility, may help us to design better information systems.

Notes

1. Phase I funding for the UCB DL came from the National Science Foundation (NSF), National Aeronautics and Space Administration (NASA), the Defense Advanced Research Projects Agency (ARPA) Digital Libraries Initiative. Currently, the UC Berkeley Digital Library Project is part of the Digital Libraries Initiative sponsored by the National Science Foundation and many others. Additional funding at Berkeley currently comes from the CNRI-sponsored D-Lib Test Suite, and the NSF-sponsored National Partnership for Advanced Computational Infrastructure (NPACI). The work described in this chapter was supported, in part, by the NSF/NASA/ARPA Digital Library Initiative under NSF IRI 94-11334.

2. In particular, this project should not be confused with the California Digital Library (⟨http://www.cdlib.org⟩), an initiative of the University of California Office of the President.

3. Much of this collection now has a separate identity as CalFlora: ⟨http://www/calflora.org⟩, which consists of the following:

• *CalFlora Species Database* Information about the 8,363 currently recognized vascular plants in California, including scientific and common names, synonymy, distribution from literature sources, legal status, wetland codes, habitat information, and more. It includes a photo of the plant, if available, from the CalPhotos California Plants and Habitats Collection.

• *California Plant Synonymy Table* Translates older scientific names to those currently recognized and finds synonyms for names in current use.

• *CalPhotos California Plants and Habitats Photographs* 31,000 images of California plants and fungi linked by taxonomic name to the CalFlora Names database.

• *CalFlora Occurrence Database* Over 850,000 records of plant observations from fifteen state, federal, and private databases.

• *GIS Viewer* Displays observation point data in conjunction with United States Geographical Survey topographical maps, geopolitical maps, relief maps, false-color satellite images, political borders, and aerial orthophotographs.

4. Interviews were conducted by the author, Mark Butler, and Lisa Schiff.

5. For example, a front-page newspaper article began as follows: "In the same 'water's for fightin'' spirit that has defined power and politics in California, East Bay Municipal Utility district directors rejected a proposal Tuesday that might have ended a long, bitter water war between Sacramento and the Bay Area" ("New Clash Surfaces" 1999, p. 1).

6. Others are interested in using the DL for publishing, as well. A recent e-mail, for example, offered to the DL an unpublished paper that the sender described as "the most important and

precious document in the world today. I am sure that my paper [will] bring a revolution in the field of water and soil management on the Earth which is the fundamental requirement of life on the planet."

7. Authors like Mario Biagioli (1999) and David J. Hess (1997) carefully sidestep the definition of science studies. Even the choice of a phrase with which to refer to this broad and diverse area of research is problematic, with important distinctions made among science studies, science and technology studies (STS), sociology of scientific knowledge (SSK), technoscience studies, and other variants.

8. ANT is most identified with the work of Latour, Callon, and Law (Callon 1986a, 1986b; Callon, Law, and Rip 1986a, 1986b; Latour 1987; Latour and Woolgar 1991; Law 1986, 1990, 1992; Law and Hassard 1999).

9. ANT has been criticized for being too agnostic about social formations such as power and gender, for paying too much attention to design and development of sociotechnical systems and leaving out use and users, for failing to consider why some networks are more enduring than others, for having a Machiavellian view of the world, and for using often warlike language (Ormrod 1995; Haraway 1997).

10. The discussion that follows draws heavily on Kaghan and Bowker (2000).

References

Agre, P. 1998. Designing Genres for New Media: Social, Economic, and Political Contexts. In S. Jones, ed., *CyberSociety 2.0: Revisiting CMC and Community*. Newberry Park, CA: Sage.

Biagioli, M., ed. 1999. *The Science Studies Reader*. New York: Routledge.

Bourdieu, P. 1977. *Outline of a Theory of Practice*. Translated by R. Nice. Cambridge: Cambridge University Press.

Callon, M. 1986a. The Sociology of an Actor-Network: The Case of the Electric Vehicle. In M. Callon, J. Law, and A. Rip, eds., *Mapping the Dynamics of Science and Technology: Sociology of Science in the Real World* (pp. 19–34). London: Macmillan Press.

Callon, M. 1986b. Some Elements of a Sociology of Translation: Domestication of the Scallops and the Fishermen of St. Brieuc Bay. In J. Law, ed., *Power, Action, and Belief: A New Sociology of Knowledge?* (pp. 196–233). London: Routledge and Kegan Paul.

Callon, M., J. Law, and A. Rip. 1986a. How to Study the Force of Science. In M. Callon, J. Law, and A. Rip, eds., *Mapping the Dynamics of Science and Technology: Sociology of Science in the Real World* (pp. 3–18). London: Macmillan Press.

Callon, M., J. Law, and A. Rip, eds. 1986b. *Mapping the Dynamics of Science and Technology: Sociology of Science in the Real World*. London: Macmillan Press.

Garfinkel, H. 1967. *Studies in Ethnomethodology*. Englewood Cliffs, NJ: Prentice-Hall.

Giddens, A. 1984. *The Constitution of Society*. Berkeley: University of California Press.

Haraway, D. 1997. *Modest_Witness@Second_Millenium. FemaleMan©_Meets_Oncomouse^{TM}: Feminism and Technoscience*. New York: Routledge.

Hess, D. J. 1997. *Science Studies: An Advanced Introduction*. New York: New York University Press.

Kaghan, W., and G. Bowker. 2001. Out of Machine Age?: Complexity, Sociotechnical Systems and Actor Network Theory. *Journal of Engineering and Technology Management*, 18, 253–269.

Latour, B. 1986. Visualization and Cognition: Thinking with Eyes and Hands. *Knowledge and Society*, 6, 1–40.

Latour, B. 1987. *Science in Action*. Cambridge, MA: Harvard University Press.

Latour, B. 1999. On Recalling ANT. In J. Law and J. Hassard, eds., *Actor Network Theory and After*. Oxford: Blackwell.

Latour, B., and S. Woolgar. 1991. *Laboratory Life: The Construction of Scientific Facts*. Princeton, NJ, Princeton University Press.

Lave, J. 1988. *Cognition in Practice: Mind, Mathematics, and Culture in Everyday Life*. Cambridge: Cambridge University Press.

Lave, J. 1993. The Practice of Learning. In S. Chaiklin and J. Lave, eds., *Understanding Practice: Perspectives on Activity and Context*. Cambridge University Press.

Lave, J., and J. Wenger. 1991. *Situated Learning: Legitimate Peripheral Participation*. Cambridge: Cambridge University Press.

Law, J. 1986. The Heterogeneity of Texts. In M. J. Callon, J. Law, and A. Rip, eds., *Mapping the Dynamics of Science and Technology: Sociology of Science in the Real World* (pp. 67–83). London: Macmillan Press.

Law, J. 1990. Technology and Heterogeneous Engineering: The Case of Portuguese Expansion. In W. Bijker et al., eds., *The Social Construction of Technological Systems: New Directions in the Sociology and History of Technology* (pp. 111–134). Cambridge, MA: MIT Press.

Law, J. 1992. Notes on the Theory of the Actor-Network: Ordering, Strategy, and Heterogenity. *Systems Practice*, 5, 379–393.

Law, J., and J. Hassard, eds. 1999. *Actor Network Theory and After*. London: Blackwell.

Livingston, E. 1987. *Making Sense of Ethnomethodology*. New York: Routledge and Kegan Paul.

New Clash Surfaces in Water War. 1999. *Oakland Tribune*, June 23, p. 1.

Ormrod, S. 1995. Feminist Sociology and Methodology: Leaky Black Boxes in Gender/Technology Relations. In K. Grint and R. Gill, eds., *The Gender-Technology Relation: Contemporary Theory and Research* (pp. 31–78). London: Taylor and Francis.

Pickering, A. 1992. From Science as Knowledge to Science as Culture. In A. Pickering, ed., *Science as Practice and Culture* (pp. 1–26). Chicago: University of Chicago Press.

Schiff, L. R., N. A. Van House, and M. H. Butler. 1997. Understanding Complex Information Environments: A Social Analysis of Watershed Planning. In R. Allen and E. Rasmussen, eds., *Digital Libraries '97: Proceedings of the Second ACM International Digital Libraries (Philadelphia, PA, July 23–26, 1997)* (pp. 161–168). New York: ACM Press.

Shapin, S. 1994. *A Social History of Truth*. Chicago: University of Chicago Press.

Shapin, S. 1995. Here and Everywhere: Sociology of Scientific Knowledge. *Annual Review of Sociology*, 21, 289–321.

Star, S. L. 1989. The Structure of Ill-Structured Solutions: Boundary Objects and Heterogeneous Distributed Problem Solving. In L. Gasser and M. Huhns, eds., *Distributed Artificial Intelligence* (vol. 2, pp. 37–54). London: Pitman.

Star, S. L. 1995. Introduction. In S. L. Star, ed., *Ecologies of Knowledge: Work and Politics in Science and Technology* (pp. 1–35). Albany: State University of New York Press.

Star, S. L., and J. Griesemer. 1989. Institutional Ecology, "Translations," and Boundary Objects: Amateurs and Professionals in Berkeley's Museum of Vertebrate Zoology, 1907–39. *Social Studies of Science*, 19(3), 387–420.

Star, S. L., and K. Ruhleder. 1996. Steps Toward an Ecology of Infrastructure: Design and Access for Large Information Spaces. *Information Systems Research*, 7(1), 111–134.

Star, S. L., and A. Strauss. 1999. Layers of Silence, Arenas of Voice: The Ecology of Visible and Invisible Work. *Computer Supported Cooperative Work*, 8(1/2), 9–30.

Van House, N. 1995. User Needs Assessment and Evaluation for the UC Berkeley Electronic Environmental Library Project. In F. M. Shipman III, R. Furuta, and D. M. Levy, eds., *Digital Libraries '95: The Second Annual Conference on the Theory and Practice of Digital Libraries (Austin, TX, June 11–13, 1995)*. College Station, TX: Hypermedia Research Laboratory, Department of Computer Science, Texas A&M University. ⟨http://csdl.tamu.edu/DL95/papers/vanhouse/vanhouse.html⟩.

Van House, N., M. Butler, V. Ogle, and L. Schiff. 1996. User-Centered Iterative Design for Digital Libraries: The Cypress Experience. *D-Lib Magazine*. ⟨http://www.dlib.org/dlib/february96/02vanhouse.html⟩.

Van House, N., M. Butler, and L. Schiff. 1998. Cooperative Knowledge Work and Practices of Trust: Sharing Environmental Planning Data Sets. In S. Poltrok and J. Grudin, eds., *CSCW '98: Proceedings of the ACM Conference on Computer Supported Cooperative Work* (pp. 335–343). New York: ACM Press.

Watson-Verran, H., and D. Turnbull. 1995. Science and Other Indigenous Knowledge Systems. In S. Jasanoff et al., eds., *Handbook of Science and Technology Studies* (pp. 115–139). Newberry Park, CA: Sage.

Weedman, J. 1998. The Structure of Incentive: Design and Client Roles in Application-Oriented Research. *Science, Technology, and Human Values*, 23(3), 315–354.

Wenger, E. 1998. *Communities of Practice: Learning, Meaning, and Identity*. New York: Cambridge University Press.

Wood, D. 1992. *The Power of Maps*. New York: Guilford Press.

12

The Flora of North America Project: Making the Case [Study] for Social Realist Theory

Mark A. Spasser

Introduction: The Case Study

The Flora of North America (FNA) project is constructing a digital library, and in this chapter, I use the term *digital library* (DL) in a relaxed and inclusive manner. I borrow from Margaret Elliott and Rob Kling's (1997, p. 1023) definition of *digital libraries* as "information systems (IS) and services that provide electronic documents—text files, digital sound, digital video—available in dynamic or archival repositories" and from J. Alfredo Sánchez's (1998, p. 1) definition of *digital library* as "a virtual space in which scholars conduct research, collaborate, and publish their work." Both definitions conceive of digital libraries as in some way integrating technology, content, and services (Bishop and Star 1996), and both imply that complex (that is, structured, irreducible, and relatively enduring) "things" are interwoven with peoples' situated choices and with the available collective resources.

Moreover, a critical and heretofore unexamined facet of DL design and use is how library content is assembled and vetted, which in turn has profound implications for subsequent DL usefulness and usability. This chapter applies a social realist evaluation framework to the FNA DL and to its context of development and use—specifically, the organizational issues (or contradictions) that made its construction and use problematic. A significant part of the present analysis focuses on the publication subsystem of the FNA DL—Collaborative Publishing Services—and on using problems related to its design and use to explain FNA not only as a functioning DL project but as an organizational form in contradiction-driven expansive development. (More will be said about the role of contradictions and about expansive development later in the chapter.)

The Flora of North America Project

Flora of North America (FNA) is a project undertaken by the community of North American botanists to provide a wide range of users (including scientific and academic entities, government agencies, private industry, and amateur enthusiasts) with authoritative information on the names, relationships, characteristics, and distributions of all plants that grow outside of cultivation in North America north of Mexico. The FNA project is gathering and making accessible, in a variety of media, scientifically authoritative and current information on the names, characteristics, relationships, and distributions of the approximately 20,700 species of vascular plants and bryophytes needed for decision making, resource management, and innovative research. Thus, FNA is intended to serve as a means of identifying plants of the region, as a means of delineating taxa and geographic areas in need of additional study, and as a systematic conspectus of the North American flora. The project first received funding in 1987 and is expected to be completed around 2009 (Magill, Barkley, Morin, Schnase, and Thiers 1999; Morin, Whetstone, Wilken, and Tomlinson 1989; Morin 1991).

The major product being produced by the Flora of North America is a comprehensive set of taxonomic treatments for all North American plant species and infraspecific taxa. These treatments are published electronically on the Web and as thirty printed volumes by Oxford University Press. FNA treatments include accepted names, synonyms, bibliographic references, keys for identification, descriptions, economic uses, conservation status, weed status, and distribution data. The nomenclatural and taxonomic data compiled for FNA form the backbone of the project and constitute a primary information product that will soon be made public. This list of plant names will become an entry point—or portal—into FNA's electronic information and other related botanical resources. Despite their synoptic format, many of the treatments present, for the first time, knowledge from a systematist's lifetime of study (Schnase et al. 1997). Figure 12.1 shows the description and key sections of the treatment for *Berberis Linnaeus*. While all treatments include georeferenced distribution maps and many include illustrations, the present study refers only to the textual portions of treatment production to simplify and focus the exposition.

FNA, then, is a compilation of the best knowledge available on the patterns of biodiversity among plants in the continental United States and Canada. In terms of the number of participants and their geographic distribution, FNA is one of the

[DESCRIPTION]

Berberis Linnaeus

3. BERBERIS Linnaeus, Sp. Pl. 1: 330. 1753; Gen. Pl. ed. 5, 153. 1754 - Barberry, Oregon-grape, berbéris, algerita [Mediaeval Latin barbaris]

Mahonia Nuttall, name conserved; Odostemon Rafinesque

Shrubs or subshrubs, evergreen or deciduous, 0.1–4.5(-8) m, glabrous or with tomentose stems. Rhizomes present or absent, short or long, not nodose. Stems branched or unbranched, monomorphic or dimorphic, i.e., all elongate or with elongate primary stems and short axillary spur shoots. Leaves alternate, sometimes leaves of elongate shoots reduced to spines and foliage leaves borne only on short shoots; foliage leaves simple or 1-odd-pinnately compound; petioles usually present. Simple leaves: blade narrowly elliptic, oblanceolate, or obovate, 1.2–7.5 cm. Compound leaves: rachis, when present, with or without swollen articulations; leaflet blades lanceolate to orbiculate, margins entire, toothed, spinose, or spinose-lobed; venation pinnate or leaflets 3–6-veined from base. Inflorescences terminal, usually racemes, rarely umbels or flowers solitary. Flowers 3-merous, 3–8 mm; bracteoles caducous, 3, scalelike; sepals falling immediately after anthesis, 6, yellow; petals 6, yellow, nectariferous; stamens 6; anthers dehiscing by valves; pollen exine punctate; ovary symmetrically club-shaped; placentation subbasal; style central. Fruits berries, spheric to cylindric-ovoid or ellipsoid, usually juicy, sometimes dry, at maturity. Seeds 1–10, tan to red-brown or black; aril absent. x = 14.

Species ca. 500 (22 in the flora): almost worldwide.

Many species of Berberis are grown as ornamental shrubs. Some species harbor the black stem-rust of wheat (Puccinia graminis Persoon); the sale or transport of susceptible or untested species is illegal in the United States and Canada. Data on susceptibility of Berberis spp. to infection by Puccinia graminis was supplied by Dr. D. L. Long, U.S. Department of Agriculture (pers. comm.).

The berries of many species are edible and frequently are used for jam and jelly.

The genus Berberis as recognized below is divided into two genera, Berberis and Mahonia, by some authors (e.g., L. Abrams 1934). Species 1–5 below represent Berberis in the narrow sense (characterized by dimorphic stems, with elongate primary stems and short axillary shoots; leaves of primary stems modified as spines; foliage leaves simple; and inflorescences usually rather lax, with acuminate bracteoles and 1–20 flowers; most species susceptible to Puccinia). Species 13–22 represent the segregate genus Mahonia (with stems never regularly dimorphic; stem spines absent; leaves pinnately compound; and inflorescences dense, with rounded or obtuse [rarely acute] bracteoles and 25–70 flowers; never susceptible to Puccinia). Species 6–12, traditionally included in Mahonia when that genus is recognized (L. Abrams 1934), are actually intermediate, resembling Berberis proper in their dimorphic stems, inflorescence structure, and susceptibility to Puccinia, and Mahonia in their spineless stems and compound leaves. Species showing different combinations of the characteristics of the two groups are found in other parts of the world (J. W. McCain and J. F. Hennen 1982; R. V. Moran 1982), so these segregate genera do not seem to be natural. Mahonia is often recognized in horticultural works, but it is seldom recognized by botanists.

SELECTED REFERENCES
Abrams, L. 1934. The mahonias of the Pacific states. Phytologia 1: 89–94. McCain, J. W. and J. F. Hennen. 1982. Is the taxonomy of Berberis and Mahonia (Berberidaceae) supported by their rust pathogens Cumminsiella santa sp. nov. and other Cumminsiella species (Uredinales)? Syst. Bot. 7: 48–59. Moran, R. V. 1982. Berberis claireae, a new species from Baja California; and why not Mahonia. Phytologia 52: 221–226.

Figure 12.1
Description and key of the Flora of North America treatment for *Berberis Linnaeus*

[KEY]

Berberis Linnaeus

1. Stems spiny; leaves simple; plants deciduous or evergreen.

    2. Plants evergreen; leaf blades thick and rigid, each margin with 2–4 teeth or shallow lobes, each tooth or lobe 1–3 mm, tipped with spine 1.2–1.6 _ 0.2–0.3 mm; stems tomentose. ..... 5. Berberis darwinii

    2. Plants deciduous; leaf blades thin and flexible, margins entire or each with 3–30 teeth, each tooth 0–1 mm, tipped with bristle 0.2–1.4 _ 0.1–0.2 mm; stems glabrous.

        3. Inflorescences of solitary flowers or umbellate; margins of leaf blade entire. ..... 4. Berberis thunbergii

        3. Inflorescences racemose; margins of leaf blade entire or toothed.

            4. Bark of 2d-year branches gray; each margin of leaf blade with (8-)16–30 teeth; racemes 10–20-flowered. ..... 3. Berberis vulgaris

            4. Bark of 2d-year branches brown, purple, or reddish; leaf blade entire or each margin with 3–12 teeth; racemes 3–15-flowered.

                5. Leaf blade oblanceolate or sometimes narrowly elliptic, apex rounded or rounded-obtuse; surfaces adaxially ± glaucous. ..... 1. Berberis canadensis

                5. Leaf blade narrowly elliptic, apex acute to obtuse or rounded; surfaces adaxially not glaucous, often shiny. ..... 2. Berberis fendleri

1. Stems not spiny; leaves compound; plants evergreen.

    6. Racemes loose (rather dense in B. harrisoniana), 1–11-flowered; bracteoles acuminate.

        7. All leaves 3-foliolate; terminal leaflet sessile.

            8. Terminal leaflet blade 0.9–2 cm wide; berries red. ..... 6. Berberis trifoliolata

            8. Terminal leaflet blade 2.2–3.2 cm wide; berries blue-black. ..... 7. Berberis harrisoniana

        7. Leaves 5-11-foliolate (sometimes a minority of leaves 3-foliolate); terminal leaflet stalked on most or all leaves.

            9. Marginal spines of leaflet blade 0.4–1.2 _ 0.1–0.15 mm.

                10. Bracteoles (at least proximal ones) leathery, spine-tipped; berries white or red, somewhat glaucous, 9–16 mm, usually hollow; c Texas. ..... 11. Berberis swaseyi

                10. Bractoles usually membranous, seldom spine-tipped; berries yellowish red to red, not glaucous, 5–6 mm, solid; s California. ..... 12. Berberis nevinii

Figure 12.1
(continued)

9. Marginal spines of leaflet blade 0.8–3 _ 0.2–0.3 mm.

    11. Berries dry, inflated, 12–18 mm. ..... 8. Berberis fremontii

    11. Berries juicy, solid, 5–8 mm.

        12. Blade of terminal leaflet mostly 2–5 times as long as wide; berries purplish red. ..... 9. Berberis haematocarpa

        12. Blade of terminal leaflet mostly 1–2.5 times as long as wide; berries yellowish red. ..... 10. Berberis higginsiae

6. Racemes dense, 25–70-flowered; bracteoles obtuse or acute.

    13. Bud scales persistent, 11–44 mm; leaflet blades 4–6-veined from base; anther filaments unappendaged.

        14. Shrubs 0.1–0.8(-2) m; teeth 6–13 per blade margin, 1–2(-3) mm, spines 0.1–0.2 mm thick; native, Pacific Coast states, B.C., and Idaho. ..... 21. Berberis nervosa

        14. Shrubs 1–2 m; teeth 2–7 per blade margin, 3–8 mm, spines 0.3–0.6 mm thick; locally naturalized, se United States. ..... 22. Berberis bealei

    13. Bud scales 2–8(-14) mm, deciduous; leaflet blades 1–3-veined from base (sometimes 1–5-veined in B. amplectens); distal end of each anther filament with pair of recurved teeth (status of this character in B. amplectens unknown).

        15. Leaflet blades abaxially smooth and somewhat shiny (outer surface of cells of abaxial epidermis of leaf plane).

            16. Blade of terminal leaflet 1.3–1.9 times as long as wide; lateral leaflet blades elliptic to ovate or broadly lanceolate. ..... 20. Berberis pinnata

            16. Blade of terminal leaflet 1.7–2.5 times as long as wide; lateral leaflet blades lance-ovate or lance-elliptic. ..... 19. Berberis aquifolium

        15. Leaflet blades abaxially papillose and very dull (outer surface of cells of abaxial epidermis of leaf strongly bulging).

            17. Leaflet blades thin and flexible; teeth 6–24 per blade margin, 0.1–0.25 mm thick; plants 0.02–0.2(-0.6) m. ..... 18. Berberis repens

            17. Leaflet blades thick and rigid; teeth 2–15 per blade margin, 0.2–0.6 mm thick; plants 0.3–2 m (0.1–0.4 m in B. pumila).

                18. Leaflet blades adaxially glossy.

                    19. Teeth 6–12 per blade margin; n California and Oregon. ..... 17. Berberis piperiana

                    19. Teeth 3–5 per blade margin; Arizona and New Mexico. ..... 16. Berberis wilcoxii

Figure 12.1
(continued)

18. Leaflet blades adaxially dull, ± glaucous.

    20. Blade margins strongly crispate, each margin with 3–8
    teeth. ..... 13. Berberis dictyota

    20. Blade margins plane to undulate or, if crispate, each
    margin with 9–15 teeth.

        21. Plants 0.2–1.2 m; each blade margin with 9–15
        teeth. ..... 14. Berberis amplectens

        21. Plants 0.1–0.4 m; each blade margin with 2–10
        teeth. ..... 15. Berberis pumila

Figure 12.1
(continued)

country's largest scientific collaborations. In sum, FNA provides the only comprehensive, scientifically authoritative treatment of all the plants of North America throughout their range. FNA thus provides a unified conspectus of the flora, is an essential tool for plant identification, and provides systematic discussions of foundational problems and promising new research and exploration.

The FNA Publishing Process: Organization, Medium, and Problems
Authors are invited by the FNA Editorial Committee to prepare treatments describing various taxa, and collections of taxonomic treatments, including distribution maps and illustrations, are then reviewed, databased, and assembled into published volumes. The FNA Editorial Committee is responsible for identifying experts, soliciting their participation, and managing the various review processes.

The project's daily activities have been coordinated at the Missouri Botanical Garden in the FNA Organizational Center. A mix of mostly paper and some electronic documents are used throughout. A total of five distinct review processes (taxonomic, regional, nomenclatural, bibliographic, and technical) that range from review of scientific content and style to evaluation of a taxon's conservation status are performed (sometimes repeatedly) on each treatment once it is submitted. In fact, approximately 100 discrete events are associated with the publication of a single manuscript, and each event must be tracked and coordinated with other events that are occurring serially and concurrently (see figure 12.2).

In figure 12.2, boxes labeled Technical Editor, Map Illustrator, Prep for OUP, Prep for WWW, Project Coordinator, and Artist all signal Organizational Center

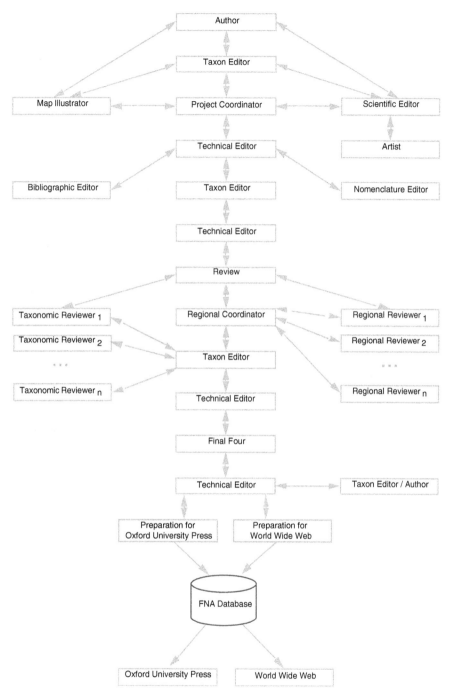

Figure 12.2
Manuscript flowchart for the Flora of North America project showing the steps involved in the edit and review process for each manuscript
Source: Adapted from Schnase et al. (1997).

activities. Boxes labeled Taxon Editor and Technical Editor, for example, refer to the same person or role but at various stages of the manuscript vetting process, demonstrating that the same person (for example, the Technical Editor in the Organizational Center) handles the manuscript multiple times. Finally, boxes to each side of the center vertical axis illustrate that steps occur concurrently as well as sequentially. For example, both the Taxon Editor and Technical Editor would see the manuscript at least four different times, while various reviewers would be reviewing the manuscript. Excluded from the figure are some of the author's functions as well as multiple back-and-forth exchanges between roles and people.

As many as 300 manuscripts can go into a single volume, and progress is coordinated across several volumes simultaneously. This means that participants in the FNA project must effectively articulate a vast number of intradocument, interdocument, intravolume, and intervolume relationships among activities and participants.

Recently, the FNA project has undergone extensive reorganization. A major strategy has evolved to establish semiautonomous editorial and service centers, the latter providing and coordinating various service activities (bibliographic, nomenclatural, artistic, GIS mapping, and portal), making it possible for editorial teams to work independently and in parallel on major taxonomic groups throughout the entire publication process, from author solicitation, through edit and review, to electronic and print publication (see figure 12.3). Authors (A_1-A_n) prepare treatments and submit manuscripts (M_1-M_n) that eventually make their way through the appropriate editorial center (each editorial center supervising the publication of multiple, or M_n, manuscripts). Arrows depict major paths of information flow between the treatment preparation and database and publishing components within the project's editorial centers (EC_1-EC_n) (adapted from Schnase et al. 1997).

The FNA project's organization can now be depicted as semiautonomous editorial centers directly communicating and coordinating their work with each other as needed. This distributed work arrangement replaces the centralized arrangement that previously prevailed.

In sum, until recently the FNA project has attempted to adapt traditional methods of small-scale print publishing to a large-scale databasing and electronic publishing effort. Volumes 1 to 3 have been published, and intensive work is well under way on volumes 4 and 22 to 24, but with over 800 participants who are scattered across North America and are involved in a decades-long effort to review hundreds of manuscripts in various stages of review by different sets of participants at any one time, traditional publishing methods have proved inadequate and inefficient. The

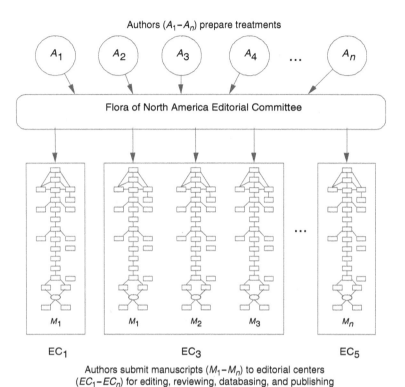

Authors (A_1–A_n) prepare treatments

Flora of North America Editorial Committee

Authors submit manuscripts (M_1–M_n) to editorial centers
(EC_1–EC_n) for editing, reviewing, databasing, and publishing

Figure 12.3
Overview of the major work processes in the Flora of North America project

present research constitutes an initial attempt to explain the expansive reorganization of the FNA project and the role of Collaborative Publishing Services in its collective reconstruction.

Collaborative Publishing Services (CPS)

In terms of hope and vision, FNA is a big project. It is a densely stratified workspace
where many collaborators process differing and sometimes conflicting agendas
across heterogeneous and (semi)autonomous information infrastructures. Because of
the huge scope of the project and the staggering number of intertask and interpersonal dependencies that must be articulated in publishing thousands of manuscripts
contributed by several hundred geographically dispersed scientists, traditional
methods of small-scale print publishing have not scaled. New tools had to be

developed to enable new work processes if FNA was not to fall even further behind its planned publication schedule (see Spasser 1998, 2000; Tomlinson, Spasser, and Schnase 1997; and Tomlinson, Spasser, Sánchez, and Schnase 1998 for research on Collaborative Publishing Services development and evaluation).

Collaborative Publishing Services (CPS) is one of a new generation of Web-based coordination environments that is being designed both to help reduce the complexity of database publishing in cases, such as FNA, where complexity arises from the inefficiencies in the publishing practices themselves and to improve the speed and quality of global scientific information gathering and the community construction of large, collaborative scientific databases. It is relatively lightweight, modular, extensible, and scalable. As a Web-based environment, it attempts to integrate communication, information sharing (through creation of a common information object repository), and coordination support features and is accessible by unmodified Web browsers across heterogeneous, autonomous, and distributed information technology infrastructures. With regard to the last, CPS helps reduce the cognitive load entailed by project participation (Tomlinson, Spasser, and Schnase 1997), as well as effectively manage projectwide collaborative load (Spasser 1998).

In developing the first version of the Web-based project coordination and publishing environment called Collaborative Publishing Services, FNA management, in consultation with CBI, brought together a multi-institutional team of sociologists and information scientists to help develop a strategy for streamlining—and, in particular, making simultaneous and parallel—the distributed operations of the FNA project. The experience has resulted in a promising new approach to large-scale project coordination. CPS provides a way of managing project information by means of dynamically constructed activity-and-information spaces, or *role-based views*. Role-based views are derived from the socially constructed roles (such as Project Editor, Taxon Editor, Taxonomic Reviewer, and Author) that already exist within the project. Through these role-based views, project participants organize information and perform the tasks required for the FNA scientific process to work. The various role-based views are delivered through dynamically constructed, personalized Web pages.

CPS accomplishes three things that are significant in terms of streamlining the publishing process:

• CPS lessens individual cognitive load by delegating information and task organizational duties to the interface. CPS does the organization "behind the scenes" and presents to the user information that is in the right place at the right time and in a form that is explicit and easier to use. CPS transforms the task that the scientist confronts by representing it in such a way that the user can readily see exactly how to perform it. This cognitive offloading results in "What you see is what you need to do."

• CPS enhances system performance by enabling massively parallel simultaneous use of a large information space via mapping of permissible views plus suites of operations onto individual and group knowledge resources and capabilities. Distributed cognitive systems like the Flora of North America achieve their computational or information-processing power by superimposing several kinds of representations or representational structures on a single framework. In our case, the framework is a single, very large information space.

• CPS structures not just the information but also the tasks. It simultaneously affords and constrains opportunities for the user to interact with the information. Concentrating information organizational complexity in this manner is a particularly efficient way to facilitate simultaneous and tailored access to, and use of, information over a single, large information space.

CPS V1.1 has been fully operational for the bryophyte component of FNA since September 1997. It was extended to the *Poaceae* group in mid-1998 and to the remaining vascular plant groups in late 1998 on a trial basis. CPS V1.3 became available projectwide to support FNA's distributed centers in early 1999. (For a detailed history of CPS development and early deployment, see Spasser 1998.)

CPS is the primary document-management system for FNA and essentially provides a Web-based, platform-independent mechanism for storing, retrieving, and tracking treatments as they make their way through the FNA publishing process. CPS allows authors, editors, and reviewers of individual families or groups of families (considered "publishing groups") to carry out their work independently. Within CPS, project participants can determine the status of a treatment (where the treatment is in the FNA process, such as "submitted," "out for taxonomic review," or "accepted") and can download copies of all documents pertinent to a treatment, including versions of the manuscript, reviews, e-mail exchanges between authors and editors, electronic copies of maps and illustrations, and so on. Access is determined by a person's role. Editors and editorial center staff, for example, can view all documents. Authors can see only their own files, and reviewers can see only designated copies of the manuscript to review.

Study Approach and Method

The Nature of Social Realist Evaluation

Realist evaluation is a relatively new evaluation paradigm (Henry, Julnes, and Mark 1998; Pawson and Tilley 1994, 1995, 1997) that posits that outcomes (outcome patterns or regularities—the things and behaviors that interest us as social scientists) follow from mechanisms (sets of internally related practices and objects) that are acting in contingently configured contexts. Unpacking the "realist" part of the phrase, we find that realist evaluation is about the real, employs a realist methodology, and has realistic outcomes as its goal. First, evaluation should concern the *real*, but this reality should be stratified and tensed, involving the interplay between the individual and institution, agency and structure, or the lifeworld and system in contingently configured circumstances over time. Social interaction creates interdependencies (Archer 1995 refers to these, when reproduced, as *situational logics*) that develop into real-world customs, rules, and divisions of labor that, in turn, condition (enable and constrain) the interests and opportunities of a given cohort of actors. These are the realities that programs, initiatives, or system implementations seek to change. The key explanatory resource for social realism is not that programs work or that computer-based information systems such as digital libraries are deployed but that such social forms constitute a spiral of new ideas and transforming social conditions and thus, when successful, introduce the appropriate ideas and opportunities to participants and users in the appropriate cultural and socio-organizational conditions. All else follows from such explanatory propositions.

Second, evaluation should follow a *realist* methodology, one that is scientific (systematic and rigorously eclectic) and strives to register the influence of the objective and of the situationally emergent. However, scientific evaluation is not method- or measurement-driven but instead suggests a more extensive role for theory in the formulation of evaluation methodology (that is, realist methodologies are theory-led).

Finally, evaluation needs to be *realistic*, which means both that it is a form of applied research pursued to inform the practice and work of designers, users, and managers and that it modestly attempts to perfect a particular method of evaluation that will work for a specific class of project in well-contextualized circumstances (that is, the scope of realistic evaluation is middle-range).

Realistic perspectives have been widely adopted in the human sciences, such as accounting (Manicas 1993), economics (Lawson 1989, 1994), education (Henry,

Julnes, and Mark 1998), history (McLennan 1981), human geography (Sayer 1985), linguistics (Pateman 1987), nursing (Ryan and Porter 1996; Wainwright 1997), psychoanalysis (Collier 1981; Will 1980), psychology (Manicas and Secord 1983), social psychology (Greenwood 1994; Harré and Secord 1972), and sociology (Archer 1995; Keat and Urry 1982; Pawson 1996; Pawson and Tilley 1997; Stones 1996). Moreover, realism has been employed by several scholars as a comprehensive philosophy of social science (Bhaskar 1986, 1989; Layder 1994; Manicas 1987; Outhwaite 1987; Pawson 1989). Finally, the scientific realist perspective is a dominant approach in philosophy of the natural sciences (Aronson, Harré, and Way 1995; Bhaskar 1978; Harré 1970, 1986). Thus, according to Andrew Sayer (1992, p. xi), "realism is a philosophy of and for the *whole* of the natural and social sciences."

Social Realism in Practice: Methodological Considerations

Because realistic evaluation is at once analytic, stratified, process oriented, and oriented toward explaining change, it builds a coherent framework based on objectivity, intentional agency, and contextual sensitivity. Interventions and systems are always embedded in a range of attitudinal, individual, institutional, and societal processes, and thus observable outcomes are always generated by a range of micro and macro forces that together produce observed situated activities. Stakeholders' capacity for choice is always conditioned (constrained and enabled) by the power and resources of their "stakeholding." Human activity must be understood in terms of its embeddedness, its location within distinct yet interdependent layers of social reality.

Social realist theory, in general, rests on two assumptions: things that are observed in the world are real, *and* those things are products of complex and contingent causal mechanisms that may not be directly accessible to us. In particular, realistic evaluation seeks to understand *for whom* and *in which circumstances* a program works through the study of contextual conditioning. *Context* refers to much more than spatial, geographical, and institutional location; it also refers to the *prior* set of social rules, norms, values, and roles and their interrelationships that condition information system usefulness and usability. A key act of design and analysis is thus to identify the people and situations for whom the system is useful or usable by drawing on success and failure rates of different subgroups of subjects within and between implementations.

Realistic evaluation is pragmatic and outcome-oriented. Social realists recognize actors as being active agents who could always do otherwise (but usually have good reasons not to). Thus, realist evaluation is sensitive to these agents' motives, objectives, and goals. Evaluators need to understand what the outcomes of an initiative are and how they are produced. Programs or systems cannot be understood as undifferentiated wholes, or as "things" because they trigger (or fire) multiple mechanisms having different effects on different subjects in different situations and so produce multiple situated outcomes. Outcomes are not inspected simply to verify whether systems are used but are analyzed to discover if conjectured mechanism or context theories are confirmed and the extent to which they transfactually apply. In effect (Pawson n.d., pp. 15–16),

> The key is to empty the notions of mechanisms, contexts and outcomes of their architectonic splendour and ... to see them as describing local resources, bounded capacities, specific choices, habitual forms of reasoning, which then act as the routine ingredients for hypothesis-making.... What is needed is a method which gets closer to the now-you-see-them-now-you-don't patterns of social activities. We need a middle-road strategy for theory-construction so that it becomes neither a jumble of ad hoc stories about why particular events are connected, nor a set of critical claims for the ubiquitous (if metaphorical) presence of master mechanisms.

Purpose(s) of the Study, or Why Study the Flora of North America Project?

The purposes of this research are threefold:

- To contextualize the use of CPS in terms of the changing organizational form of the FNA project,
- To study the organizational issues that made the adoption of CPS as well as the construction of the FNA DL problematic, and
- To explain, in social realist terms, the reorganizational pressures faced by the FNA project and the role of CPS in its organizational reconfiguration.

Thus, the overarching focus of this research is to identify the underlying mechanisms generative of the interorganizational dysfunction that has been observed.

A Minor Digression on Activity Theory

Because social realism is a mechanics of explanation (supplying only a scaffolding for hypothesis generation and an ontological chassis for collection of evidence) (Pawson n.d.), it makes no assumptions about the content or substance of social reality. Thus, scientific realism is substantively neutral; it ushers in or presumes no specific theory of social life. Accordingly, realism has been conjoined with structuration theory (Pawson 1996; Pawson and Tilley 1997), with Habermasian critical theory (Morrow and Brown 1994), and sketchily but suggestively with activity

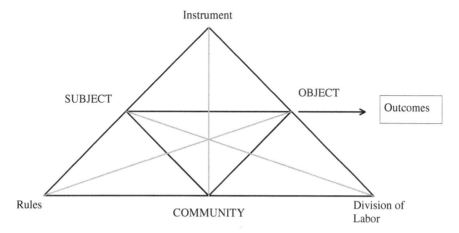

Figure 12.4
Yrjö Engeström's basic model of a human activity system

theory (Hjørland 1997). Building on Birger Hjørland's work, I propose that realism provides a solid foundation for activity theoretic analyses of social life in general and for work and technology in particular, while activity theory provides a conceptually and substantively rich vocabulary for explanatory reasoning about technologically mediated social praxis.

Activity theory is a philosophical and cross-disciplinary framework for studying different forms of human practices and social life in a multilevel, stratified manner developmentally in time and through space. Activity theorists (see, for example, Engeström 1987, 1990, 1991) consider the activity (or activity system) to be the minimally meaningful unit of study. An activity system involves an activity undertaken by a human actor (either individual or collective) motivated toward or by an object (in the sense of an objective), mediated by artifacts and tools, and conditioned by emergent community structural or institutional properties (such as rules, conventions, and roles and divisions of labor).

Yrjö Engeström's (1987) basic model of an activity is depicted in figure 12.4. An activity system is a triply mediated, outcome-oriented unit of analysis. Instruments (resources) mediate the relationship between subject and object; social rules and regulations mediate interpersonal relationships; and roles (division of labor) condition the relationship between a community of practice and its negotiated object. This chapter builds on Engeström's seminal work, and a variant of his basic model of an activity system figures prominently in subsequent analysis. Summarizing the major tenets of activity theory, all human activity is at once artifactually mediated, pragmatic (or objectively motivated), situated, provisional (historically developing or evolving), and more or less contested (Blackler 1995).

Data-Collection Methods

Activity theory is employed as a middle-range realist theory to initiate, provisionally order, and govern the ongoing conduct of data collection and analysis. Data have been collected from several sources, employing a variety of techniques, such as documentary analysis and extended participant-observation. The idea is to take as many "cuts" at the data from as many angles as is feasible to maximize the strength, density, and validity of theoretical ideas that emerge from data collection and analysis.

Such an approach automatically contributes to multiple triangulations, which strengthens the multiperspectival validity of the work's conceptual framework and findings (Yeung 1997). Specifically, in terms of the present research, triangulation across sources and techniques is especially beneficial to theory generation, as the analytic strategy provides multiple perspectives on an issue, supplies more (and higher-quality) information on emerging concepts, allows for cross-checking of assumptions and conjectures, and yields stronger substantiation of constructs (Layder 1994, 1998; Orlikowski 1993; Yeung 1997; Yin 1994).

A source of data especially important to the present research is the intensive participation of the researcher in the ongoing activities of the FNA project. As the bioinformatics coordinator and CPS analyst, I have had access to the inner workings of project management as well as to the informal and more or less spontaneous activities in the Organizational Center (OC) at the Missouri Botanical Garden since May 1997. This immersion in the FNA project as both a participant and an observer provides a rich source of data, as well as a deeply grounded context for the analysis of other sources of data. Similar to that described in Torbjorn Näslund (1997, p. 173), "my role can be characterized as a *participant observer* of issues closely related to usability, and as an *observer participant* regarding other issues in the project." Both role variants facilitated the sort of unstructured interactive interviewing that makes possible the study of individual in their contingently configured casual contexts. In this way, data collection for research and for formative system evaluation happily coincided.

Moreover, a large and *living* corpus of such written materials as organizational documentation and communications (such as minutes from committee meetings)

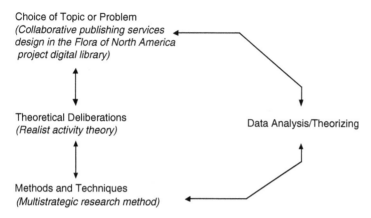

Figure 12.5
Realist case study research methodology
Source: Adapted from Layder (1998). Italicized elements refer to its application in the present research.

about the development and current use of written coordinative mechanisms and the design of CPS were analyzed to help further contextualize and situate the participant-observation. In addition, notes from selected interviews, use sessions, and meetings were content analyzed. Another important source of data was feedback from users as they learned to use CPS over time. Communication with the analyst, the FNA bryological community's primary point of contact with the system development effort, was facilitated by Mailto links on each CPS Web page. Input from CPS users was important because their questions and comments were used both to contextualize system use and, as a filtering mechanism, to elicit extended feedback from selected users. Feedback from users about their difficulties or problems using CPS provides protocol-analytic data to the extent that users were encouraged to reconstruct what they did and saw and even to anticipate what they planned to do.

Case studies allow researchers to trace phenomena over time, which is essential if the interplay of (inter)action and its causally efficacious conditions are to be effectively separated and analyzed. The realist case study approach employed in the present research is depicted in figure 12.5. Inspired by key realist and activity theoretic principles, the case study approach employed in this research is deliberately sensitive to and equally privileges the contextual conditions of behavior. Moreover, by exhaustively specifying and contextualizing the findings (or outcomes), previously or emergently developed theory can be used as a revisable template for constantly comparing the results of other social realist case studies.

Study Findings and Analysis

The approach to understanding the evolution of the Flora of North America project that is summarized here is based on realist activity theory (discussed earlier in the chapter) and has been developed during an eighteen-month, ongoing study of the FNA project. Because detailed presentation of the empirical work is beyond the scope of this chapter, aspects of it are used to illustrate the application and development of the framework used to analyze the FNA project. Thus, while this presentation can only suggest ways in which the theoretical approach developed herein can be used to evaluate trajectories of DL development, in future work this theoretical framework will be explored as a more general conceptual framework for framing DL evaluation.

The present work focuses on describing how attempts to resolve interorganizational problems (contradictions) among the participants in the FNA project have collectively led to its expansive development (Engeström 1991) and reconceptualization, which in turn, at least potentially, provides a more hospitable environment for the development and deployment of CPS as a publishing coordination environment and viably sustainable boundary infrastructure (Bowker and Star 1999). A couple of the steps that Engeström includes in cycles of expansive development will be the focus of future work. An additional way in which this presentation simplifies actual cycles of expansive development is by suggesting that the steps progress in a linear fashion with no overlap between them. In fact, reality as lived and experienced is never so tidy. The cycle of expansive development proceeded as described below.

Ethnography of Trouble: Describing and Explaining the Initial Situation

In figure 12.6, FNA depicts an activity network whose primary communities of activity are the Center for Botanical Informatics (CBI); Flora of North America Organization (FNA O), which represents those responsible for the execution, planning, and oversight of the entire project; and the Flora of North America Organizational Center (FNA OC), the workaday, administrative center of project coordination physically located at the Missouri Botanical Garden in St. Louis. Since its initial funding in 1989, FNA OC personnel have been the only full-time paid staff in the FNA project (more about this below).

Figure 12.6 depicts the FNA activity network as not only multiply mediated but multiply stratified. Material mediation is embedded in sociocultural mediation.

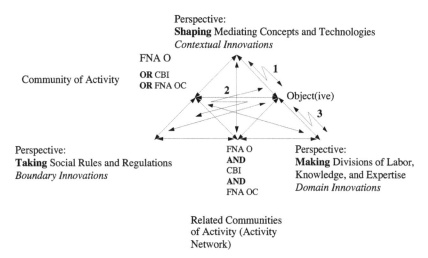

Figure 12.6
An activity network within the Flora of North America project
Source: Adapted from Blackler, Crump, and McDonald (2000).

Communities of activity are embedded in activity networks (networks of related activity systems). And the material, technological, and socio-organizational infrastructures of work-related social interaction (object-oriented resources, rules, and roles derived from the relatively constrained objective(s) of work) generalize to processes of perspective shaping, taking, and making when activity systems are properly located in the networks of related communities of activity in which they customarily operate. The numbered lightning bolts indicate "contradiction sets" that organize the following discussion.

This approach has been applied to identify and understand key differences between the three communities of activity that are vital to the project's future viability. Factors that help account for differences include approaches to work adopted by communities of activity, management and leadership styles, and the ways that communities of activity define themselves in relation to others in the context of the project as a whole. These factors are mediated by conceptual and material artifacts, tools, and instruments (contextual resources), formal and informal social rules and regulations (boundary formation and transgression—that is, innovation), and divisions of labor, expertise, and knowledge or know-how (domain-specific innovations).

As Frank Blackler, Norman Crump, and Seonaidh McDonald (2000, p. 18) explain, with regard to their two-year study of an international supplier of high-tech electro-optical devices to the military:

The infrastructure and priorities that make it possible for particular expert communities to focus on shared goals, to develop an identity, and to act competently can also act as barriers to close collaboration across different communities. Interactions across activity networks are easier to manage in times of relative stability when systems can be developed to minimise conflict or misunderstandings, and more difficult to achieve in times of uncertainty when priorities, methods and group identities may need to be re-examined.

For example, both CBI and the FNA OC are physical places with colocated members, and each community of activity has formed a strong sense of shared identity and goals that have interfered with their ability to cooperate to develop an electronic tool to facilitate the work of the project. Moreover, while CBI functions as a source of specialized consultants who are charged by the FNA O to help administer (Zuboff 1989) the project, members of the FNA OC are the only full-time paid staff of the FNA project. Unlike both CBI and FNA OC personnel, members of FNA O are in effect volunteers who are geographically dispersed and who often serve multiple roles for the project.

Given their differing and inevitably conflicting roles and their equally divergent conception of the objects of their work, the project has been riddled with destabilizing, structurally embedded contradictions from its initial funding. The FNA OC considered their primary objective to be the timely production of the print *Flora of North America* (Flora of North America Editorial Committee, 1993–1997). CBI was primarily concerned with the design, development, and deployment of CPS as the Web-based coordination environment for the construction of the FNA digital library—the electronic *Flora of North America*. Finally, the FNA O's main concern was the overall identity of the project and its continued funding, which, until very recently, required them to fund FNA OC operations with monies ostensibly earmarked for FNA informatics development.

The FNA OC has never been involved in decisions concerning CBI's role as the project's informatics consultant; these decisions have been made by those in the FNA O who were responsible for managing the project. Accordingly, the FNA OC became extremely competitive with, and suspicious of, CBI's efforts to develop an electronic infrastructure to enable the FNA O to adaptively construct its floristic DL and until recently successfully lobbied for substantial internal funds to enable it to continue its work. Thus, it is easy to see (even in this brief sketch) how differences between the communities of activity that constitute the FNA activity network (in terms of both identity and objectives) generate profound difficulties and threaten the viability and continuity of the project in its entirety. In other words, community of activity boundary-spanning activities that would lead to projectwide cooperation and overall identity construction and reinforcement have been severely undermined by "pathogens" resident in the very structure or configuration of the FNA network.

Analysis of Contradictions: Historically and Empirically

As can be seen in figure 12.6, three sets of contradictions are identified (by three numbered two-headed lightning bolts). Each focuses on and emanates from differing conceptions of the objects of work within and between communities of activity, and these communities collectively have been fueled by cyclical yet ongoing funding pressures. Not only do each of the communities of activity have different conceptions of the same object, but in some cases, the objects of their work themselves actually differ. For example, the object of work for CBI is CPS, for the FNA OC the print *FNA*, and for the FNA O a fully and renewably funded project (which because of funding pressures and conflicting work exigencies is being transformed from the paper-based production of *FNA* as a traditional multivolume reference set to the construction of a floristic DL, of which the reference set is one product).

Referring to figure 12.6, the focus of contradiction set 1 is on the tension between the object of work and mediating instruments, which jointly shape the context of work. For example, the model of FNA as a floristic DL (a conceptual mediating instrument) used by CBI to guide its work on CPS (its object of activity) is vastly different from and in conflict with the model of FNA as a print reference set used by FNA OC to guide its work on the print *FNA*, both of which differ from the management and funding strategies used by FNA O to shape the FNA project into a fundable entity.

In terms of facilitating projectwide coordination strategies and developing some sense of overall identity, contradictions emanating from differing objectives are profoundly problematic. For example, the contextual resources employed by CBI reinforced its domain-specific (informatics consulting) orientation and approach— its sense of technical expertise. However, by restrictively employing the traditional tools of paper-based publishing, based on its conceptual model of small-scale print publishing, the FNA OC developed and reinforced yet another domain-specific orientation and approach. The result of these radically different contexts was to create rather than cross boundaries, rendering the development of either participative (interaction-related) or reificative (artifact-related) connections problematic if not impossible (Wenger 1998). In terms of the latter, differing investments in domain-specific perspective-making activities prevented the development of coordinating boundary objects (Star 1989; Star and Griesemer 1989).

Because each community of activity—CBI, the FNA OC, and the FNA O—construes its purpose and place in the overall project differently, additional tensions ramify throughout the FNA activity network (see lightning bolts/contradiction sets 2 and 3 in figure 12.6). For example, FNA OC construed its object as the publisher-directed publication of the print *FNA* and thus required all manuscripts to be processed locally and physically (according to a set of elaborate rules and procedures) by its staff to ensure print publication standards were being met. This clearly conflicted with the conception of *FNA* as first and foremost a digital library, of which one revenue-generating product is the print *FNA* multivolume reference set. Moreover, the role of the FNA OC as the centralized, projectwide center of administrative coordination conflicted with the development and especially the implementation of CPS as a means to facilitate the fundamentally distributed nature of the project's work and, even more fundamentally, with the conception of *FNA* as a digital library and as a fundable test bed for studying and supporting large-scale and highly distributed knowledge work.

Until recently, the FNA activity network could be diagrammatically characterized as in figure 12.7. The network was deeply fragmented because of differing and incompatible objects, identities, and differing contexts of work. Instead of providing a common or shared object of interest and work, CPS became an arena of conflict. As Richard J. Boland Jr. and Ramakrishnan V. Tenkasi (1995, p. 362) astutely note, "Boundary objects can, of course, be a center of intense conflict as easily as one of cooperative effort. Creating and reshaping boundary objects is an exercise of power that can be collaborative or unilateral." In terms of the analysis above, such a situation is likely to develop in the absence of a collectively held vision of future direction. Not surprisingly, the outcome has been a complex project whose work is behind schedule and whose very existence is threatened by votes of no confidence on the part of benefactors and funding agencies.

Ethnography of Transformation: Describing and Explaining the New Situation
Recently, the FNA project has undergone a dramatic change in its structure to resolve the contradictions described above (as well as many others) under renewed and intense funding pressures. The current funding for the project ended in Decem-

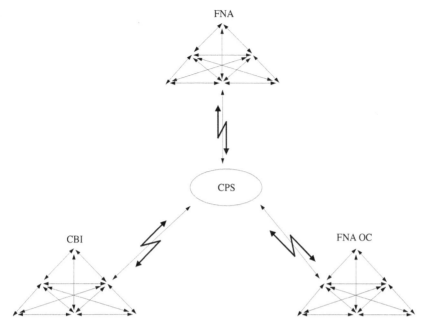

Figure 12.7
Contradictions in an activity network within the Flora of North America project

ber 1998, which forced project management to confront several particularly inter-
necine contradictions—*FNA* as primarily a print reference set or a digital library;
the adequacy of small-scale print publication practices or the necessity for implementing Web-based database publishing and DL construction techniques; and the
FNA project as a centrally administered activity system or fundamentally distributed
activity network (in the parlance of realist activity theory). The way to resolve these
contradictions has come in the form of a newly co-constructed organizational form
and context. As can be seen in figure 12.8, the FNA project has been radically
reconstrued as a decentralized network of semiautonomous editorial centers (EC)
whose work will be at least partially coordinated by a newly designed CPS. The
FNA OC has been reconfigured as one of five editorial centers (it will be the site for
a couple service centers, as well). In effect, CPS has been upwardly (or forwardly)
contextualized by creating a new, more inclusive perspective for the future of the
entire FNA activity network: from the FNA project as centralized structure centered
in the FNA OC to FNA as network of semiautonomous editorial centers; from print
FNA multivolume reference set to *FNA* as a floristic digital library; and from CPS as

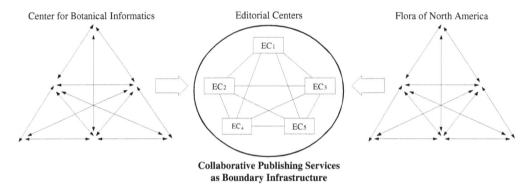

Figure 12.8
The Flora of North America project as a distributed network of semiautonomous publishing centers

optional tool or application to CPS as an integral working coordinative boundary infrastructure serving multiple communities of activity, including the newly reconstituted FNA OC, simultaneously.

However, apparent success may be short-lived. As Engeström (1990, p. 193) reminds us, a "new descriptive 'what' artifact [such as FNA as a floristic digital library] may provide for temporary involvement and discussion—but it *alone* is not likely to achieve much change in the practical actions of the community members. Correspondingly, a new algorithm or a new explanatory model [such as CPS] *alone* will probably remain a curiosity, no matter how obvious its need may be to the researcher [such as CBI]." Neither involves a needed change in perspective, and thus both only downwardly contextualize. Then again, when the FNA project was explicitly reconceptualized as the highly distributed project it always had been (acknowledging its true nature, what about the project makes it what it is) and when *FNA* was formally recognized as a floristic DL, a *prospective*, perspective-changing, upwardly contextualizing artifact was co-constructed. It was only then that project participants could finally agree that CPS will function as the accepted environment for *FNA* DL construction.

Finally, however, for the social realist such transformative activity always occurs in some extant context and thus begins from some definite (if unspecified) somewhere. In the case of the FNA project, the new vision of it as a distributed network of semiautonomous centers was enacted on a small scale from the beginning of this research by bryologists whose lead editor is based at the New York Botanical Garden (NYBG). Given that they study lower, nonvascular plants (as opposed to the rest of the FNA project, which covers the vascular flora, the higher plants, of North America), the bryology community of activity was structured differently from the start. They were included in the FNA project several years after the project was initially funded, only after strenuous lobbying that a flora of North America could not be considered complete if the bryoflora were omitted.

While they were begrudgingly admitted to the project and are represented on projectwide committees, the bryologists have formed an especially tightly integrated community of activity, often holding their own editorial meetings and communicating intensively among themselves. And crucially for the present work, they have, for the most part, chosen not to work with the FNA OC and instead to use earlier versions of CPS to coordinate the publication of their FNA contributions. While there have been problems with these earlier versions of CPS, the bryology community of activity has worked (not always harmoniously) with the CBI informatics team to iteratively develop CPS and, in the process, have modeled the new perspective of the FNA project as a truly distributed network and also as a more adaptive and constructive approach to information infrastructure design, development, and possible deployment. In fact, the transformative potential for the project as a whole was prefigured and at least partially modeled by the autonomous emergence shortly after this research began of the bryoflora center at the New York Botanical Garden.

As the example of the bryology community of activity illustrates, the FNA project is both highly stratified and relationally differentiated (Stones 1996). Social systems (activity systems or communities of activity) do not exist in a vacuum but are always in contact with the differentiated powers and liabilities of the objects, processes, people, and practices in other systems. Indeed, "for the standard view of science, the world is a determined concatenation of contingent events; for the realist it is a contingent concatenation of real structures. *And this difference is monumental*" (Layder 1990, p. 14).

The difference was certainly monumental for the bryology group at NYBG, which in turn made a monumental difference to the development of the FNA project as a whole. In the open, stratified, and differentiated world that is the FNA project, its development could never have been predicted because the consequences of the interaction between the unique configuration of structural processes (such as the necessary independence of the bryology community of activity) and such contingently intervening factors as funding deadlines and national research priorities can never be captured by simple empiricist logic (for the realist, human beings are spontaneous loci of causality—generators of social behaviors rather than simple transmitters of external constraints). The ontology of any nontrivially complex activity network is simply too richly variegated. However, what has happened can be causally explained (or evaluated), which is the vital and generative province of realist social science.

Conclusions

The Flora of North America project and collaborative publishing services are emerging and evolving. Drawing conclusions about the ultimate form of the former or the usefulness and productive integration of the latter are thus impossible. However, while the new funding cycle that triggered the crystallization of the FNA project as a network of semiautonomous centers has not yet begun, the vision described in the ethnography of transformation in this chapter is guiding all current project developments, and it is highly likely that the trajectories outlined above are at minimum directionally accurate.

However, while conclusions are hard to come by, the analysis in the present work has demonstrated the utility of a realist activity theory conceptual framework for ecologically evaluating the socio-organizational and institutional embeddedness of computer-based information systems, such as digital libraries, and explaining the contextual conditioning of our experience of them. Such a framework includes a concern with intrinsic (or generative) causation (such as latent systemic contradictions); a stratified and relationally differentiated view of social life (such as material and socio-organizational mediation—activity systems related by shared objects and identities into activity networks); and cycles of expansive development, learning, and object co-construction (in terms of the temporal patterning and interplay among structural conditioning, social interaction, and structural reproduction, elaboration, or transformation).

For the social realist, all valid social scientific research (and even more so for the evaluation of social systems, of which DLs are paradigms) must inhabit the middle range somewhere between the Scylla of fables of continuous, unconditioned, and unfettered local construction and the Charybdis of grand narratives of the inexorable, depeopled, and totalizing march of master mechanisms. To paraphrase Karl Marx's celebrated dictum, people make history but not in circumstances of their choosing. Instead, the social realist adopts a research strategy of appropriate form and scope for leading and, perhaps even more important, federating (cumulating) empirical inquiry. This type of empirical inquiry leads to the development of a generic fuselage of concepts, which in turn provides a scaffolding for middle-range hypothesis generation and an ontological chassis for evidence collection (Pawson n.d.). The social realist "moves from one case to another, not because *they* are descriptively similar, but because [*she* or *he* has] ideas that can encompass them both (Pawson and Tilley 1997, p. 119). It is axiomatic for the social realist who is evaluating social systems, such as DLs, that any useful insights provided by an evaluation depend on the clarity and acuteness of its explanatory vision and means.

So let us end where we began, with the assertion that this research would be real, realist, and realistic. First, and obviously, this research deals with the real activity network that is the FNA project. Additionally, the organizational incoherencies and the interorganizational conflicts and contradictions have been all too real. Second, its methodology is realist in that thoroughly contextualized generative mechanisms have been identified. Last, this research has been realistic; its applied focus is on better understanding those contradictions to improve the fit between CPS, the Web-based coordination environment, and the organizational form of the FNA project as a truly distributed Web-enabled activity network.

Finally, while the conceptual foundations of the framework outlined in this work have been widely accepted throughout the natural and social sciences, social realism has yet to find acceptance in the social informatics, computer-supported cooperative work, and human-computer/computer-human interaction communities (all of which include at least aspects of research on DLs). Thus, to the extent that the analysis is valid and compelling, it is hoped that the ideas contained herein may find greater acceptance in better understanding DL evaluation in its tensed, dialectical interplay with design, development, deployment, and use.

Postscript

As of this writing, the fate of the FNA project has taken yet another unanticipated turn. The NSF grant that outlines the new organizational form of the project has been withdrawn. It was decided that too much of the project is incomplete and that not enough was done for the previous grant. FNA leaders were told not to apply for funding until at least 60 percent of the project was complete (estimates are that about a quarter of the project is now finished). The FNA OC is being subsidized by the Missouri Botanical Garden, CBI has been eliminated, and CPS continues to be used by the bryology group. Thus, whether FNA truly becomes the distributed network of editorial and service centers outlined in the grant application, whether CPS can be successfully adopted now that CBI is out of the picture, and whether the fit between FNA's organizational structure and state-of-the-art information technologies can be improved so that the former can effectively capitalized on the availability of the latter are all unclear and to be determined.

Acknowledgment

This research was conducted while the author was at the Center for Botanical Informatics, LLC, Missouri Botanical Garden, St. Louis, Missouri, and was supported in part by grants from the Andrew W. Mellon Foundation and the National Science Foundation (DEB-9505383).

References

Archer, M. S. 1995. *Realist Social Theory: The Morphogenetic Approach*. Cambridge: Cambridge University Press.

Aronson, J. L., R. Harré, and E. C. Way. 1995. *Realism Rescued: How Scientific Progress Is Possible*. Chicago: Open Court.

Bhaskar, R. 1978. *A Realist Theory of Science* (2nd ed.). Brighton, UK: Harvester.

Bhaskar, R. 1986. *Scientific Realism and Human Emancipation*. London: Verso.

Bhaskar, R. 1989. *The Possibility of Naturalism: A Philosophical Critique of the Contemporary Human Sciences* (2nd ed.). Brighton, UK: Harvester Wheatsheaf.

Bishop, A. P., and S. L. Star. 1996. Social Informatics for Digital Library Use and Infrastructure. In M. E. Williams, ed., *Annual Review of Information Science and Technology* (vol. 31, pp. 301–401). Medford, NJ: Information Today.

Blackler, F. 1995. Knowledge, Knowledge Work and Organizations: An Overview and Interpretation. *Organization Studies*, 16(6), 1021–1046.

Blackler, F., N. Crump, and S. McDonald. 2000. Organizing Processes in Complex Activity Networks. *Organisation: The Interdisciplinary Journal of Organisation, Theory and Society*, 7(2), 277–300.

Boland, R. J., Jr., and R. V. Tenkasi. 1995. Perspective Making and Perspective Taking in Communities of Knowing. *Organization Science*, 6(4), 350–372.

Bowker, G. C., and S. L. Star. 1999. *Sorting Things Out: Classification and Practice*. Cambridge, MA: MIT Press.

Collier, A. 1981. Scientific Realism in the Human World: The Case of Psychoanalysis. *Radical Philosophy*, (29), 8–18.

Elliott, M., and R. Kling. 1997. Organizational Usability of Digital Libraries: Case Study in Legal Research in Civil and Criminal Courts. *Journal of the American Society for Information Science*, 48(11), 1023–1035.

Engeström, Y. 1987. *Learning by Expanding: An Activity-Theoretical Approach to Developmental Research*. Helsinki: Orienta-Konsultit Oy.

Engeström, Y. 1990. *Learning, Working, and Imagining: Twelve Studies in Activity Theory*. Helsinki: Orienta-Konsultit Oy.

Engeström, Y. 1991. Developmental Work Research: Reconstructing Expertise through Expansive Learning. In M. I. Nurminen and G. R. S. Weir, eds., *Human Jobs and Computer Interfaces* (pp. 265–290). Amsterdam: Elsevier Science.

Flora of North America Editorial Committee. 1993–1997. *Flora of North America: North of Mexico* (Vols. I–III). New York: Oxford University Press.

Greenwood, J. D. 1994. *Realism, Identity, and Emotion: Reclaiming Social Psychology*. Thousand Oaks, CA: Sage.

Harré, R. 1970. *The Principles of Scientific Thinking*. London: Macmillan.

Harré, R. 1986. *The Varieties of Realism: A Rationale for the Natural Sciences*. Oxford: Blackwell Scientific.

Harré, R., and P. F. Secord. 1972. *The Explanation of Social Behaviour*. Oxford: Blackwell Scientific.

Henry, G. T., G. Julnes, and M. M. Mark, eds. 1998. *Realist Evaluation: An Emerging Theory in Support of Practice*. New Directions for Evaluation No. 78. San Francisco: Jossey-Bass.

Hjørland, B. 1997. *Information Seeking and Subject Representation: An Activity-Theoretical Approach to Information Science*. Westport, CT: Greenwood.

Keat, R., and J. Urry. 1982. *Social Theory as Science* (2nd ed.). London: Routledge & Kegan Paul.

Lawson, T. 1989. Abstraction, Tendencies and Stylised Facts: A Realist Approach to Economic Analysis. *Cambridge Journal of Economics*, 13(1), 59–79.

Lawson, T. 1994. A Realist Theory for Economics. In R. E. Backhouse, ed., *New Directions in Economic Methodology* (pp. 257–285). London: Routledge.

Layder, D. 1990. *The Realist Image in Social Science*. London: Macmillan.

Layder, D. 1994. *Understanding Social Theory*. London: Sage.

Layder, D. 1998. *Sociological Practice: Linking Theory and Social Research*. London: Sage.

Magill, R. E., T. M. Barkley, N. R. Morin, J. L. Schnase, and B. M. Thiers. 1999. *Flora of North America: A Framework of the Construction and Long-Term Maintenance of North American Botanical Knowledge*. Unpublished manuscript.

Manicas, P. T. 1987. *A History and Philosophy of the Social Sciences*. Oxford: Blackwell Science.

Manicas, P. T. 1993. Accounting as a Human Science. *Accounting, Organizations and Society*, 18(2/3), 147–161.

Manicas, P. T., and P. F. Secord. 1983. Implications for Psychology of the New Philosophy of Science. *American Psychologist*, 38(4), 399–413.

McLennan, G. 1981. *Marxism and the Methodologies of History*. London: Verso.

Morin, N. R. 1991. Beyond the Hardcopy: Databasing Flora of North America Information. In E. C. Dudley, ed., *The Unity of Evolutionary Biology (Proceedings of the International Congress for Systematic and Evolutionary Biology IV)* (Vol. II, pp. 973–980). Portland, OR: Dioscorides Press.

Morin, N. R., R. D. Whetstone, D. Wilken, and K. L. Tomlinson, eds. 1989. *Floristics for the Twenty-first Century: Proceedings of the Workshop Sponsored by the American Society of Plant Taxonomists and the Flora of North America Project*. St. Louis: Missouri Botanical Garden.

Morrow, R. A., and D. D. Brown. 1994. *Critical Theory and Methodology*. Thousand Oaks, CA: Sage.

Näslund, T. 1997. Computers in Context—But Which Context? In M. Kyng and L. Mathiassen, eds., *Computers and Design in Context* (pp. 171–200). Cambridge, MA: MIT Press.

Orlikowski, W. J. 1993. CASE Tools as Organizational Change: Investigating Incremental and Radical Changes in Systems Development. *MIS Quarterly*, 17(3), 309–340.

Outhwaite, W. 1987. *New Philosophies of Social Science: Realism, Hermeneutics and Critical Theory*. Basingstoke, UK: Macmillan.

Pateman, T. 1987. *Language in Mind and Language in Society*. Oxford: Clarendon Press.

Pawson, R. 1989. *A Measure for Measures: A Manifesto for Empirical Sociology*. London: Routledge.

Pawson, R. 1996. Theorizing the Interview. *British Journal of Sociology*, 47(2), 295–314.

Pawson, R. No date. *Middle-Range Realism*. Unpublished manuscript.

Pawson, R., and N. Tilley. 1994. What Works in Evaluation Research? *British Journal of Criminology*, 34(3), 291–306.

Pawson, R., and N. Tilley. 1995. Whither (European) Evaluation Methodology? *Knowledge and Policy: The International Journal of Knowledge Transfer and Utilization*, 8(3), 20–33.

Pawson, R., and N. Tilley. 1997. *Realistic Evaluation*. London: Sage.

Ryan, S., and S. Porter. 1996. Breaking the Boundaries between Nursing and Sociology: A Critical Realist Ethnography of the Theory-Practice Gap. *Journal of Advanced Nursing*, 24(2), 413–420.

Sánchez, J. A. 1998. Improving the Collaborative Use of Information Spaces by Enhancing Group Awareness. Position paper for the Workshop on Collaborative Information Seeking at the 1998 Conference on Computer Supported Cooperative Work, Seattle, WA, November 1998. ⟨http://ict.pue.udlap.mx/pubs/cscw98.html⟩.

Sayer, A. 1985. Realism in Geography. In R. J. Johnston, ed., *The Future of Geography* (pp. 159–173). London: Methuen.

Sayer, A. 1992. *Method in Social Science: A Realist Approach* (2nd ed.). London: Routledge.

Schnase, J. L., D. L. Kama, K. L. Tomlinson, J. A. Sánchez, E. L. Cunnius, and N. R. Morin. 1997. The Flora of North America Digital Library: A Case Study in Biodiversity Database Publishing. *Journal of Network and Computer Applications*, 20(1), 87–103.

Spasser, M. A. 1998. Computational Workspace Coordination: Design-in-Use of Collaborative Publishing Services for Computer-Mediated Cooperative Publishing. Ph.D. dissertation, School of Library and Information Science, University of Illinois, Urbana.

Spasser, M. A. 2000. Articulating Collaborative Activity: Design-in-Use of Collaborative Publishing Services in the Flora of North America Project. *Scandinavian Journal of Information Systems*, 12(1), 149–172.

Star, S. L. 1989. The Structure of Ill-Structured Solutions: Boundary Objects and Heterogeneous Distributed Problem Solving. In M. Huhs and L. Gasser, eds., *Readings in Distributed Artificial Intelligence 3* (pp. 37–54). Menlo Park, CA: Morgan Kaufmann.

Star, S. L., and J. R. Griesemer. 1989. Institutional Ecology, "Translations" and Boundary Objects: Amateurs and Professionals in Berkeley's Museum of Vertebrate Zoology, 1907–39. *Social Studies of Science*, 19(3), 387–420.

Stones, R. 1996. *Sociological Reasoning: Towards a Past-Modern Sociology*. New York: St. Martin's Press.

Tomlinson, K. L., M. A. Spasser, J. A. Sánchez, and J. L. Schnase. 1998. Managing Cognitive Overload in the Flora of North America Project. In R. H. Sprague Jr., ed., *HICSS-31: Proceedings of the Hawaii International Conference on Systems Sciences*, Vol. 2, *Digital Documents Track* (pp. 296–304). Los Alamitos, CA: IEEE Computer Society Press.

Tomlinson, K. L., M. A. Spasser, and J. L. Schnase. 1997. Flora of North America: A Distributed Cognitive System. In S. M. Lobodzinski and I. Tomek, eds., *Proceedings of WebNet '97: World Conference of the WWW, Internet, and Intranet* (pp. 918–920). Charlottesville: Association for the Advancement of Computing in Education. Available from AACE at ⟨http://www.aace.org⟩.

Wainwright, S. P. 1997. A New Paradigm for Nursing: The Potential of Realism. *Journal of Advanced Nursing*, 26(6), 1262–1271.

Wenger, E. 1998. *Communities of Practice: Learning, Meaning, and Identity*. Cambridge: Cambridge University Press.

Will, D. 1980. Psychoanalysis as a Human Science. *British Journal of Medical Psychology*, 53(3), 201–211.

Yeung, W. H. 1997. Critical Realism and Realist Research in Human Geography: A Method or a Philosophy in Search of a Method? *Progress in Human Geography*, 21(1), 51–74.

Yin, R. K. 1994. *Case Study Research: Design and Methods* (2nd ed.). Thousand Oaks, CA: Sage.

Zuboff, S. 1989. *In the Age of the Smart Machine: The Future of Work and Power*. New York: Basic Books.

Contributors

Philip E. Agre (⟨http://dlis.gseis.ucla.edu/people/pagre/⟩) is an Associate Professor of Information Studies at the University of California, Los Angeles. His research interests include: information technology and institutional change, technology and privacy, genre theory, linguistic aspects of computing, and Internet culture. Recent publications include: "Writing and Representation" (in Michael Mateas and Phoebe Sengers, eds., *Narrative Intelligence*, John Benjamins, 2003); "Real-time Politics: The Internet and the Political Process" (*The Information Society*, 18(5) [2002], 311–331); and *Computation and Human Experience* (Cambridge University Press, 1997).

Imani Bazzell (⟨ibazzell@parkland.edu⟩) is Founder and Director of SisterNet, a local network of African American women committed to the physical, emotional, intellectual, and spiritual health of black women. She has worked as a community educator and organizer in the areas of racial justice, gender justice, healthcare access, educational reform, and leadership development for the past 25 years. In addition to her responsibilities as a grant coordinator at Parkland College, she serves as an independent consultant with public schools, colleges and universities, unions, nonprofits, state and international agencies, and community-based organizations to promote organizational development and social justice.

Ann Peterson Bishop (⟨abishop@uiuc.edu⟩) is an Associate Professor in the Graduate School of Library and Information Science, University of Illinois at Urbana–Champaign. She is also the cofounder of Prairienet, the community network serving East Central Illinois. She focuses her research and teaching on community informatics, inquiry-based learning and system design, and social justice in the information professions. Bishop's research has been funded by the National Science Foundation, the U.S. Department of Commerce, the W. K. Kellogg Foundation, and the U.S. Institute of Museum and Library Services. She is currently working on the Paseo Boricua Community Library Project, a neighborhood effort based in the Puerto Rican Cultural Center in Chicago's Humboldt Park (⟨http://www.prairienet.org/pbclp⟩).

Christine L. Borgman (⟨http://is.gseis.ucla.edu/cborgman/⟩), Professor, holds the Presidential Chair in Information Studies at the University of California, Los Angeles. Her teaching and research interests include digital libraries, human-computer interaction, electronic publishing, information seeking behavior, scholarly communication, and bibliometrics. She is the author of *From Gutenberg to the Global Information Infrastructure: Access to Information in a Networked World* (MIT Press, 2000). Borgman's current research addresses the use of digital libraries in high school and undergraduate education. She has lectured or conducted research in more than twenty countries.

Geoffrey C. Bowker (⟨http://weber.uscd.edu/~gbowker⟩) is Professor and Chair of the Department of Communication, University of California, San Diego. He studies social and organizational aspects of the development of very large scale infrastructures. With Leigh Star he has recently completed a book on the history and sociology of medical classifications (*Sorting Things Out: Classification and Its Consequences*, published by the MIT Press in 1999). He is conducting a number of projects to analyze the development of scientific cyberinfrastructure, with particular relationship to ecology and geoscience. His next book, *Memory Practices in the Sciences*, about archival practices in the sciences over the past 200 years, will appear in 2004. He is on the steering committee of the University of California Digital Cultures project.

Barbara P. Buttenfield (⟨babs@colorado.edu⟩) is Professor of Geography at the University of Colorado in Boulder and Director of the Meridian Lab, a research facility focusing on visualization and modeling of geographic information and technology. Her research interests lie in data delivery on the Internet, information management in hostile environments, and design and usability of visualization tools for environmental modeling. She was a Co-Principal Investigator for the Alexandria Digital Library Project (1994–1998), in collaboration with the University of California–Santa Barbara, and led the User Interface Evaluation team during its early years. Dr. Buttenfield is a past president of the American Cartographic Association and a fellow of the American Congress on Surveying and Mapping (ACSM). She was a member of the National Research Council's Panel on Distributed Geolibraries (1997–1999).

Anita Komlodi (⟨komlodi@umbc.edu⟩) is an Assistant Professor in the Department of Information Systems at the University of Maryland–Baltimore County. Her research interests include: human-computer interaction, information visualization, international user interface design, information storage and retrieval, and legal informatics. Recent publications include: "The Role of Interaction Histories in Mental Model Building and Knowledge Sharing in the Legal Domain" (*Journal of Universal Computer Science* 8(5) [2002], 557–566); and, with Dagobert Soergel, "Attorneys Interacting with Legal Information Systems: Tools for Mental Model Building and Task Integration" (in Elaine Toms, ed., *2002 Proceedings of the 65th Annual Meeting of the American Society for Information Science and Technology*).

David M. Levy (⟨dmlevy@u.washington.edu⟩) is a Professor in the Information School of the University of Washington. He holds a Ph.D. from Stanford University in computer science (1979) and a diploma in calligraphy and bookbinding from the Roehampton Institute, London (1982). Prior to joining the Information School faculty, he was a researcher at the Xerox Palo Alto Research Center (PARC). His book, *Scrolling Forward: Making Sense of Documents in the Digital Age*, was published in November, 2001.

Clifford Lynch (⟨clifford@cni.org⟩) is the Executive Director of the Coalition for Networked Information. Prior to joining CNI, Lynch spent eighteen years at the University of California Office of the President, the last ten as Director of Library Automation. Lynch, who holds a Ph.D. in Computer Science from the University of California–Berkeley, is an Adjunct Professor at Berkeley's School of Information Management and Systems. He is a past president of the American Society for Information Science and a fellow of the American Association for the Advancement of Science and the National Information Standards Organization. Lynch currently serves on the Internet 2 Applications Council and the National Digital Preservation Strategy Advisory Board of the Library of Congress; he was a member of the National Research Council committees that published *The Digital Dilemma: Intellectual Property in the Information Infrastructure* and *Broadband: Bringing Home the Bits* and now serves on the NRC's committee on digital archiving and the National Archives and Records Administration.

Gary Marchionini (⟨march@ils.unc.edu⟩; ⟨http://www.ils.unc.edu/~march⟩) is the Cary C. Boshamer Professor of Information Science in the School of Information and Library Science at the University of North Carolina–Chapel Hill. His Ph.D. is from Wayne State University in mathematics education with an emphasis on educational computing. His research interests are in information seeking in electronic environments, digital libraries, human-computer interaction, digital government, and information technology policy. He has had grants or contracts from the National Science Foundation, U.S. Department of Education, Council on Library Resources, the National Library of Medicine, the Library of Congress, the W. K. Kellogg Foundation, and NASA, among others. He was the Conference Chair for ACM Digital Library '96 Conference and program chair for ACM-IEEE Joint Conference on Digital Libraries in 2002. He is editor-in-chief for ACM Transactions on Information Systems and serves on the editorial boards of a dozen scholarly journals. He has published more than one hundred articles, chapters, and conference papers in the information science, computer science, and education literatures. He founded the Interaction Design Laboratory at UNC–CH.

Catherine C. Marshall (⟨http://www.csdl.tamu.edu/~marshall⟩) is a Senior Researcher at Microsoft Corporation. She was a longtime member of the research staff at Xerox PARC and is an affiliate of the Center for the Study of Digital Libraries at Texas A&M University. Her research interests include applying field methods to study reading and other types of intellectual work and designing and evaluating new digital library technologies.

Bharat Mehra (⟨http://www.students.uiuc.edu/~b-mehra1⟩) is a doctoral candidate in the Graduate School of Library and Information Science, University of Illinois at Urbana–Champaign. His research focuses broadly on social aspects in library and information science and representation of human factors in socio-technical research. Specifically, he is interested in community networking and cultural informatics as related to social justice/social equity issues for minority and marginalized populations. He has also been involved in the use of emerging visualization technologies in education and environmental informatics. Mehra's work has incorporated socially grounded methods for "mapping" human factors in human-computer interactions, user-centered design, participatory action research, and formative and summative evaluations of technology based on users' needs and expectations.

Bonnie A. Nardi (⟨bonnie@darrouzet-nardi.net⟩; ⟨http://www.darrouzet-nardi.net/bonnie⟩) is an Associate Professor in the School of Information and Computer Science, University of California–Irvine. She is the author of *A Small Matter of Programming: Perspectives on End User Computing* (MIT Press, 1993), *Context and Consciousness: Activity Theory and Human Computer Interaction* (MIT Press, 1996), and, with Vicki O'Day, *Information Ecologies: Using Technology with Heart* (MIT Press, 1999).

Laura J. Neumann is a Usability Engineer at Microsoft Corporation. She has a Ph.D. in Library and Information Science from the University of Illinois and has a background in sociology and cultural anthropology. Her past research examines how people use, organize, find, and gather information, organize their work space, and carry out their work. She first pursued these interests studying scientists and engineers and then extended her research to humanities scholars in her dissertation. Currently, she is applying this research perspective to focus on finding solutions to real-world problems in the information technology industry.

Vicki O'Day (⟨http://www.calterra.com/people/vicki.html⟩) is studying anthropology at University of California–Santa Cruz. Before returning to student life, she worked in several research labs in Silicon Valley, where she designed software and studied technology use in offices, schools, and libraries. Her earlier work addressed problems in information access, collaborative work, and online communities for senior citizens and children. She is currently studying the uses of computational materials and ideas in biological research, particularly in the context of research into age-related genes and new models of human aging. She is the author (with Bonnie A. Nardi) of *Information Ecologies: Using Technology with Heart* (MIT Press, 1999).

Catherine Plaisant (⟨plaisant@cs.umd.edu⟩; ⟨http://www.cs.umd.edu/hcil/members/cplaisant/⟩) is Associate Research Scientist and Associate Director of the Human-Computer Interaction Laboratory of the University of Maryland Institute for Advanced Computer Studies. She earned a Doctorat d'Ingenieur degree in France in 1982 and has eighteen years of experience in developing and evaluating user interfaces. In 1988 she joined Professor Ben Shneiderman at the Human-Computer Interaction Laboratory. Her research has focused on a variety of topics focusing around the design and evaluation of visual interfaces for public information systems and data visualization.

Bruce R. Schatz (⟨schatz@uiuc.edu⟩; ⟨http://www.canis.uiuc.edu⟩) is Professor in the Graduate School of Library and Information Science at the University of Illinois at Urbana–Champaign. In this role, he was the Principal Investigator of the DLI-1 project that built a federated collection of 100,000 journal articles served to 2,500 users around campus. He is also a Senior Research Scientist at the National Center for Supercomputing Applications (NCSA), where he served as scientific advisor for information systems during the development of Mosaic, and formerly the Director of the Digital Library Research Program in the University Library. He works on information infrastructure and digital libraries and has published major reviews in these areas in *Computer and Science*. He also works on biomedical informatics, constructing analysis environments for the post-genome era, and healthcare infrastructure, constructing population monitors for chronic illness.

Cynthia Smith has worked for the Decatur School District in Illinois for many years. She received her B.A. in Child, Family and Community Services at the University of Illinois in Springfield and her master's degree in Social Work at the University of Illinois at Urbana-Champaign. She currently works as a school social worker.

Mark A. Spasser (⟨mas1200@bjc.org⟩) is Chief, Library & Information Services and Associate Professor at the Jewish Hospital College of Nursing and Allied Health in St. Louis, Missouri. He received his Ph.D. from University of Illinois, where he studied the development-through-use of Web-based publishing tools. His publications and interests focus on the application of such theoretical approaches as activity and realist social theories to better understand social informatic relationships and interactions among interdependent communities of professional practice.

Susan Leigh Star ("Leigh") is a Professor of Communication at the University of California–San Diego, where she is also faculty in the Department of Sociology and in the Science Studies

Program. Prior to joining the faculty at UCSD, Leigh was Professor of Information Science at the Graduate School of Library and Information Science at the University of Illinois at Urbana–Champaign. For many years she has collaborated with computer and information scientists, with whom she has studied work, practice, organizations, scientific communities and their decisions, and the social/moral aspects of information infrastructure. She originally trained as an ethnographer and grounded theorist (with Anselm Strauss), and received her Ph.D in sociology of science and medicine from the University of California–San Francisco. She is a feminist activist, poet, and social theorist, in addition to a troubler of categories.

Nancy A. Van House (⟨http://www.sims.Berkeley.edu/~vanhouse⟩) is a Professor in the School of Information Management and Systems, University of California–Berkeley. Before that she was a Professor in the School of Library and Information Studies at UC–Berkeley, from which she received a Ph.D. Her current research is concerned with epistemic communities, especially their practices of trust and authority, and the articulation between communities' knowledge practices and their information artifacts. She has published extensively in these areas, as well as on digital libraries and the evaluation of libraries and information services.

Index

www.ingramcontent.com/pod-product-compliance
Lightning Source LLC
Chambersburg PA
CBHW080151060326
40689CB00018B/3937